Modeling Fixed Income Securities and Interest Rate Options

Third Edition

CHAPMAN & HALL/CRC
Financial Mathematics Series

Series Editors

M.A.H. Dempster
Centre for Financial Research
Department of Pure Mathematics and Statistics
University of Cambridge

Dilip B. Madan
Robert H. Smith School of Business
University of Maryland

Rama Cont
Department of Mathematics
Imperial College

Aims and scope:

The field of financial mathematics forms an ever-expanding slice of the financial sector. This series aims to capture new developments and summarize what is known over the whole spectrum of this field. It will include a broad range of textbooks, reference works and handbooks that are meant to appeal to both academics and practitioners. The inclusion of numerical code and concrete real-world examples is highly encouraged.

For more information about this series please visit: *https://www.crcpress.com/Chapman-and-HallCRC-Financial-Mathematics-Series/book-series/CHFINANCMTH*

Modeling Fixed Income Securities and Interest Rate Options

Third Edition

Robert A. Jarrow

CRC Press
Taylor & Francis Group
Boca Raton London New York

CRC Press is an imprint of the
Taylor & Francis Group, an **Informa** business

A CHAPMAN & HALL BOOK

CRC Press
Taylor & Francis Group
6000 Broken Sound Parkway NW, Suite 300
Boca Raton, FL 33487-2742

First issued in paperback 2022

ISBN 13: 978-1-03-247526-4 (pbk)
ISBN 13: 978-1-138-36099-0 (hbk)

DOI: 10.1201/9780429432842

**Visit the Taylor & Francis Web site at
http://www.taylorandfrancis.com**

**and the CRC Press Web site at
http://www.crcpress.com**

This book is dedicated to my wife Gail.

Contents

Section II   **Theory**

Chapter 4 ▪ The Term Structure of Interest Rates   41

Chapter 5 ▪ The Evolution of the Term Structure of Interest Rates   57

Preface to the Third Edition

THIS BOOK IS ENTITLED *Modeling Fixed-Income Securities and Interest Rate Options*. Its primary purpose is to teach students the basics of fixed-income securities, but not in the fashion of traditional courses and texts in this area. Traditional fixed-income courses and texts emphasize institutional details, with theories included in an ad hoc fashion and only occasionally. In contrast, this book teaches the basics of fixed-income securities from a unified theoretical framework. The framework is that of the arbitrage-free pricing and complete markets methodology. This textbook is therefore more abstract than traditional textbooks in this area. This is the reason for the word *modeling* in the title. It is the hope (and belief) of this author, however, that this material is the approach of the future.

As a secondary purpose, this textbook explains the arbitrage-free term structure models used for pricing interest rate derivatives with particular emphasis on the Heath-Jarrow-Morton model and its applications. It is designed to make the Heath-Jarrow-Morton model accessible to MBAs and advanced undergraduates, with a minimum of course prerequisites. The course prerequisites are just some familiarity with high school algebra and mathematical reasoning. This textbook has already been used multiple times for an MBA class at Cornell's Johnson Graduate School of Management on fixed-income securities with no prerequisites other than a basic core finance course and a core quantitative methods course.

Contrary to what a quick skimming of this text might suggest, this book is designed for an MBA elective. Each chapter's material is introduced through examples. The examples themselves are designed to illustrate the key concepts. The formal and more general presentation of the same material follows after the examples. This organization facilitates various levels of presentation. For an MBA elective, only the examples should be discussed in class. The formal presentation is left for background reading. For the Ph.D. level, the reverse ordering is appropriate.

This edition differs from the second only in that it is updated to remove typos, to polish the prose, and to reflect new insights from the literature and changes in current market structures.

This manuscript's organization and content were greatly influenced by the comments from reviewers, friends, and students. Thanks are especially extended to Arkadev Chatterjea, Joseph Cherian, Steve Choi, Raoul Davie, William Dimm, Blair Kanbar, Heedong Kim, David Heath, David Lando, Bill Margrabe, Sam Priyadarshi, Tal Schwartz, Ahmet Senoglu, Stuart Turnbull, Don van Deventer, and Yildiray Yildirim.

I

Introduction

CHAPTER 1

Introduction

1.1 THE APPROACH

This book studies an approach for understanding and analyzing fixed income securities and interest rate options that has revolutionized Wall Street. The approach is to apply the tools of derivatives pricing and hedging to understand the *risk management* of fixed income securities. This is in contrast to the traditional focus of textbooks in this area. Traditional textbooks concentrate on the institutional setting of fixed income markets, which consists of descriptions of the various markets and the instruments that trade within – e.g., the key players, conventions in quoting prices, and contract specifics. Risk management – the pricing and hedging of fixed income securities – is, at best, only an afterthought. Traditional fixed income textbooks often introduce pricing theories in an ad hoc and inconsistent fashion. Today, with the advent of sophisticated and readily available computer technology that can be applied to study these securities, the traditional approach is now outdated. Although institutional details are still important, their importance is now secondary to risk management considerations.

This book provides a self-contained study of this new approach for pricing and hedging fixed income securities and interest rate options. This new approach is based on the Heath, Jarrow, and Morton (HJM) model, an interest rate derivatives pricing model, which is used extensively in the industry. The HJM model was developed by Heath, Jarrow, and Morton in a sequence of papers.[*] It was motivated by the earlier work of Ho and Lee[†]

---

[*] See Heath, Jarrow, Morton [2,3,4]. Also see Jarrow [6] for a review of the history behind the development of the HJM model.

[†] See Ho and Lee [5].

on this same topic, and by the martingale pricing methods of Harrison and Pliska.*

The HJM model is presented using the standard binomial model so often used to analyze the pricing and hedging of equity options. The standard binomial model is chosen because of its mathematical simplicity, in comparison to the more complex stochastic calculus techniques. In addition, this is with little loss of generality, as the alternative and more complex stochastic calculus techniques still need to be numerically implemented on a computer, and a standard approach for implementing these techniques on the computer is to use the binomial model. So, from an implementation point of view, the binomial model is all one needs to master.

This flexible, new approach to fixed income securities has revolutionized the industry. The HJM model is employed by commercial and investment banks to price and hedge numerous types of fixed income securities and interest rate options. One reason for its extensive use is that this model provides a "lego" building block technology. With minor modifications, this technology can be easily extended (built upon) to handle the pricing of other more complex instruments, e.g., foreign currency derivatives, commodity derivatives, and credit derivatives. These extensions will also be illustrated, albeit briefly, in this book.

1.2 MOTIVATION

Examining three figures can motivate the subject matter of this textbook. The first, Figure 1.1, contains the graph of a yield curve for Treasury securities. For now, do not worry about the exact definition of a Treasury security or its yield. These will be discussed more fully, later on in the text. Here, it suffices to understand that Treasury securities are bonds issued by the U.S. government. A bond is an IOU issued for borrowing a stated amount of dollars (e.g., $10,000), called the principal, for a fixed period of time (e.g., 5 years), called the bond's maturity. Interest is paid regularly (often semi-annually) on this IOU. A Treasury bond's *yield* can be thought of as the interest earned per year from buying and holding the bond until its maturity.

In Figure 1.1 we see a Treasury yield curve on December 12, 2018. The *y*-axis units are percentages per year. The *y*-axis starts at 0 percent. The *x*-axis gives the different bonds' maturities in years. The maturities run

---

* See Harrison and Pliska [1].

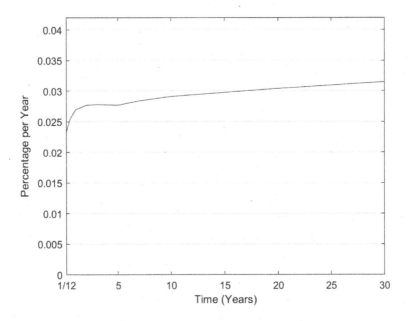

FIGURE 1.1   Treasury Bond Yields on December 12, 2018.

from 1 month to 30 years. As depicted, the yield curve is upward sloping, indicating that it costs more to borrow for 30 years than it does for 5 years or for 1 year. The longer the borrowing horizon, the larger the interest rate charged per year. This increasing interest rate reflects the different risks of the longer borrowing horizons and the market's perception of how short-term interest rates will change in the future.

The Treasury yield curve changes through time. This change is illus-trated in Figure 1.2, which repeats Figure 1.1, but graphs it at different dates over the time period from January 2014 through January 2019. As seen, the yield curve's shape and height differ at different times over this 5-year period. Sometimes the yield curve is nearly flat (see January 2019). The shape of the yield curve is referred to as *the term structure of interest rates*. Its fluctuation through time is called *the evolution* of the term struc-ture of interest rates. This evolution is random because it is not predict-able in advance. Forecasting the evolution of the term structure of interest rates well is crucial for risk management procedures. Models for the evo-lution of the term structure of interest rates will form a big part of this book's content.

Interest rate options or derivatives are financial contracts whose cash flows depend, contractually, on the Treasury yield curve as it evolves

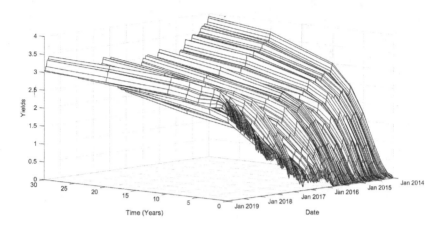

FIGURE 1.2   Yields from January 2014 to December 2018.

through time. For example, one such interest rate derivative is a financial contract that pays its owner on the second anniversary date of the contract (its maturity date), $10,000 times the difference between the 10-year and 5-year yields observed on the maturity date, but only if this difference is positive. This is a *call option* on an interest rate spread. To understand how to price this spread option, one needs to understand (or model) how the term structure of interest rates evolves through time. The better the model (or forecast) of the term structure evolution, the better the pricing model will be. After all, the expectation of the profits or losses from the spread option depends on what the shape of the term structure looks like in 2 years.

A key theme in this book, therefore, is how to model the evolution of the term structure of interest rates illustrated in Figure 1.2. Much of our emphasis will be on developing this structure. Then, given this evolution, the second theme is how to price and hedge the interest rate derivatives written on it. This is a nontrivial exercise and reflects the remaining focus of the book. This topic is referred to as risk management.

A key characteristic of U.S. Treasury securities is that they are considered to be default-free. That is, an IOU issued by the U.S. government is considered to be safe, with the receipt of the interest and principal payments considered a sure bet. Of course, not all IOUs are so safe. Corporations and government municipalities also borrow by issuing bonds. These corporate and municipal loans can default and have done so in the past. The interest and principal owed on these loans may not be paid in full if the corporation or municipality defaults prior to the

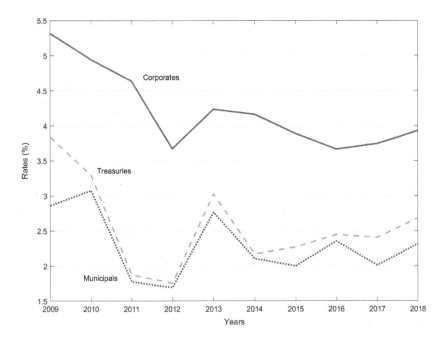

FIGURE 1.3 Various Bond Yields.

maturity date of the borrowing. This different default risk generates different interest rates for different borrowers, reflecting a credit risk spread, as illustrated in Figure 1.3.

Figure 1.3 plots the average yields per year on Treasury, corporate, and municipal bonds over the years 2009 through 2018. The y-axis units are percentages per year, and it starts at 1.5 percent. The x-axis units are time measured in years from 2009 to 2018. As seen, the cost for borrowing is higher for corporates than it is for Treasuries. Surprisingly, however, the cost for borrowing is higher for Treasuries than it is for municipals. The corporate--Treasury spread (positive) is due to credit risk. This credit spread represents the additional compensation required in the market for bearing the risk of default. The Treasury--municipal spread (negative) is due to both default risk and the differential tax treatments on Treasuries versus municipals. Treasuries and corporates are taxed similarly at the federal level. So, tax differences do not influence their spread. In contrast, U.S. Treasury bond income is taxed at the federal level, while municipal bond income is not. This differential tax treatment influences the spread. As seen, the tax benefit of holding municipals dominates the credit risk involved, making the municipal borrowing rate less than the Treasury rate. Understanding the different spreads between various fixed income securities is an important aspect of fixed income markets.

Unfortunately, for brevity, this book concentrates almost exclusively on the Treasury curve (default-free borrowing), and it ignores both credit risk and taxation. Nonetheless, this is not too big an omission. It turns out that if one understands how to price interest rate derivatives issued against Treasuries, then extending this technology to derivatives issued against corporates or municipals is (conceptually) straightforward. This is due to the "lego" building block aspect of the HJM technology. Consequently, most of the economics of fixed income markets can be understood by considering only Treasury markets.

This claim can be intuitively understood by studying the similarity in the evolutions of the three different yields – Treasuries, corporates, municipals – as depicted in Figure 1.3. The movements in these three rates are highly correlated, almost moving in a parallel fashion. As one rate increases, so do the remaining two. As one declines, so do the others, although the magnitudes of the changes may differ. A good model for one rate, with some obvious adjustments, therefore, will also be a good model for the others. This implies that the same pricing technology should also apply. And, it does. Therefore, the same risk management practices can be utilized for all three types of fixed income securities. Consequently, mastering risk management for Treasury securities is almost sufficient for mastering the risk management techniques for the rest. The last chapter in the book will briefly solidify this intuition.

1.3 THE METHODOLOGY

As mentioned earlier, we use the standard binomial option pricing methodology to study the pricing and hedging of fixed income securities and interest rate options. The binomial approach is easy to understand, and it is widely used in practice. It underlies the famous Black-Scholes-Merton option pricing model.* As shown later, the only difference between the application of these techniques to fixed income securities and those used for pricing equity options is in the construction of a more complex binomial tree. To understand this difference, let us briefly review the standard binomial pricing model.

A diagram with nodes and *two* branches emanating from each node on the tree is called a *binomial* tree (see Figure 1.4). The nodes occur at

---

* An excellent reference for the Black-Scholes-Merton option pricing model is Jarrow and Chatterjea [7].

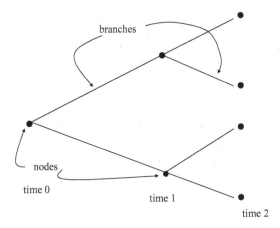

FIGURE 1.4   A Binomial Tree.

particular times, e.g., times 0, 1, 2, and so forth. The nodes represent the state of the economy at these various times. The first node at time 0 represents the present (today's) state of the economy. The future states possible at times 1 and 2 are random. The branches represent the paths for reaching these nodes, called histories.

For risk management purposes, the relevant state of the economy at each node is characterized by the prices of some set of securities. For equity options, a single security's price is usually provided at each node, the price of the underlying equity. For fixed income securities, an array (a vector) of bond prices is considered at each node of the tree, rather than the single equity's price (a scalar) as in the case of equity options. Otherwise, the two trees are identical. Hence, term structure modeling is just the multidimensional extension of the standard equity pricing model. Unfortunately, this multidimensional extension to a vector of prices at each node of the tree is nontrivial. The complexity comes in ensuring that the given evolution of prices is arbitrage-free.

By *arbitrage-free* we mean that the prices in the tree are such that there are no trading strategies possible using the different bonds that can generate profits with no initial investment and with no risk of a loss. For an equity pricing model, constructing an arbitrage-free binomial tree is trivial. For fixed income pricing models, however, constructing an arbitrage-free binomial tree is quite complex. Quantifying the conditions under which this construction is possible is, in fact, the key contribution of the HJM model to the literature. Once this arbitrage-free construction is accomplished, the standard binomial pricing arguments apply.

The standard binomial pricing argument determines the arbitrage-free price of a traded derivative security, in a frictionless and competitive market, where there is no counterparty risk (these terms are explained later in the text). It does this by constructing a *synthetic* derivative security, represented by a portfolio of some underlying assets (bonds) and a riskless money market account. This portfolio is often modified, rebalanced, through time in a dynamic fashion. This portfolio's (the synthetic derivative security's) cash flows are constructed to exactly match those of the traded derivative security, but the synthetic derivative security's cost of construction is known. This cost of construction represents the "fair" or "theoretical" value of the derivative security.

Indeed, if the traded derivative security's price differs from the theoretical value, an arbitrage opportunity is implied. For example, suppose the theoretical price is lower than the traded price. Then, sell the traded derivative (a cash inflow) and construct it synthetically (a cash outflow). The difference in cash flows is positive because the traded price is above the cost of construction. One pockets this initial dollar difference, and there is no future liability, as the cash flows are equal and opposite from thereon.

The identification of such arbitrage opportunities is one of the most popular uses of these techniques in practice. Other uses are in the area of risk management, and these uses will become more apparent as the pricing and hedging techniques are mastered.

1.4 AN OVERVIEW

This book is divided into four parts. Part I is an introduction to the material studied in this book. Part II describes the economic theory underlying the HJM model. Part III applies the theory to specific applications, while Part IV studies implementation and estimation issues, as well as extensions of the HJM model to foreign currency derivatives, credit derivatives, and commodity derivatives.

Part I, the introduction, consists of three chapters. Chapter 1 is the introduction, the current chapter. Chapter 2 gives the institutional description of the fixed income security and interest rate derivative markets. Included are Treasury securities, repo markets, Forward Rate Agreements (FRAs), Treasury futures, caps, floors, swaps, swaptions, and other exotics. This presentation is brief with additional references provided. Chapter 3 describes the classical approach to fixed income security risk management. The classical approach concentrates on yields, duration, and convexity. The classical approach is studied for two reasons. First, it is still in use today, although its use is diminishing as time progresses. Second, and perhaps most important, understanding the limitations of these classical techniques provides the motivation for the HJM model presented in Part II.

Part II presents the HJM model's theoretical framework. Chapter 4 introduces the notation, terminology, and assumptions used in the remainder of the text. The traded securities are discussed, where, for the sake of understanding, many of the actually traded financial securities are simplified. Chapter 5 sets up the binomial tree for the entire term structure of zero-coupon bonds. A one-factor model is emphasized in Chapter 5 and throughout the remainder of the text. Markets with two or more factors are also considered. Chapter 6 describes the traditional expectations hypothesis, which, after a curious transformation, becomes essential to the pricing methodology used in subsequent chapters. Chapter 7 introduces the basic tools of analysis: trading strategies, arbitrage opportunities, and complete markets. Chapter 8 illustrates the use of these tools for zero-coupon bond trading strategies with an example. Chapter 9 provides the formal theory underlying Chapter 8. Finally, Chapter 10 discusses option-pricing theory in the context of the term structure of interest rates.

Part III consists of various applications of this pricing and hedging technology. Chapter 11 studies the pricing of coupon bonds. Chapter 12 investigates the pricing of options on bonds (both European and American). Chapter 13 studies the pricing and hedging of forwards, futures, and options on futures. Chapter 14 analyzes swaps, caps, floors, and swaptions. Chapter 15 prices various interest rate exotic options including digitals, range notes, and index-amortizing swaps. Index-amortizing swaps are considered to be one of the most complex interest rate derivatives traded. They will be seen as just a special, albeit more complex, example of our pricing technology.

Part IV investigates implementation and estimation issues. Chapter 16 shows that the model of Parts I–III can serve as a discrete time approximation to the continuous time HJM model. Both the continuous time HJM model and its limits are studied herein. Chapter 17 investigates the statistical estimation of the inputs: forward rate curves and volatility functions. Empirical estimation of these inputs is also provided in this chapter. Chapter 18 completes the book with a discussion of various extensions and a list of suggested references. The extensions are to foreign currency derivatives, credit derivatives, and commodity derivatives.

REFERENCES

1. Harrison, J. M., and S. Pliska, 1981. "Martingales and Stochastic Integrals in the Theory of Continuous Trading." *Stochastic Processes and Their Applications* 11, 215–260.
2. Heath, D., R. Jarrow, and A. Morton, 1990. "Bond Pricing and the Term Structure of Interest Rates: A Discrete Time Approximation." *Journal of Financial and Quantitative Analysis* 25 (4), 419–440.
3. Heath, D., R. Jarrow, and A. Morton, 1990. "Contingent Claim Valuation with a Random Evolution of Interest Rates." *Review of Futures Markets* 9 (1), 54–76.
4. Heath, D., R. Jarrow, and A. Morton, 1992. "Bond Pricing and the Term Structure of Interest Rates: A New Methodology for Contingent Claims Valuation." *Econometrica* 60 (1), 77–105.
5. Ho, T. S., and S. Lee, 1986. "Term Structure Movements and Pricing Interest Rate Contingent Claims." *Journal of Finance* 41, 1011–1028.
6. Jarrow, R., 1997. "The HJM Model: Past, Present, Future." *Journal of Financial Engineering* 6 (4), 269–280.
7. Jarrow, R., and A. Chatterjea, 2019. *An Introduction to Derivative Securities, Financial Markets, and Risk Management,* 2nd edition, World Scientific Press.

CHAPTER 2

Traded Securities

THIS CHAPTER PROVIDES THE institutional background on the financial securities studied in this book. These securities include Treasury bonds; notes; bills; Treasury futures; Eurodollar deposits; Eurodollar futures; and various interest rate options like swaps, caps, floors, and swaptions. Although this material has also been applied to price and hedge mortgage-backed securities, corporate debt, municipal bonds, and agency securities, this later set of securities contains either prepayment risk (mortgages) or credit risk (corporate, municipal, agency debt). Incorporating these additional risks requires an extension of the methods studied in this book. A brief discussion of these extensions is contained in Chapter 18.

The presentation of the institutional material in this chapter is brief, as the emphasis of this book is on models and not institutional considerations. In subsequent chapters, some of the institutional features of these financial contracts are simplified. This is done to facilitate understanding. An example is Treasury futures contracts. These contracts are quite complex, containing numerous embedded delivery options. These delivery options are simplified in Chapter 13. Fortunately, these more complicated contractual provisions can be easily mastered once the simplified versions of the contracts studied in this book are well understood.

2.1 TREASURY SECURITIES

U.S. Treasury securities (bonds, notes, floating rate notes, and bills) are debt obligations issued by the U.S. government, and their payment (coupons plus principal) is guaranteed by the taxing authority of the United States.

As such, they are generally considered to be default-free. These securities are used to finance the cumulative U.S. government spending deficits.

Treasury securities are issued in three basic types: *(i)* coupon-bearing instruments paying interest every 6 months with a principal amount (or face value) paid at maturity, called *coupon bonds*, *(ii)* discount securities bearing no coupons and paying only a principal amount at maturity, called *zero-coupon bonds*, and *(iii) floating rate notes* paying a changing (floating) interest rate quarterly that is indexed to the 13-week Treasury bill rate and a principal amount at maturity. By historic convention, the Treasury issues all securities with maturities of 1 year or less as zero-coupon bonds, and all securities with maturities greater than a year as coupon or interest-paying notes and bonds.

The zero-coupon Treasury securities are called *bills*. The coupon and interest-paying Treasury securities are called *notes* or *bonds* depending on whether the maturity at issuance is from 2 to 10 years or greater than 10 years, respectively. The bonds, notes, floating rate notes, and bills currently issued can be found on www.treasurydirect.gov. The minimum purchase size of a Treasury security is $100.

In addition to bills, notes, floating rate notes, and bonds, Treasury securities called STRIPS (separate trading of registered interest and principal of securities) are traded. Treasury STRIPS have been traded since August 1985. These are securities issued against Treasury bonds and notes that provide just the coupon interest payment at a particular date or just the principal payment of the Treasury bond or note. STRIPS are created by financial institutions via the commercial book-entry system. They are, in effect, *synthetically* created zero-coupon bonds.

New Treasury securities are issued in an *auction market* on a regular basis. The auction uses competitive bids, although noncompetitive bids are accepted and tendered at the average yield of the competitive bids (see www.treasurydirect.gov for additional details). The auction market is called the *primary market*. The Treasury auction mechanism is, in fact, an important component in the determination of secondary market Treasury security prices.

Those Treasury security issues of a particular time to maturity (3 months, 6 months, 1 year, 2 years, 3 years, 5 years, 7 years, 10 years and 30 years) that are the most recently auctioned are called *on-the-run*. Those Treasury security issues with similar maturities but that were offered in previous auctions are called *off-the-run*. On-the-run Treasury securities are the most liquid of the traded Treasuries, being held in government security dealer inventories. These securities are often used as benchmarks against which other security prices are quoted; for example, agency bonds are quoted as a spread to Treasuries.

2.2 TREASURY SECURITY MARKETS

The majority of Treasury securities are traded in the *over-the-counter market* (the New York Stock Exchange lists some issues, but the trading volume is small.) An over-the-counter market is a "screen-based" or "phone-based" market between investment banks, commercial banks, and government bond dealers. Bid/ask quotes are regularly provided in this market, and trades are executed without an official exchange location.

Treasury notes (including floating rate) and bonds are quoted in units of one 32d on a 100-dollar par basis. The *bid price* is that price the government dealer is willing to purchase the issue for, and the *ask price* is that price they are willing to sell the issue for. The difference between the bid and ask price is called the *bid-ask spread*. The bid-ask spread is the compensation that the government security dealers earn for making a market in this particular security.

The *ask yield* corresponds to the internal rate of return or, equivalently, the holding period return on the bond. It is computed according to the following equation:

$$\text{ask price} = \sum_{t=1}^{2T} \frac{\text{coupon}/2}{\left(1+\text{ask yield}/2\right)^{t-\tau}} + \frac{\text{face value}}{\left(1+\text{ask yield}/2\right)^{2T-\tau}}$$

where*

T = the number of years to maturity from the settlement date, and

$$\tau = \frac{\text{number of days from settlement}}{\text{number of days in the coupon period}}$$

The ask yield is that rate which when discounted on a semi-annual basis equates the present value of the coupon and principal payments to the ask price. Coupon rates are quoted on a per year basis. As coupon payments are paid semi-annually, the yearly coupon payment must be divided by two.

---

* The days to maturity are measured from the settlement day. The settlement day is the day when the actual cash flows are exchanged for the Treasury security transaction. Settlement for institutional traders in Treasury securities is usually the next business day.

Ask yields are also quoted on a per year basis, hence, the division by two in the numerator as well. Semi-annual compounding of interest is used. This explains the number of compounding periods as 2T.

Bid and asked quotes are given without *accrued interest*. The price the buyer actually pays (or receives) when transacting in the bonds or notes is the quoted price *plus accrued interest*, which is calculated as follows:

$$\text{accrued interest} = \left(\frac{\text{coupon}}{2} \right) \tau.$$

Treasury STRIPS are quoted on the same basis as notes and bonds – in units of 32nds on a 100-par basis.

Finally, we revisit Treasury bills. Treasury bills are quoted on a different basis than either Treasury bonds, notes, or STRIPS. Treasury bills' bid/ask quotes are given in percentages and not dollar prices. The quotes are provided on a *banker's-discount* basis. The *banker's-discount yield* (ask or bid) is calculated as follows:

$$\text{Banker's discount yield} = \left(\frac{\text{face value} - \text{price}}{\text{face value}} \right)$$
$$\times \left(\frac{360}{\text{number of days from settlement until maturity}} \right).$$

The banker's discount yield is just a convenient statistic for quoting prices. It does not correspond to either the internal rate of return or, equivalently, the holding period return on the Treasury bill. In contrast, the ask yield is intended to be such a measure of the return on the Treasury bill. Therefore, it is calculated differently as follows:

$$\text{Ask yield} = \left(\frac{\text{face value} - \text{ask price}}{\text{ask price}} \right)$$
$$\times \left(\frac{365}{\text{number of days from settlement until maturity}} \right).$$

The difference between the ask yield and the banker's discount yield is two-fold. First, the denominator in the banker's discount yield is the face value, while in the ask yield it is the ask price. Second, the banker's discount yield annualizes on a 360-day basis, while the ask yield annualizes on a 365-day basis. The *ask yield* is meant to be comparable to the ask yield on a Treasury bond, note, or STRIP.

For the subsequent analysis we are primarily interested in the prices (ask/bid) quoted for the various Treasury securities. The banker's discount yield is meaningful only to the extent that it is the necessary input to a transformation (given above) that generates the Treasury bill's price. We will have no use for it otherwise.

Another class of traded Treasury securities is inflation-indexed bonds (called TIPs).* These bonds pay fixed dollar coupon payments semi-annually, where the final principal is adjusted to make the dollar coupon payment as a percentage of the principal equal to the "real" or "inflation adjusted" rate of return on these bonds. The principal adjustment is based on a consumer price index.

2.3 REPO MARKETS

In addition to trading in the Treasury securities themselves, there is an active repurchase agreement market for Treasury securities, both over-night and term (more than 1 day). Repurchase agreements (repos) are the key vehicle by which government dealers finance their huge inventories of Treasury securities (see Sundaresan [3]).

From the dealer's perspective, a *repo*, or *repurchase agreement*, is a transaction in which a dealer sells a Treasury security to an investor and simultaneously promises to buy it back at a fixed future date and at a fixed future price. This transaction is diagrammed in Figure 2.1. The price at which the Treasury security is repurchased is larger than the selling price,

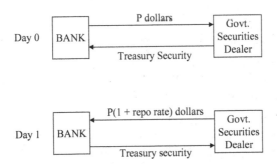

FIGURE 2.1   A Typical Repurchase Agreement.

---

* For an analysis of TIPs using an HJM model see Jarrow and Yildirim [2].

the difference being interest, called the *repo rate*. The investor in the above transaction is said to have entered a *reverse repo*. A repurchase agreement is equivalent to the dealer's borrowing funds from the investor and using the Treasury security as collateral.

Alternatively, a repurchase agreement is simply a forward contract on a Treasury security, having an initial value equal to zero (the Treasury security is exchanged for its fair value in cash), and with a forward price equal to the repurchase price. The repo rate is then nothing more than a specific maturity, *forward interest rate*, as subsequently defined in Chapter 4.

There are repo rates for general collateral (any Treasury security), and there are often special repo rates for a particular Treasury security. These special repo rates are lower than the general collateral rates, and they occur when there is an unusual demand for the particular underlying bond (or note). In essence, these special repo rates represent lowered borrowing rates available only to the holder of a particular Treasury security – a convenience yield (see Sundaresan [3] for related discussion and evidence).

2.4 TREASURY FUTURES MARKETS

In summary, there are basically three markets in which to buy or sell a Treasury security: *(i)* the spot market, *(ii)* the forward market, and *(iii)* the futures market.

The first method by which to buy or sell a Treasury security is in the *spot* or *cash* market for immediate delivery. These are the primary and secondary markets discussed above. The price at which the exchange takes place is called the *spot* or *cash* price.

The second method is by entering a *forward contract*. A forward contract is an agreement made in the present (today) between a buyer and seller of a commodity to exchange the commodity for cash at a predetermined future date, but at a price agreed upon today. The agreed-upon price is called the *forward price*. For Treasury securities, the repo market is an example of a forward market.

Finally, the third method is by entering a *futures contract*. A *futures contract* is a standardized financial security, issued by an organized exchange, for future purchase/sale of a commodity at a predetermined future date and at an agreed-upon price. The agreed-upon price is called the *futures price*, and it is paid via a sequence of unequal and random "daily installments" over the contract's life. These "daily installments" were instituted by the futures exchange to guarantee that the purchaser/seller of a futures contract would fulfill his or her obligations. It is this daily settlement procedure that differentiates forward contracts from futures contracts and forward prices from futures prices.

Daily settlement is called *marking to market*. The procedure is as follows. When a buyer/seller of a futures contract opens a position, a margin account is required. The margin account is usually an interest-earning account consisting of an initial cash deposit. The margin account's magnitude is set by the exchange so as to ensure (with reasonable probability) that the buyer/seller's obligation of the futures contract will be fulfilled. Marking to market occurs when the daily change to the futures price is added or subtracted from the margin account. For example, the purchaser of a futures contract would have the margin account's balance decreased if the futures price falls, increased if the futures price rises, and unchanged otherwise. This adjustment resets the value of the futures contract to zero at the end of each trading day. Marking to market is discussed in greater detail in Chapter 4.

Futures contracts on Treasury securities trade on various exchanges, and they are standardized in terms of the contract size, the delivery months, the deliverable securities, and the delivery procedure. In fact, for Treasury notes and bonds this contract specification is quite complicated, involving various embedded options known as *(i)* the delivery option, *(ii)* the wildcard option, and *(iii)* the quality option. These delivery options are included to increase the deliverable supply of the underlying Treasury securities (both across time and in quantity) in order to reduce the possibility of manipulation. For the details of the contract specifications and an explanation of these embedded options see Jarrow and Chatterjea [1].

2.5 INTEREST RATE DERIVATIVES ON TREASURIES

The phrase *interest rate derivatives* is a catchall that includes all derivatives, not already considered, whose payoffs depend on the evolution of the Treasury zero-coupon bond price curve or, as it is more commonly called, the *Treasury term structure of interest rates*.

Two types of interest rate derivatives traded on various Treasuries are *call* and *put options*. A *call option* is a financial security that gives its owner the right to buy the underlying Treasury security or Treasury futures at a fixed price on or before a fixed date. The fixed price is called the *exercise* or *strike* price of the option. The fixed date is called the *maturity* or *expiration date* of the option. A *put option* gives the owner the right to sell the underlying Treasury security or Treasury futures at a fixed price on or before a fixed date. The owner does not have to purchase (the call) or sell (the put) the options if he or she does not want to. This is why they are called "options." Options provide opportunities to hedge and/or speculate on the future evolution of Treasury rates with a minimum initial payment, called the option's *premium*. Understanding these contracts is an essential part of this text. Call and put options are discussed in Chapters 4 and 12.

2.6 EURODOLLAR SPOT, FORWARD, AND FUTURES MARKETS

Eurodollar deposits are dollar deposits held in banks outside the United States (and therefore exempt from Federal Reserve regulations). An index of rates for interbank lending of Eurodollars for various maturities is known as LIBOR (London Interbank Offered Rates). These are the spot or cash markets in Eurodollars. These markets are most active for transactions of less than a year.

Figure 2.2 contains the LIBOR rates for Wednesday, January 2, 2019, as reported in the *Wall Street Journal* on Thursday, January 3. The 1-month LIBOR rate is 2.50713 percent, the 3-month rate is 2.79388 percent, the 6-month rate is 2.87394 percent, and the 1-year rate is 3.00200 percent.

Forward contracts in LIBOR trade in the over-the-counter market. These are called forward rate agreements (FRAs). An FRA is a forward contract written on LIBOR that requires a cash payment at the delivery date equal to the realized LIBOR rate on the delivery date less the pre-specified forward rate times some principal dollar amount. FRAs differ with respect to the maturity of the LIBOR rate (usually 1 month to 1 year) and the delivery date of the contract (usually 4 months to 18 months; see Jarrow and Chatterjea [1] for more explanation).

There is also an active market in Eurodollar futures contracts. They are written on different maturity Eurodollar rates (e.g., 3 months). Eurodollar futures contracts are settled in cash on the delivery date. Delivery dates

---

Wednesday, January 2, 2019

one month .0250713
three months .0279388
six months .0287394
one year .0300200

FIGURE 2.2   LIBOR Rates.

(**As Reported in the *Wall Street Journal* on Thursday, January 3, 2019.**)

of the contracts range from 1 month to 10 years. These contracts are quite active because they are the main instruments used to hedge more complex interest rate derivatives based on LIBOR.

2.7 INTEREST RATE DERIVATIVES ON LIBOR

There are active markets in interest rate derivatives on LIBOR – both exchange-traded and over-the-counter. In the over-the-counter market are traded swaps, floors, caps, swaptions, and more exotic interest rate derivatives, like digitals and index-amortizing swaps.

Briefly, an *interest rate swap* is an agreement between two counterparties to exchange fixed rate payments for floating rate payments (based on LIBOR) at regular time intervals on some principal amount for a given period of time (maturity of the contract). A *cap* is a sophisticated type of call option on LIBOR rates. A *floor* is the corresponding sophisticated type of put option on LIBOR rates. A *swaption* is a call or put option on a swap. Although complex sounding, swaps, caps, floors, and swaptions are very practical interest rate derivatives, used to hedge (provide insurance for) and to speculate on future interest rate movements. These instruments are common tools for corporate debt management. These contracts are discussed in great detail in Chapter 14. Interest rate exotic options are discussed in Chapter 15.

REFERENCES

1. Jarrow, R., and A. Chatterjea, 2019. *An Introduction to Derivative Securities, Financial Markets, and Risk Management*, 2nd edition, World Scientific Press.
2. Jarrow, R., and Y. Yildirim, 2003. "Pricing Treasury Inflation Protected Securities and Related Derivatives Using an HJM Model." *Journal of Financial and Quantitative Analysis* 38 (2), 337–358.
3. Sundaresan, S., 2009. *Fixed Income Markets and Their Derivatives*, 3rd edition, Academic Press.

CHAPTER 3

The Classical Approach

3.1 MOTIVATION

The chapter studies the classical approach to pricing fixed income securities. Unfortunately, for the classical approach, pricing fixed income securities means just the pricing of coupon-bearing bonds with no embedded options. This is due to the limitation of the tools and techniques involved. The classical approach to coupon bond pricing uses only the notion of an internal rate of return and simple calculus to generate the traditional risk management parameters – yields, duration, modified duration, and convexity. The classical approach uses little, if any, economic theory. As such, one should not anticipate that significant insights into the subject can be obtained via the classical approach.

We present the classical approach in this book for two reasons. One, these tools are still used by many fixed income traders today (although these tools are slowly being replaced by the techniques presented in subsequent chapters). So, any serious student of finance needs to understand the classical techniques. Two, understanding the limitations of these classical techniques motivates the approach used in the subsequent chapters. Indeed, one perspective on the HJM model is that it is precisely that generalization of the classical techniques needed to overcome the limitations of duration and convexity hedging.

3.2 COUPON BONDS

As mentioned earlier, the classical approach to fixed income analysis concentrates on studying coupon bonds. Although the techniques are often (and incorrectly) applied to Treasury, corporate, and municipal bonds, we will concentrate only on Treasury bonds in this section. The U.S. government (Treasury) issues coupon-bearing notes and bonds (see Chapter 2). A coupon-bearing bond is a loan for a fixed amount of dollars (e.g., $10,000), called the principal or face value. The loan extends for a fixed time period, called the life or maturity of the bond (usually 5, 10, 20, or 30 years). Over its life, the issuer is required to make periodic interest payments (usually semi-annually) on the loan's principal. These interest payments are called coupons. Hence, the term a coupon bond. The U.S. government also issues short-term bonds (less than a year in maturity) with no coupons, called Treasury bills. These Treasury bills are called zero-coupon bonds. The same techniques apply to both coupon and zero-coupon bonds.

A callable coupon bond is a coupon bond that can be repurchased by the issuer, at predetermined times and prices, prior to its maturity. This repurchase provision is labeled a "call" provision. For example, a typical callable coupon bond includes a provision where the issuer can repurchase the bond at face value (the predetermined price), anytime during the last 5 years of its life (the predetermined times). This section studies only non-callable coupon bonds. Callable coupon bonds are studied in a subsequent chapter. Unfortunately, the techniques of the classical approach cannot be used to price callable coupon bonds. This is a serious limitation of the classical approach that is overcome by the HJM model introduced later in the book.

To discuss these techniques, we need to introduce some notation. First, consider a time horizon of length T, divided into unit subintervals with dates $0, 1, 2, \ldots, T-1, T$. This is a discrete trading model.

We define a *coupon bond* with principal L, coupons C, and maturity T to be a financial security that is entitled (contractually) to receive a sequence of future coupon payments of C dollars at times $1, \ldots, T$ with a principal repayment of L dollars at time T. The times $1, 2, \ldots, T$ represent payment dates on the coupon bond. The coupon rate (one plus a percentage) on the bond per period is given by $c = 1 + C/L$. This cash flow sequence is illustrated in Figure 3.1. For simplicity, we assume that the bond makes a coupon payment every period. Relaxing this assumption to have coupon payments only every so often is straightforward and left to the reader as an exercise.

We let $\mathcal{B}(t)$ denote the price of this coupon bond at time t. By construction, the price of the coupon bond is forward-looking. It represents

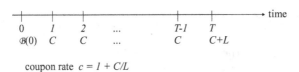

coupon rate $c = 1 + C/L$

FIGURE 3.1 Cash Flows to a Coupon Bond with Price $\mathcal{B}(0)$, Principal L, Coupons C, and Maturity T.

the present value of all the future cash flows (those cash flows received after time t). For example, if a coupon payment C occurs at time t, then $\mathcal{B}(t)$ does not include this coupon payment, i.e., it is the bond's price *ex-coupon*.

Recall from Chapter 2 that the market convention in buying and selling bonds is to pay a quoted price plus accrued interest (because we are in frictionless markets, the ask price equals the bid price, which equals the quoted price). We next discuss how $\mathcal{B}(t)$ relates to the bond's quoted price. The relation is straightforward. The price paid for the coupon bond must equal $\mathcal{B}(t)$. That is,

$$\text{quoted price} + \text{accrued interest} = \mathcal{B}(t).$$

Thus,

$$\text{quoted price} = \mathcal{B}(t) - \text{accrued interest}.$$

For example, on a coupon-payment date t (the ex-date), accrued interest equals zero, so the quoted price equals $\mathcal{B}(t)$. For a date t halfway between the coupon payment dates, the accrued interest equals $C/2$, so the quoted price $= \mathcal{B}(t) - C/2$. Of course, if the date t lies somewhere else between the coupon payment dates, then the accrued interest calculation must be modified accordingly (in a prorated fashion).*

We point out that a zero-coupon bond paying a dollar at time T is a special case of this definition. It is the case in which the coupon payment is set identically to zero ($C=0$) and the principal is set equal to unity ($L=1$). Zero-coupon bonds play an important role in the subsequent theory.

---

* For example, if date t occurs one-third of the way to the next coupon payment date, then the accrued interest is $C/3$.

3.3 THE BOND'S YIELD, DURATION, MODIFIED DURATION, AND CONVEXITY

This section defines the bond's yield, duration, modified duration, convexity and explores their relationship to the bond's price.

The *bond's yield*, $Y(t)$, is defined as one plus its percentage internal rate of return; i.e., $Y(t)$ is defined by the expression

$$\mathcal{B}(t) = \sum_{i=1}^{T-t} C/Y(t)^i + L/Y(t)^{T-t} \text{ with } Y(t) > 0. \tag{3.1}$$

As a convention in this book, for simplicity of notation and exposition, *all rates will be denoted as one plus a percentage* (these are sometimes called dollar returns). This convention greatly simplifies all of the subsequent formulas. This is why the denominator in expression (3.1) is given by $Y(t) > 0$ and not by $(1 + y(t))$ with $y(t) > -1$.

The rate $Y(t)$ is the internal rate of return on the bond because it is that rate which equates the cost of the bond $(\mathcal{B}(t))$ to the present value of all its future cash flows (the right side of expression (3.1)). To see this, note that the term inside the summation on the right side of expression (3.1), $(C/Y(t)^i)$, represents the present value of a coupon payment of C dollars received at time $t + i$. All such coupon payments are included in the summation. The last term $(L/Y(t)^{T-t})$ is the present value of the principal repayment at time T.

Example: Computing a Bond's Yield

Let the time step be years. Consider a coupon bond with maturity $T = 10$ years, paying a coupon of $C = 5$ dollars per year on a face value of $L = 100$ dollars. At time $t = 0$ the bond is currently trading *at par* or, equivalently, its current price is $\mathcal{B}(t) = 100$ dollars. The yield is determined by solving expression (3.1), i.e.,

$$100 = \sum_{i=1}^{10} 5/Y(0)^i + 100/Y(0)^{10}.$$

This expression must be solved numerically (on a computer). Such a computation shows that the bond's yield is $Y(0) = 1.05$.

This example illustrates a general result. When the bond trades at par, it will always be true that the yield equals one plus the coupon rate:

$$Y(0) = 1 + (C/L).$$

Suppose instead that the bond's price $\mathcal{B}(t) = 90$ dollars. In this case the bond's yield is determined by solving the modified equation:

$$90 = \sum_{i=1}^{10} 5/Y(0)^i + 100/Y(0)^{10}.$$

The solution (determined on a computer) is $Y(0) = 1.0638$. We see that the bond's yield increases as the bond's price declines, everything else constant.

Finally, it is instructive to compute the yield on a zero-coupon bond. Let the zero-coupon bond have a maturity of $T = 5$ years and a face value of $L = 1$ dollar. Let the bond's price be $\mathcal{B}(t) = .85$ dollars. The zero-coupon bond's yield solves the simplified expression,

$$.85 = 1/Y(0)^5.$$

Equivalently, $Y(0) = (1/.85)^{(1/5)}$.

This expression also needs to be solved on a computer. The solution is $Y(t) = 1.0330$.

This completes the example. ☐

Bond markets usually trade bonds by quoting yields. This is because a bond's yield is a "standardized" price, roughly comparable across different bonds with different coupon rates and maturities. Unfortunately, this standardization is not exact. Comparing bonds based solely on yields has serious problems. This is because the yield is not a good measure of the bond's expected yearly return. Implicit in this use of the yield as a measure of return is an unrealistic reinvestment rate assumption. A complete discussion of these problems is outside the scope of this text and is left to independent reading. We point out in passing, however, that the problems that occur with using yields to compare different bonds are equivalent to the well-known problems that occur with using the internal rate of return to choose different investment projects (see Brealey and Myers [1]).

Risk management is concerned with understanding and managing the exposure of a bond portfolio's value to the evolution of (changes in) the underlying coupon bond prices through time. The coupon bond's price, as given by expression (3.1), is a function of its yield. So, we can understand the stochastic evolution of a bond portfolio's value by understanding the stochastic evolution of each of the underlying bonds' yields. This is the direction to which we now turn.

Consider the bond given in Figure 3.1. Suppose that the bond's yield shifts from $Y(t)$ to $Y(t) + \Delta$. A yield change of size Δ. Let $\Delta \mathcal{B}(Y) \equiv \mathcal{B}(Y(t)+\Delta) - \mathcal{B}(Y(t))$ represent the change in the value of the coupon bond's price* given a change in the yield of size Δ. Our task is to understand this bond price change in terms of simple expressions involving Δ and Δ raised to a power.

A Taylor series expansion[†] of the coupon bond's price gives us such an expression:

$$\Delta \mathcal{B}(Y) = (\partial \mathcal{B}(Y)/\partial Y)\Delta + (\partial^2 \mathcal{B}(Y)/\partial Y^2)\Delta^2/2 + \cdots \qquad (3.2)$$

The change in the bond's price is seen to equal a polynomial function of the change in the bond's yield Δ. Thus, understanding the change in the bond's price involves understanding the first and second derivatives of the bond's price with respect to the bond's yield. We next study these two derivatives.

A straightforward computation shows that the first derivative is:

$$\partial \mathcal{B}(Y)/\partial Y = -D\mathcal{B}(Y)/Y < 0 \qquad (3.3)$$

where

$$D \equiv \left[\sum_{i=1}^{T-t} iC/Y^i + (T-t)L/Y^{T-t} \right] / \mathcal{B}(Y).$$

Expression (3.3) gives an expression for the first derivative of the bond's price with respect to the bond's yield. First, we see that this expression is negative, that is, as the bond's yield increases, the bond's price declines. This makes sense because as the yield rises, the present value of all the future cash payments to the bond declines. The quantity D in expression (3.3) is called the bond's *duration*. It has two interesting economic interpretations.

---

* The astute reader will notice that we have chosen to keep t fixed and only change the yield. This is done to keep the mathematics simple.

† Given a function $f(x)$ that is infinitely differentiable, the Taylor series expansion of $f(x)$ around (x_0) is $f(x) = f(x_0) + \dfrac{\partial f(x_0)}{\partial x}(x - x_0) + \dfrac{1}{2}\dfrac{\partial^2 f(x_0)}{\partial x^2}(x - x_0)^2 + \cdots$

The first economic interpretation is obtained by performing some simple algebra on expression (3.3). One obtains:

$$D = -\frac{[\partial \mathcal{B}(Y)/\partial Y]Y}{\mathcal{B}(Y)} \approx -\frac{\partial \mathcal{B}(Y)/\mathcal{B}(Y)}{\partial Y/Y}. \tag{3.4}$$

Duration is seen to be minus the percentage change in the bond's price for a given percentage change in the bond's yield. Thus, duration is a measure of the sensitivity of the coupon bond's price to percentage changes in the bond's yield. It is a "risk measure" for the bond. Shortly, we will see how this risk measure can be used to do risk management.

To get the second economic interpretation, we can rewrite the definition of D.

$$D = \sum_{i=1}^{T-t} w_i i \tag{3.5}$$

where $w_i = C/Y^i \mathcal{B}(Y)$ for $i = 1, 2, \ldots, T - t - 1$, $w_{T-t} = (C+L)/Y^{T-t}\mathcal{B}(Y)$,

and $\sum_{i=1}^{T-t} w_i = 1$.

Expression (3.5) shows that duration can also be interpreted as the *average life* of the bond. This is because it represents a weighted average of the time periods where the bond's cash flows are received. The weights (w_i) correspond to the percentage of the bond's price that the ith period cash flow represents. Of course, the weights sum to one.

This interpretation of duration is the one most often used when comparing bond portfolios. For example, a bond portfolio with a 5 percent yield and a duration of 10 (years) is sometimes (incorrectly) viewed as similar, in terms of risk, to a single bond with a yield of 5 percent and a maturity of 10 years. The reason why this comparison is incorrect is explained later in the text.

Example: Computing a Zero-Coupon Bond's Duration

We first consider computing the duration of a zero-coupon bond.

Consider a zero-coupon bond with maturity $T = 5$ years, a face value of $L = 1$ dollar, and a price of $\mathcal{B}(t) = .85$ dollars. Its yield is computed via expression (3.1) and equals $Y(0) = 1.0330$.

Using expression (3.3), the duration is

$$D = \left[(5)1/(1.0330)^5 \right]/.85 = 5(.85)/.85 = 5.$$

This zero-coupon bond's duration is 5 years. This equals the zero-coupon bond's time to maturity!

This completes the example. □

The fact that a zero-coupon bond's duration is its time to maturity is a general result. Indeed, consider a zero-coupon bond with maturity T, face value L, and price $\mathcal{B}(t)$. Its yield is determined from expression (3.1) as $Y(t)$. Substituting these symbols into the expression for the bond's duration gives

$$D = \left[(T-t)L/Y(t)^{T-t} \right]/\mathcal{B}(Y).$$

But, the definition of the bond's yield is

$$\mathcal{B}(t) = L/Y(t)^{T-t}.$$

Substitution and simplification give the desired result:

$$D = (T-t).$$

Example: Computing a Coupon Bond's Duration

We now compute a coupon bond's duration.

Consider a coupon bond with maturity $T=10$ years, paying a coupon of $C=5$ dollars per year on a face value of $L=100$ dollars. Suppose that the bond is currently trading at par so that its current price is $\mathcal{B}(t)=100$ dollars. We know that the bond's yield is $Y(0)=1.05$. Substituting this information into expression (3.3) gives:

$$D = \left[\sum_{i=1}^{10} i5/(1.05)^i + (10)100/(1.05)^{10} \right]/100 = 8.1078.$$

The bond's duration is 8.1078 years, or slightly more than three-fourths of its remaining time to maturity. This completes the example. □

Expression (3.3) is often written alternatively as

$$\partial \mathcal{B}(Y)/\partial Y = -D_M \mathcal{B}(Y) \qquad (3.6)$$

where $D_M = D/Y \approx -[\partial \mathcal{B}(Y)/\partial Y]/\mathcal{B}(Y)$.

The quantity D_M in expression (3.6) is called *modified duration*. It has the interpretation of being the percentage change in the bond's price for a unit change in the yield. Its usage is justified by the simpler form that the first derivative takes in expression (3.6) versus expression (3.3).

Example: Computing a Bond's Modified Duration

First, we compute the modified duration for a zero-coupon bond. Consider a zero-coupon bond with maturity T, face value L, and price $\mathcal{B}(t)$. We know that the zero-coupon bond's duration is its time to maturity, or $D=(T-t)$. Hence, a zero-coupon bond's modified duration is

$$D_M = D/Y = (T-t)/Y.$$

For a specific example, let the zero-coupon bond have maturity $T=5$ years, a face value of $L=1$ dollar, and a price of $\mathcal{B}(t) = .85$ dollars. Its yield is computed via expression (3.1) and equals $Y(0) = 1.0330$. Substitution into the above expression gives

$$D_M = 5/(1.0330) = 4.8401.$$

Next, we compute modified duration for a coupon-bearing bond. Consider the coupon bond in the previous example for the computation of duration. The coupon bond has maturity $T=10$ years and pays a coupon of $C=5$ dollars per year on a face value of $L=100$ dollars. At time $t=0$ the bond is currently trading at par, so its current price is $\mathcal{B}(t) = 100$ dollars. We know, in this case, that the bond's yield is $Y(0) = 1.05$. The duration was computed to be $D=8.1078$. Then,

$$D_M = D/Y = 8.1078/1.05 = 7.7217.$$

This completes the example. ☐

Finally, we next compute the second derivative of the bond's price with respect to its yield. This computation is given in expression (3.7).

$$\partial^2 \mathcal{B}(Y)/\partial Y^2 = C_v \mathcal{B}(Y) > 0 \tag{3.7}$$

where $C_v \equiv \left[\sum_{i=1}^{T-t} i(i+1)C/Y^{i+2} + (T-t)(T-t+1)L/Y^{T-t+2} \right]/\mathcal{B}(Y).$

The second derivative of the bond's price with respect to the bond's yield is always positive. It is written in expression (3.7) as a product of C_v times the bond's price. The quantity C_v is called the bond's *convexity*. It represents a normalized second derivative, where the normalization is the bond's price. This normalization allows a comparison of convexities across different bonds.

Example: Computation of a Bond's Convexity

We now compute a coupon-bearing bond's convexity. Consider a coupon bond with maturity $T = 10$ years, paying a coupon of $C = 5$ dollars per year on a face value of $L = 100$ dollars. Suppose that the bond is currently trading at par so $\mathcal{B}(t) = 100$ dollars. The bond's yield is $Y(0) = 1.05$. Substituting into expression (3.7) gives

$$C_v = \left[\sum_{i=1}^{10} i(i+1)5/(1.05)^{i+2} + (10)(11)100/(1.05)^{12} \right]/100 = 74.9977.$$

This completes the example. □

Using these expressions, we can now give a simpler form for the change in the bond's price given a change in the bond's yield. Substitution of expressions (3.6) and (3.7) into (3.2) gives

$$\Delta\mathcal{B}(Y) = -D_M \mathcal{B}(Y)\Delta + C_v \mathcal{B}(Y)\Delta^2/2 + \cdots. \tag{3.8}$$

That is, the change in the bond's price due to a change in the bond's yield is seen to be composed of two parts. The first component is a negative change that is proportional to the bond's modified duration. The second component is a positive change, related to the bond's convexity. The next section of this chapter shows how these quantities are useful for managing the risk in holding a portfolio of coupon bonds.

3.4 RISK MANAGEMENT

This section of the chapter discusses the use of modified duration and convexity for hedging the interest rate risk in a portfolio of coupon bonds.

Consider a portfolio consisting of two coupon bonds, one denoted with a subscript a and the other with a subscript b. Let the value of the portfolio be denoted

$$V(Y_a, Y_b) = n_a \mathcal{B}_a(Y_a) + n_b \mathcal{B}_b(Y_b) = \mathcal{B}_a(Y_a) + n_b \mathcal{B}_b(Y_b) \qquad (3.9)$$

where $n_a = 1$ is the number of shares of bond a and n_b is the number of shares of bond b.

For simplicity, we have set the holdings in bond a equal to 1 unit. The idea is that we have just purchased bond a. It was undervalued, and we want to lock in the mispricing profit. To do this, we want to hedge its interest rate risk. Useful in this regard will be expression (3.8).

The portfolio's value changes when yields change. Given a shift in the yields of bond a by Δ_a and bond b by Δ_b, the change in the value of the portfolio is

$$\Delta V(Y_a, Y_b) = \Delta \mathcal{B}_a(Y_a) + n_b \Delta \mathcal{B}_b(Y_b). \qquad (3.10)$$

Note that the yield on bond a, Y_a, can differ from the yield on bond b, Y_b. Similarly, the changes in the yields on both bonds can also differ; i.e., $\Delta_a \neq \Delta_b$ is possible.

Substitution of expression (3.8) into (3.10) yields an equivalent expression in terms of modified duration and convexity:

$$\Delta V(Y_a, Y_b) = -D_{M,a} \mathcal{B}_a(Y_a) \Delta_a - n_b D_{M,b} \mathcal{B}_b(Y_b) \Delta_b + C_{v,a} \mathcal{B}_a(Y_a) \Delta_a^2 / 2$$
$$+ n_b C_{v,b} \mathcal{B}_b(Y_b) \Delta_b^2 / 2 + \cdots \qquad (3.11)$$

For small changes in yields, i.e., $\Delta_a \approx 0$ and $\Delta_b \approx 0$ (e.g., .0001 or 1 basis point), then the squared changes of yields are orders of magnitude smaller than the change itself, i.e., $\Delta_a^2 \ll \Delta_a$ and $\Delta_b^2 \ll \Delta_b$ (e.g., (.0001) (.0001) = .00000001 ≪ .0001). In this case, the terms in expression (3.11) involving convexity are small and can be ignored. Then,

$$\Delta V(Y_a, Y_b) \approx -D_{M,a}\mathcal{B}_a(Y_a)\Delta_a - n_b D_{M,b}\mathcal{B}_b(Y_b)\Delta_b. \qquad (3.12)$$

Expression (3.12) gives us a very simple expression for the change in the value of a bond portfolio when yields change.

A *hedged portfolio* is one where $\Delta V(Y_a, Y_b) = 0$, i.e., the value of the portfolio is invariant to changes in the underlying bond's yields. This is because in a hedged portfolio, bond a's price changes are exactly offset by changes in bond b's price. Such a portfolio is said to be riskless, because its value does not change when the underlying term structure changes.* We next study the choices of n_b necessary to obtain such a hedged portfolio.

To obtain a hedged portfolio, imagine for the moment that $\Delta_a = \Delta_b = \Delta$. Then, we can write

$$\Delta V(Y_a, Y_b) \approx -[D_{M,a}\mathcal{B}_a(Y_a) + n_b D_{M,b}\mathcal{B}_b(Y_b)]\Delta.$$

This will be zero if the term preceding Δ is zero, i.e., $D_{M,a}\mathcal{B}_a(Y_a) + n_b D_{M,b}\mathcal{B}_b(Y_b) = 0$. Solving for n_b gives $n_b = -D_{M,a}\mathcal{B}_a(Y_a)/D_{M,b}\mathcal{B}_b(Y_b)$. This simple calculation motivates the following definition.

The portfolio is said to be (*modified*) *duration hedged* if

$$n_b = -D_{M,a}\mathcal{B}_a(Y_a)/D_{M,b}\mathcal{B}_b(Y_b). \qquad (3.13)$$

For a (modified) duration hedged portfolio, substitution of (3.13) into expression (3.12) gives that

$$\Delta V(Y_a, Y_b) \approx -D_{M,a}\mathcal{B}_a(Y_a)[\Delta_a - \Delta_b].$$

Thus, we see that for a (modified) duration hedged portfolio,

$$\Delta V(Y_a, Y_b) \approx 0 \text{ if and only if } \Delta_a = \Delta_b. \qquad (3.14)$$

This says that a bond portfolio is hedged against random changes in yields using modified duration hedging if and only if changes in bond yields are identical for all bonds included in the portfolio.

---

* This assumes that the change in time is zero. If time changes as well, from t to $t + \delta$, then a riskless portfolio of positive investment will earn the riskless rate over this time period, i.e., $\Delta V \approx rV\delta$ where r is the riskless rate. We will show later that r is called the spot rate of interest.

Extending this result to arbitrary bond portfolios, duration hedged bond portfolios will be hedged against random changes in yields if and only if changes in bond yields are identical for all bonds. However, because zero-coupon bonds are possible for inclusion in such portfolios, modified duration hedging provides a hedged portfolio only if there is a parallel shift in the zero-coupon bond's yield curve.

Unfortunately, parallel shifts in the zero-coupon bond's yield curve are unlikely events. Glancing back at Figure 1.2 in Chapter 1, we see that yield curve changes are almost never parallel. In fact, historically yield curves have shifted through time in quite a complex fashion. This empirical observation indicates that modified duration hedging does not provide a perfect hedge. Hence, the duration-hedged portfolio is not expected to have a zero change in value. This is a significant limitation of using the classical duration hedging approach for fixed income risk management.

A better procedure for hedging the bond portfolio is to choose the number of shares of bond b, n_b, such that the portfolio is hedged for the actual shift in the yield curve (not just parallel shifts). But, yield curve shifts are random. This implies that we should use a model that hedges the evolution for yield changes that are expected to occur based on historical experience. This is the motivation behind the HJM model presented in subsequent chapters.

Example: Modified Duration Hedging

This example illustrates a modified duration hedge.

Suppose our portfolio consists of a single coupon-bearing bond with maturity $T_a = 10$ years, paying a coupon of $C_a - 5$ dollars per year on a face value of $L_a = 100$ dollars. The bond is currently trading at par so that its current price is $\mathcal{B}_a(0) = 100$ dollars.

We want to apply a modified duration hedge using another coupon-bearing bond with maturity $T_b = 5$ years, paying a coupon of $C_b = 6$ dollars per year on a face value of $L_b = 100$ dollars. Let the bond's price be trading above par at $\mathcal{B}_b(0) = 103$ dollars.

Computing the bond yields and modified durations gives:

$$Y_a(0) = 1.05, \quad D_{M,a} = 7.7217, \quad Y_b(0) = 1.0530, \quad D_{M,b} = 4.2487.$$

The modified duration neutral hedge in bond b is:

$$n_b = -(7.7217)100/(4.2487)103 = -1.7645.$$

That is, we need to short 1.7645 units of the second coupon bond. This completes the example. □

In addition, there is an additional limitation of classical duration hedging. As mentioned previously, even under parallel shifts, modified duration hedging is only valid given small changes in yields. Remember that expression (3.12) was only an approximation because we omitted the terms involving convexity. For small changes in yields, $\Delta_a^2 \ll \Delta_a$ and $\Delta_b^2 \ll \Delta_b$. But, for large changes in yields, $\Delta_a^2 \approx \Delta_a$ and $\Delta_b^2 \approx \Delta_b$ (e.g., if $\Delta_a = 1$, i.e., 100 basis points, then $\Delta_a^2 = 1$ so $\Delta_a = \Delta_a^2$). In this case, the second-order terms are as important as the first-order terms.

For this reason, convexity hedging can also be performed. One can hedge a bond portfolio with respect to both modified duration and convexity. We now discuss convexity hedging. Here, at least three bonds are needed in the portfolio. The second bond is needed to hedge duration, and the third bond is needed to hedge convexity.

Consider three coupon bonds with indexes a, b, and c. The portfolio value is

$$V(Y_a, Y_b, Y_c) = \mathcal{B}_a(Y_a) + n_b \mathcal{B}_b(Y_b) + n_c \mathcal{B}_c(Y_c). \tag{3.15}$$

Given a change in yields, the change in the portfolio's value is

$$\Delta V(Y_a, Y_b, Y_c) = -D_{M,a}\mathcal{B}_a(Y_a)\Delta_a - n_b D_{M,b}\mathcal{B}_b(Y_b)\Delta_b - n_c D_{M,c}\mathcal{B}_c(Y_b)\Delta_c$$

$$+ C_{v,a}\mathcal{B}_a(Y_a)\Delta_a^2/2 + n_b C_{v,b}\mathcal{B}_b(Y_b)\Delta_b^2/2 \tag{3.16}$$

$$+ n_c C_{v,c}\mathcal{B}_c(Y_c)\Delta_c^2/2 + \cdots.$$

To make this portfolio delta and convexity hedged, choose n_b and n_c such that the following terms involving modified duration and convexity are zero, i.e.,

$$D_{M,a}\mathcal{B}_a(Y_a) + n_b D_{M,b}\mathcal{B}_b(Y_b) + n_c D_{M,c}\mathcal{B}_c(Y_c) = 0 \text{ (modified duration hedged)}$$

$$C_{v,a}\mathcal{B}_a(Y_a) + n_b C_{v,b}\mathcal{B}_b(Y_b) + n_c C_{v,c}\mathcal{B}_c(Y_c) = 0 \text{ (convexity hedged)}.$$

$$\tag{3.17}$$

As before, this system gives a hedged portfolio if and only if there is a parallel shift in yields, i.e., $\Delta_a = \Delta_b = \Delta_c$. Standard linear algebra can be used to solve this system of two equations in two unknowns.

Example: Duration and Convexity Hedging

This example illustrates a modified duration and convexity hedge. We continue with the previous example.

Suppose our portfolio consists of a single coupon-bearing bond with maturity $T_a = 10$ years, paying a coupon of $C_a = 5$ dollars per year on a face value of $L_a = 100$ dollars. The bond is currently trading at par so that its current price is $\mathcal{B}_a(0) = 100$ dollars.

We want to apply a modified duration and convexity hedge using two additional bonds.

The first is a coupon-bearing bond with maturity $T_b = 5$ years, paying a coupon of $C_b = 6$ dollars per year on a face value of $L_b = 100$ dollars. Let the bond's price be trading above par at $\mathcal{B}_b(0) = 103$ dollars. The second is a coupon-bearing bond with maturity $T_c = 15$ years, paying a coupon of $C_c = 4$ dollars per year on a face value of $L_c = 100$ dollars. Let the bond's price be $\mathcal{B}_c(0) = 94$ dollars.

Computing the bond yields and modified durations gives:

$$Y_a(0) = 1.05, \quad D_{M,a} = 7.7217, \quad Y_b(0) = 1.0530, \quad D_{M,b} = 4.2487,$$

$$Y_c(0) = 1.0456, \quad D_{M,c} = 10.9393.$$

The convexities are: $C_{v,a} = 74.9977$, $C_{v,b} = 23.2842$, $C_{v,c} = 151.6231$. The system of equations (3.17) becomes

$$(7.7217)100 + n_b(4.2487)103 + n_c(10.9393)94 = 0$$

$$(74.9977)100 + n_b(23.2842)103 + n_c(151.6231)94 = 0.$$

The solution is: $n_b = -.8734 \quad n_c = -.3792$.

Both additional bonds need to be shorted to create the appropriate hedge.

This completes the example. $\quad\square$

In summary, there are two problems with classical modified duration hedging: first, it is valid only for small shifts in the yield curve, and second, it is valid only for parallel shifts in the yield curve. The convexity adjustment to the modified duration hedge is useful only for reducing the first of these biases, i.e., that due to small changes in yields. It does not remove the nonparallel shift bias. Historical term structure evolutions are inconsistent with parallel shifts in the yield curve, making modified duration hedging imprecise. For this reason, we do not recommend the use of the classical approach to risk management. Instead, we recommend the procedure discussed in the remainder of the book.

REFERENCE

1. Brealey, R., and S. Myers, 1999. *Principals of Corporate Finance*, 6th edition, McGraw-Hill Book Company, New York.

II

Theory

CHAPTER 4

The Term Structure
of Interest Rates

THIS CHAPTER PRESENTS THE preliminaries of the model. As is true of all models, the model selected for presentation is an abstraction from reality. It is an abstraction because it is a simplification. It is simplified in order to facilitate understanding and analysis. Of course, the hope is that the simplified model is still a good approximation to the actual economy. The approximation needs to be good enough to provide both *(i)* accurate valuation of interest rate options and *(ii)* accurate synthetic construction of the cash flows to interest rate options using the traded Treasury or Eurodollar securities. How accurate the valuation and the synthetic construction need to be is determined in each application. Our experience, however, is that this model and its extensions are accurate enough to have proven quite useful in the actual trading and risk management of fixed income securities and interest rate options.

To maintain a sense of realism, all of the examples presented in this chapter (and in the remainder of the book) are based on term structure parameters, calibrated to actual market rates, at the writing of the first edition. The discrete time period selected corresponds to 6 months. It is useful to keep this statement in mind as one reads the remainder of the book. Although simple in appearance, the examples provided are based on data obtained from real markets. Generalized slightly, at one time, these examples were used to price traded interest rate derivatives.

4.1 THE ECONOMY

We consider a *frictionless, competitive,* and *discrete trading* economy. By frictionless we mean that there are no transaction costs in buying and selling financial securities, there are no bid/ask spreads, there are no restrictions on trade (legal or otherwise) such as margin requirements or short sale restrictions, and there are no taxes.

The frictionless markets assumption can be justified on two grounds. First, very large institutional traders approximate frictionless markets since their transaction costs are minimal. If these traders determine prices, then this model may approximate actual pricing and hedging well. The second argument is that understanding frictionless markets is a necessary prelude to understanding friction-filled markets. Only by understanding the ideal case can we hope to understand the more complicated friction-filled economy. Both arguments have merit, and either provides us with sufficient motivation to continue with the analysis.

The markets are assumed to be *competitive*; i.e., each trader believes that she can buy/sell as many shares of a traded security as she desires without influencing its price. This implies, of course, that the market for any financial security is perfectly (infinitely) liquid. This is an idealization of actual security markets. It is more nearly satisfied by large volume trading on organized exchanges than it is in the over-the-counter markets. Nonetheless, it is a reasonable starting hypothesis, from which we will proceed. The modification of the subsequent theory for a relaxation of this competitive market assumption is a fruitful area of research. (For some work along these lines, see Back [1], Gastineau and Jarrow [2], Jarrow [3, 4].)

Last, we consider a *discrete trading* economy with trading dates {0, 1, 2, ... , τ}. This assumption is not very restrictive because it is a reasonable approximation to actual security markets, especially if τ is large and the time interval between trading periods is small. The alternative, continuous trading, provides similar results but with significantly more complicated mathematics.

4.2 THE TRADED SECURITIES

Traded in this economy are zero-coupon bonds of all maturities {0, ... , τ} and a *money market account*. The price of a zero-coupon bond at time t that pays a sure dollar at time $T \geq t$ is denoted by $P(t,T)$. All zero-coupon bonds are assumed to be default-free and have strictly positive prices.

Table 4.1 provides three different hypothetical zero-coupon bond price curves. Panel A gives these prices for a flat term structure, panel B is for a downward sloping term structure, and panel C is for an upward sloping term structure. The numbers are constructed so that each time period corresponds to half a year.

The money market account is constructed to always represent an investment portfolio in the shortest maturity zero-coupon bond. It is initialized at time 0 with a dollar investment, and its time t value is denoted by $B(t)$. Thus, by convention, $B(0) = 1$. To construct this account, at time 0 this dollar is invested in the zero-coupon bond that matures at time 1 (i.e., $1/P(0,1)$ units are purchased). This position is held until time 1. At time 1, the initial dollar plus interest earned is reinvested into the zero-coupon bond that matures at time 2. This is the shortest maturity zero-coupon

TABLE 4.1   Hypothetical Zero-Coupon Bond Prices, Forward Rates, and Yields

	Time to Maturity (T)	Zero-Coupon Bond Prices $P(O,T)$	Forward Rates $f(O,T)$	Yields $y(O,T)$
Panel A: Flat Term	0	1	1.02	1.02
Structure	1	.980392	1.02	1.02
	2	.961168	1.02	1.02
	3	.942322	1.02	1.02
	4	.923845	1.02	1.02
	5	.905730	1.02	1.02
	6	.887971	1.02	1.02
	7	.870560	1.02	1.02
	8	.853490	1.02	1.02
	9	.836755		
Panel B:	0	1	1.024431	1.024431
Downward	1	.976151	1.023342	1.023886
Sloping Term	2	.953885	1.022701	1.023491
Structure	3	.932711	1.022319	1.023198
	4	.912347	1.022025	1.022963
	5	.892686	1.021794	1.022768
	6	.873645	1.021627	1.022605
	7	.855150	1.021544	1.022472
	8	.837115	1.020748	1.022281
	9	.820099		
Panel C: Upward	0	1	1.016027	1.016027
Sloping Term	1	.984225	1.016939	1.016483
Structure	2	.967831	1.017498	1.016821
	3	.951187	1.017836	1.017075
	4	.934518	1.018102	1.017280
	5	.917901	1.018312	1.017452
	6	.901395	1.018465	1.017597
	7	.885052	1.018542	1.017715
	8	.868939	1.019267	1.017887
	9	.852514		

bond at time 1. So, interest is earned on interest in this fund. This process continues at times 2, 3, ... The value of the money market account at any time is just the accumulated value of this fund. More will be said about the money market account after the introduction of various interest rates.

4.3 INTEREST RATES

Markets tend to quote bond prices using interest rates. This is because interest rates summarize very complex cash flow patterns into a single number that is (roughly) comparable across different financial securities. In this text, various different interest rates will play a significant role in the analysis. This section defines the most important of these interest rates: yields, forward rates, and spot rates. As a convention in this book, for simplicity of notation and exposition, *all rates will be denoted as one plus a percentage* (these are sometimes called dollar returns), implying that all rates will be strictly positive. This convention greatly simplifies all the subsequent formulas. It has already been employed in Chapter 3.

The *yield* at time t on a T-maturity zero-coupon bond, denoted by $y(t,T)$, is defined by expression (4.1):

$$y(t,T) \equiv \left[\frac{1}{P(t,T)} \right]^{1/(T-t)}.$$ (4.1)

The yield is one plus the percentage return earned per period by holding the T-maturity bond from time t until its maturity and, therefore, $y(t,T) > 0$.* The yield is often called the *holding period return*. Alternatively written, expression (4.1) implies

$$P(t,T) = \frac{1}{[y(t,T)]^{(T-t)}}.$$ (4.2)

In this expression, we see that the yield is also the internal rate of return on the zero-coupon bond. It is that single rate compounded for $(T-t)$ periods that discounts the dollar payoff at time T to the present value of the zero-coupon bond, its price.

The time t *forward rate* for the period $[T, T+1]$, denoted by $f(t,T) > 0$, is defined by

---

* In more standard notation, one would write $1 + \hat{y}(t,T) = \left[1/P(t,T) \right]^{1/(T-t)}$. The difference between these two expressions is slight. Here $\hat{y}(t,T) > -1$.

$$f(t,T) \equiv \frac{P(t,T)}{P(t,T+1)}. \tag{4.3}$$

The forward rate can be understood from two perspectives. First, looking at expression (4.3), the only difference between $P(t,T)$ and $P(t,T+1)$ is that $P(t,T+1)$ earns interest for one more time period, the period $[T,T+1]$. Taking the ratio as in expression (4.3) isolates the implicit rate earned on the longer maturity bond over this last time period. This implicit rate is the forward rate.

The second interpretation of the forward rate is that it corresponds to the rate that one can contract at time t for a riskless loan over the time period $[T,T+1]$.

To see this interpretation, we construct a portfolio of zero-coupon bonds that creates a dollar loan over $[T,T+1]$ that is riskless. The interest rate on this loan is shown to be the forward rate of interest. With this goal in mind, consider forming the following portfolio at time t: *(i)* buy one zero-coupon bond maturing at time T and *(ii)* sell the quantity $[P(t,T)/P(t,T+1)]$ of zero-coupon bonds maturing at time $T+1$. Hold each zero-coupon bond until maturity. The cash flows to this portfolio are given in Table 4.2. The initial cash flow from forming this portfolio at time t is zero. Indeed, the cash outflow is $P(t,T)$ dollars, but the cash inflow is $[P(t,T)P(t,T+1)]P(t,T+1) = P(t,T)$ dollars. These net to zero. The first cash flow of one dollar occurs at time T. It is an inflow. In addition, there is a cash outflow of $P(t,T)/P(t,T+1)$ dollars at time $T+1$.

This pattern of cash flows is the same as that obtained from a dollar loan over $[T,T+1]$, contracted at time t. The implicit rate on this loan is $P(t,T)/P(t,T+1)$, the forward rate. This completes the argument.

From expression (4.3) we can derive an expression for the bond's price in terms of the various maturity forward rates:

TABLE 4.2   Cash Flows to a Portfolio Generating a Borrowing Rate Equal to the Time t Forward Rate for Date $T, f(t,T)$

Time	T	T	$T+1$
Buy bond with maturity T	$-P(t,T)$	$+1$	
Sell $[P(t,T)/P(t,T+1)]$ bonds with maturity $T+1$	$+\dfrac{P(t,T)}{P(t,T+1)}P(t,T+1)$		$-\dfrac{P(t,T)}{P(t,T+1)}$
Total Cash Flow	0	$+1$	$-\dfrac{P(t,T)}{P(t,T+1)}$

$$P(t,T) = \frac{1}{\displaystyle\prod_{j=t}^{T-1} f(t,j)}.$$

(4.4)

Derivation of Expression (4.4)

Step 1.

$$f(t,t) = P(t,t)/P(t,t+1)$$

$$= 1/P(t,t+1)$$

Since $P(t,t) = 1$. So,

$$P(t,t+1) = 1/f(t,t).$$

Step 2. Next,

$$f(t,t+1) = P(t,t+1)/P(t,t+2).$$

So,

$$P(t,t+2) = P(t,t+1)/f(t,t+1).$$

Substitution yields

$$P(t,t+2) = 1/f(t,t)f(t,t+1).$$

Continuing, we get

$$P(t,t+j) = 1/f(t,t)f(t,t+1)f(t,t+2)\ldots f(t,t+j-1).$$

This completes the proof. □

Expression (4.4) shows that the bond's price is equal to a dollar received at time T and discounted to the present by the different maturity forward rates.*

---

* The symbol $\prod_{j=t}^{T-1} f(t,j) = f(t,t)f(t,t+1)\cdots f(t,T-1)$ means the result obtained from multiplying together the terms arising when the index j runs from t to $T-1$.

Example: Computing Forward Rates and
Bond Prices Using Table 4.1

In Table 4.1, panel A gives the forward rates for a flat term structure, panel B for a downward sloping term structure, and panel C for an upward sloping term structure. The shape of the term structure is defined by the slope of the graph of the forward rate versus time to maturity. To illustrate the use of the definitions, from panel A we have, in symbols,

$$f(0,3) = P(0,3)/P(0,4)$$

or, in numbers,

$$1.02 = 0.942322/0.923845.$$

For this example the forward rate is $f(0,3) = 1.02$, and it is the rate one can contract at time 0 for borrowing starting at time 3 and ending at time 4.

Conversely, given the forward rates, we can determine the bond prices. Using expression (4.4), the calculation is

$$P(0,3) = 1/f(0,0)f(0,1)f(0,2)$$

or, in numbers,

$$0.942322 = 1/(1.02)(1.02)(1.02).$$

This shows that the 3-period zero-coupon bond's price is equal to a discounted dollar at 2 percent for three periods. This completes the example. □

Last, the *spot rate*, denoted by $r(t)$, is defined as the rate contracted at time t on a 1-period riskless loan starting immediately. By definition, therefore,

$$r(t) \equiv f(t,t). \tag{4.5}$$

From Table 4.1 we see that the spot rate $r(0) = f(0,0)$ is 1.02 for the flat term structure, 1.024431 for the downward sloping term structure, and 1.016027 for the upward sloping term structure.

Alternatively, using expressions (4.1) and (4.3), the spot rate is seen to be the holding period return on the shortest maturity bond, i.e.,

$$r(t) = y(t, t+1). \tag{4.6}$$

We can now clarify the return earned on the money market account. By construction,

$$B(t) = B(t-1)r(t-1) = \prod_{j=0}^{t-1} r(j). \tag{4.7}$$

The money market account invests its entire fund in the shortest maturity zero-coupon bond each period, thereby earning the spot rate over each subsequent period. From Table 4.1 we see that for the flat term structure in panel A, $B(1) = B(0)r(0) = 1.02$. The money market account earns 2 percent over the first interval. The data in Table 4.1 does not enable us to determine the money market account's value at any future date beyond time 1 because the future spot rates are not available.

4.4 FORWARD PRICES

This section introduces forward contracts and forward prices. Forward contracts are one of the oldest forms of financial contracts in existence, having been used over 4,000 years ago in the Roman Empire and Greece.

A *forward contract* is a financial security obligating the purchaser to buy a commodity at a prespecified price (determined at the time the contract is written) and at a prespecified date. At the time the contract is initiated, by convention, no cash changes hands. The contract has zero value. The prespecified purchase price is called the *forward price*. The prespecified date is called the *delivery* or *expiration* date. Forward contracts are written on many different commodities. In this book, we are only interested in studying forward contracts on zero-coupon bonds. Nonetheless, the mathematics for the other commodities is almost identical to that presented below.

There are three dates of importance for forward contracts on zero-coupon bonds: the date the contract is written (t), the date the zero-coupon bond is purchased or delivered (T_1), and the maturity date of the zero-coupon bond (T_2). The dates must necessarily line up as $t \le T_1 \le T_2$. We denote the time t forward price of a contract with expiration date T_1 on the T_2-maturity zero-coupon bond as $F(t, T_1; T_2)$.

By definition, the forward price on a contract with immediate delivery is the *spot price* of the zero-coupon bond, i.e.,

$$F(T_1, T_1 : T_2) = P(T_1, T_2). \qquad (4.8)$$

The payoff to the forward contract on the delivery date is

$$P(T_1, T_2) - F(t, T_1 : T_2). \qquad (4.9)$$

This is the value of the T_2-maturity zero-coupon bond on the delivery date less the agreed-upon forward price. Expression (4.9) is called the *boundary condition* or *payoff* to the forward contract.

For subsequent usage, we illustrate the cash flow to the forward contract at expiration using a *payoff diagram*. A payoff diagram is a graph of the payoff to a financial instrument, as a function of its underlying commodity's price at the delivery (or expiration) date. The payoff diagram for the forward contract is shown in Figure 4.1. On the x-axis is the price of the underlying commodity, the zero-coupon bond. On the y-axis is the payoff to the forward contract. As shown, the payoffs are positive when the zero-coupon bond's price is greater than the forward price, negative otherwise. This payoff is linear in the underlying zero-coupon bond's price. Furthermore, it has unlimited gains and unlimited loss potential.

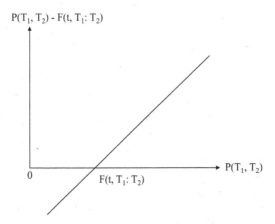

FIGURE 4.1   Payoff Diagram for a Forward Contract with Delivery Date T_1 on a T_2-Maturity Zero-Coupon Bond.

4.5 FUTURES PRICES

Futures contracts are a more recent development than are forward contracts, having been introduced to financial markets sometime during the late 19th and early 20th centuries. Futures contracts can be thought of as standardized "forward" contracts, constructed to trade on organized exchanges with a minimum of counter party risk. To minimize counter party risk, futures contracts have daily cash flows, unlike forward contracts. The differences between futures contracts and forward contracts are now detailed.

A futures contract,* like a forward contract, is an agreement to purchase a commodity at a prespecified date, called the *delivery* or *expiration* date, and for a given price, called the *futures price*. The futures price is paid via a sequence of random and unequal installments over the contract's life. At the time the contract is initiated, by convention, no cash changes hands. The contract has zero value at initiation. A cash payment, however, is made at the end of each trading interval, and it is equal to the change in the futures price over that interval. This cash payment resets the value of the futures contract to zero and is called *marking-to-market*.

To see how this process works, let us denote the time t futures price on a contract with delivery date T_1 on a T_2-maturity zero-coupon bond as $\mathcal{F}(t, T_1 : T_2)$. As before, these dates must line up as $t \leq T_1 \leq T_2$. By definition, the futures price for a contract with immediate delivery of the T_2-maturity zero-coupon bond is the bond's price, i.e.,

$$\mathcal{F}(T_1, T_1 : T_2) = P(T_1, T_2). \tag{4.10}$$

The cash flow to the futures contract at time $t+1$ is the change in the value of the futures contract over the preceding period $[t, t+1]$, i.e.,

$$\mathcal{F}(t+1, T_1 : T_2) - \mathcal{F}(t, T_1 : T_2). \tag{4.11}$$

This payment occurs at the end of every period over the futures contract's life.

Table 4.3 compares the cash flows of a forward contract and a futures contract. Let us first consider the forward contract. The forward contract has a cash flow only at the expiration date T_1. The forward contract's cash flow at delivery is equal to the zero-coupon bond's price less the forward

---

* This section discusses a hypothetical futures contract, devoid of the imbedded options associated with actual market traded futures contracts; see Chapter 2.

TABLE 4.3 Cash Flow Comparison of a Forward and Futures Contract

Time	Forward Contract	Futures Contract
T	0	0
$t+1$	0	$\mathcal{F}(t+1,T_1:T_2)-\mathcal{F}(t,T_1:T_2)$
$t+2$	0	$\mathcal{F}(t+2,T_1:T_2)-\mathcal{F}(t+1,T_1:T_2)$
$\vdots$	$\vdots$	$\vdots$
T_1-1	0	$\mathcal{F}(T_1-1,T_1:T_2)-\mathcal{F}(T_1-2,T_1:T_2)$
T_1	$P(T_1,T_2)-F(t,T_1:T_2)$	$P(T_1,T_2)-\mathcal{F}(T_1-1,T_1:T_2)$
Sum	$P(T_1,T_2)-F(t,T_1:T_2)$	$P(T_1,T_2)-\mathcal{F}(t,T_1:T_2)$

price. Next, consider the futures contract. The futures contract has a cash flow at the end of each intermediate trading date equal to the change in the futures price on that day. This is distinct from the forward contract.

Next, look at the sum of the cash flows to the two contracts. The total (undiscounted) sum of all the cash flows paid to each contract is similar. It is the spot price of the T_2-maturity bond less the forward price (futures price) for the forward contract (futures contract).

The difference between the two contracts, therefore, is solely in the timing of the cash flows. The forward contract receives one cash flow at the end of the contract, and the futures contract receives a cash flow at the end of each trading date. This cash payments to the futures contract are random.

There is a heuristic argument that helps form intuition for the relation between forward and futures prices. Consider a forward contract with delivery date T_1 on the T_2-maturity bond with forward price $F(t,T_1:T_2)$. This contract will be our standard for comparison. It has no cash flow prior to the maturity date. To make the comparison equal, any cash flow to the futures contract must be reinvested (if positive) or borrowed (if negative) so that the futures contract plus investment/borrowing is also zero.

Next, let us decide whether a long position in a forward contract is preferred to a long position in a futures contract with delivery date T_1 on the same T_2-maturity bond. If the forward contract is preferred, then the futures price should be less than the forward price. Why? Because investors would be less willing to enter into a futures contract, and to induce one to do it (with zero cash exchanging hands), the futures price must be less than the forward price, i.e.,

$$\mathcal{F}(t,T_1:T_2) < F(t,T_1:T_2).$$

We claim that the forward contract on a zero-coupon bond is preferred to a futures contract. To see this, suppose the spot rate increases, then:

(i) the zero-coupon bond price falls,

(ii) the current futures price falls,

(iii) the change in the futures price is negative, thus

(iv) the futures contract has a negative cash outflow.

But, to get this cash to cover the futures contract's loss, we need to borrow, and spot rates are high. This is a negative compared to the forward contract that has no cash flow and is implicitly borrowing at the lower interest rate, determined before the spot rate increased.

Next, consider the case where spot rates fall, then:

(i) the zero-coupon bond price rises,

(ii) the current futures price rises,

(iii) the change in the futures price is positive, thus

(iv) the futures contract has a positive cash inflow.

But, after getting this cash profit, we need to invest it and spot rates are low. This is a negative compared to the forward contract that has no cash flow and is implicitly investing at the higher interest rate, determined before the spot rate decreased.

In both cases the forward contract is preferred to the futures contract. This follows because of the negative correlation between futures prices and interest rates assumed in the above argument. The argument is only heuristic because we have not proven the relations: *(i)* implies *(ii)* implies *(iii)* above. The formal proof of these relations requires the full power of the theory presented in this book, and thus awaits a subsequent chapter.

4.6 OPTION CONTRACTS

This section discusses four types of option contracts: calls and puts of either the European or American type. A good reference for this material is Jarrow and Chatterjea [5].

4.6.1 Definitions

This section gives the definitions of the various option contracts.

A *call option* of the *European* type is a financial security that gives its owner the right (but not the obligation) to purchase (call) a commodity at a prespecified price (determined at the time the contract is written) and at a predetermined date. The prespecified price is called the *strike price* or *exercise price*. The predetermined date is called the *maturity date* or *expiration date*.

The difference between a call option of the European type and a forward contract on the same commodity with the same delivery date is that the call option gives the right to purchase, but not the obligation. This means that the owner of the call does not have to purchase the underlying commodity if he or she doesn't want to. In contrast, the forward contract's holder (long position) must purchase the commodity at the forward price at the delivery date. This is a crucial distinction, as we will see below.

A call option of the *American type* is identical to the European call except that it allows the purchase decision to be made at any time from the date the contract is written through the maturity date. This added flexibility makes the American-type call option at least as valuable as its European counterpart.

A *put option* of the *European* type is identical to the European call except that it gives the right to sell (put) the commodity.

A put option of the *American* type is identical to the European put except that it allows the sell decision to be made at any time from the date the contract is written through the maturity date. Again, this added flexibility makes the American type put option at least as valuable as its European counterpart.

Option contracts are written on many commodities. Examples include zero-coupon bonds, coupon bonds, futures contracts, interest rates, and swaps. All of these options will be discussed in this book. But, at the start, to understand these contracts in more detail, we concentrate only on options written on zero-coupon bonds.

4.6.2 Payoff Diagrams

This section studies European call and put options using payoff diagrams. For this analysis, we need some notation. Let the underlying zero-coupon bond have maturity date T_2. Its time t price is denoted by $P(t, T_2)$.

Consider a European call option with strike price K and maturity date $T_1 \le T_2$ written on this zero-coupon bond. Its time t price is denoted by $C(t,T_1,K:T_2)$.

At maturity its payoff or boundary condition is:

$$C(T_1,T_1,K:T_2) = \max[P(T_1,T_2) - K, 0]. \tag{4.12}$$

If at maturity, the zero-coupon bond's price exceeds the strike price K, then the call option is said to be *in-the-money*. In this circumstance, the holder of the call will exercise their right (option) to purchase the underlying bond for K dollars. The value of the call is then the value of the underlying zero-coupon bond less the purchase price, $P(T_1,T_2) - K$. Alternatively, if at maturity the zero-coupon bond's price is less than the strike price, the call option is *out-of-the-money*. In this circumstance, the holder of the call does not exercise the option to purchase, and the option expires worthless.

This payoff is graphed in Figure 4.2. On the x-axis is the price of the underlying zero-coupon bond at maturity. We see that the call has a zero payoff for zero-coupon bond prices below the strike K. At K, the payoff increases dollar for dollar with any increase in the price of the zero-coupon bond. The payoff to a call is *nonlinear*. The loss to the call is bounded below by zero. The gains are unlimited. This is in contrast to the payoffs to the forward contract (see Figure 4.1). The payoffs to a forward contract are unlimited both above and below.

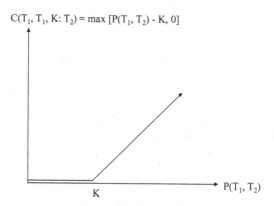

FIGURE 4.2 Payoff Diagram for a European Call Option on the T_2-Maturity Zero-Coupon Bond with Strike K and Expiration Date T_1.

Next, consider a European put option with strike K and maturity date $T_1 \leq T_2$ written on this same zero-coupon bond. Let its time t price be denoted by $\mathcal{P}(t, T_1, K : T_2)$.

At maturity its payoff or boundary condition is:

$$\mathcal{P}(T_1, T_1, K : T_2) = \max[K - P(T_1, T_2), 0]. \tag{4.13}$$

If at maturity, the zero-coupon bond's price is less than the strike price K, the put option is *in-the-money*. In this circumstance, the holder of the put will exercise their right (option) to sell the underlying bond for K dollars. The value of the put is then K dollars less the value of the underlying, $K - P(T_1, T_2)$. Alternatively, if at maturity the zero-coupon bond's price exceeds the strike price, the put option is *out-of-the-money*. In this circumstance, the holder of the put does not exercise the option to sell, and the option expires worthless.

This payoff is graphed in Figure 4.3. On the x-axis is the price of the underlying zero-coupon bond at maturity. We see that the put has positive payoff for zero-coupon bond prices below the strike K. Above K, the payoff on the put is zero. Similar to the call, the payoff to the put is *nonlinear*, and the loss to the put is bounded below by zero. In contrast to the call, the put's gains are bounded above by the strike price K. This simple observation shows that a call option is not equivalent to shorting a put. The relationship, called put-call parity, is more complex, and its explanation awaits a subsequent chapter.

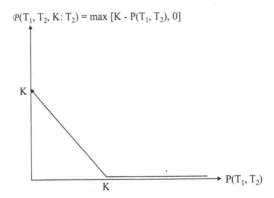

$\mathcal{P}(T_1, T_2, K : T_2) = \max[K - P(T_1, T_2), 0]$

FIGURE 4.3 Payoff Diagram for a European Put Option on the T_2-Maturity Zero-Coupon Bond with Strike K and Expiration Date T_1.

4.7 SUMMARY

The basic assumptions and traded securities introduced in this chapter are used throughout the remainder of the book. The assumptions are that the economy is frictionless and competitive and has only discrete trading.

The traded securities are zero-coupon bonds of all maturities ($P(t,T)$) and a money market account ($B(t)$). From these traded securities, we define the bond's yield ($y(t,T)$), forward rates ($f(t,T)$), and spot rates ($r(t)$). By convention, all these rates are given as one plus a percentage. The relations among these rates are provided.

Last, forward contracts, futures contracts, and options on zero-coupon bonds have been defined in this chapter. The forward price of a contract with delivery date time T_1 on a zero-coupon bond maturing at time T_2 is denoted by $F(t,T_1;T_2)$. The futures price of a contract with delivery date time T_1 on a zero-coupon bond maturing at time T_2 is denoted by $\mathcal{F}(t,T_1;T_2)$. The difference between futures and forward contracts is that futures contracts are marked to market, whereas forward contracts are not.

Call and put options of the European and American type were defined. A European call option gives its holder the right to purchase the underlying zero-coupon bond at a strike price on the maturity date of the option. The European put is the right to sell. American-type options can be exercised any time over their life. It was argued that calls, puts, and forward contracts are quite different, and their relationship awaits the derivation of a subsequent result known as put-call parity.

REFERENCES

1. Back, K., 1993. "Asymmetric Information and Options." *Review of Financial Studies* 6 (3), 435–472.
2. Gastineau, G., and R. Jarrow, 1991. "Large Trader Impact and Market Regulation." *Financial Analysts Journal* 47 (4), 40–51.
3. Jarrow, R., 1992. "Market Manipulation, Bubbles, Corners and Short Squeezes." *Journal of Financial and Quantitative Analysis* 27 (3), 311–336.
4. Jarrow, R., 1994. "Derivative Security Markets, Market Manipulation, and Option Pricing Theory." *Journal of Financial and Quantitative Analysis* 29 (2), 241–261.
5. Jarrow, R., and A. Chatterjea, 2019. *An Introduction to Derivative Securities, Financial Markets, and Risk Management*, 2nd edition, World Scientific Press.

The Evolution of the Term Structure of Interest Rates

5.1 MOTIVATION

Arbitrage pricing theory is often called a *relative pricing theory* because it takes the prices of a primary set of traded assets as given, as well as their stochastic evolution, and then it prices a secondary set of traded assets. It prices a secondary traded asset by constructing a portfolio of the primary assets, dynamically rebalanced across time, such that the portfolio's cash flows and value replicate the cash flows and value of the secondary traded asset. This portfolio is called the "synthetically created traded asset." To prevent riskless profit opportunities, or *arbitrage*, the cost of this replicating portfolio and the price of the secondary traded asset must be identical. Of course, the cost of the replicating portfolio is known, since the prices of the primary traded assets are given. The given stochastic evolution of the primary traded assets determines whether this synthetic replication is possible (whether markets are complete).

For this text, the primary set of traded assets consists of the zero-coupon bonds and the money market account. The secondary set of traded assets is either *(i)* other zero-coupon bonds (Chapter 8–9) or *(ii)* interest rate options (Chapters 11–15). The basic inputs to this relative pricing theory are the prices of the primary set of zero-coupon bonds and their stochastic evolution through time.

This chapter introduces the stochastic evolution for the zero-coupon bond price curve. The stochastic structure is introduced sequentially, starting with a one-factor model, then presenting a two-factor model, and so forth. Starting with the simplest stochastic structure (one factor) facilitates understanding. After this one-factor model is mastered, additional factors can be added in a straightforward fashion. Chapter 7 formalizes the meaning of arbitrage opportunities and market completeness.

This section motivates the discrete time processes constructed in this chapter. Using historic zero-coupon bond price observations, we can compute a time series history of forward rates (across all maturities). Such a historical forward rate curve evolution is illustrated in Figure 5.1. Figure 5.1 gives the historic evolution of the forward rate curve from January 2007 to December 2018 in weekly observation intervals.

From this evolution, we can generate histograms for changes in a particular maturity forward rate as illustrated in Figure 5.2. These histograms provide an estimate of the true underlying distributions, shown in Figure 5.2 as the continuous curve. For each histogram, there are a mean (expected value) and a standard deviation. These are estimates of the true underlying distribution's mean and standard deviation. Although we do not present it graphically, from these forward rate observations we can also compute the joint histogram

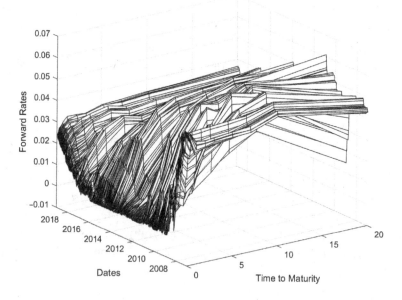

FIGURE 5.1   Forward Rate Curve Evolutions over January 2007–December 2018.

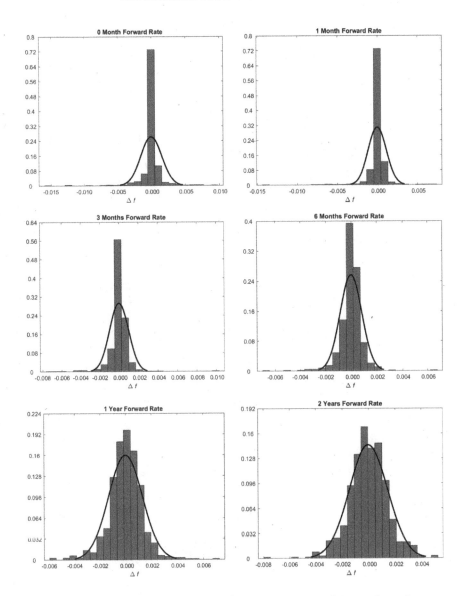

FIGURE 5.2 Histogram of Weekly Changes in Forward Rates from January 2007–December 2018.

(multidimensional distribution) and the covariance matrix of changes in forward rates.

What we want to do in this chapter is to build a model for the evolution of forward rates such that the model provides a reasonable approximation to the true underlying joint distribution. The model needs to be simple enough to be computable, but rich enough to be realistic.

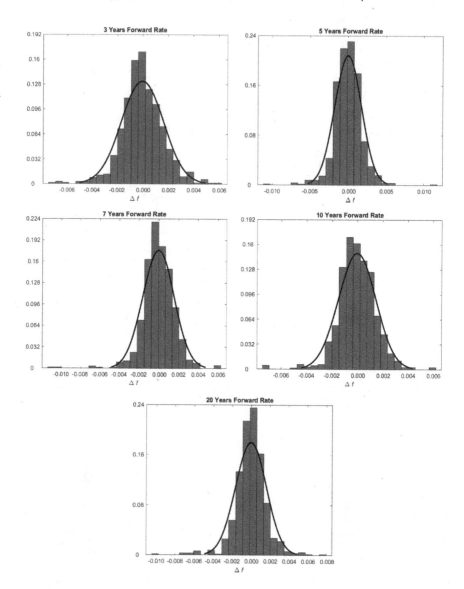

FIGURE 5.2 (Continued)

The binomial (and its generalization – the multinomial) model provides such an approach.

To understand how this works, in its simplest form, we build a binomial tree such that over a month, the distribution for changes in forward rates induced by the binomial tree approximates the true underlying distribution. This is done via a two-step process. First, the mean and standard

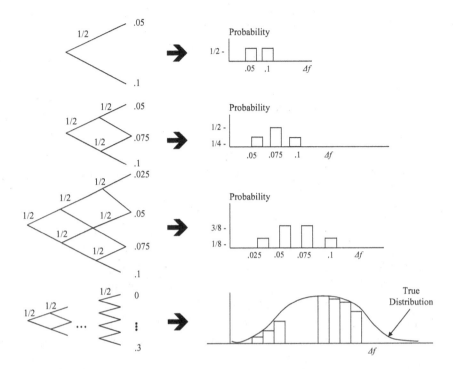

FIGURE 5.3   Example of a Binomial Approximation to the True Distribution.

deviations for the binomial model are chosen to match the mean and standard deviations of the histograms. Second, the number of intervals in the binomial tree over a month is chosen to be large enough so that the binomial model's distribution provides a reasonable approximation to the true underlying distribution.

This process is illustrated in Figure 5.3. Figure 5.3 shows that as the number of intervals in the binomial tree increases, the distribution implied by the binomial model becomes closer and closer to the true underlying distribution. The details of this procedure are discussed more precisely in Chapter 16. The next sections in this chapter demonstrate how to construct such a binomial (or multinomial) tree.

5.2 THE ONE-FACTOR ECONOMY

This section presents the one-factor binomial tree. We first describe the state space process, which characterizes the "randomness" in the economy. Based on this characterization, we then describe the bond price process, the forward rate process, and the spot rate process.

5.2.1 The State Space Process

The uncertainty in the economy is best visualized by utilizing a *tree diagram* as given in the following example.

Example: One-Factor State Space Process

An example of a one-factor state space tree diagram is given in Figure 5.4. The tree starts at time 0 and terminates at time 3. The states are indicated by a sequence of *u*'s (up) and *d*'s (down). Branches connect the various states. Above each branch is a probability. The probabilities across branches at a node sum to one.

The initial probability of jumping up is (3/4), and the initial probability of jumping down is (1/4). If a *u* occurs, at the next node in the tree, the probabilities of jumping up and down remain the same (3/4) and (1/4), respectively. This pattern repeats itself throughout the tree.

At this time, the up state and the down state have no economic interpretation. They are just used as "place-holders" for an economic state, e.g., "good" or "bad." Later in this chapter, when we introduce the bond price process, these states will take on an economic meaning.

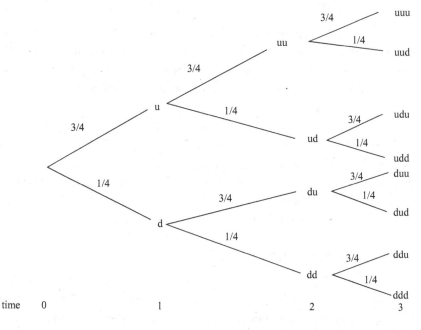

FIGURE 5.4 An Example of a One-Factor State Space Tree Diagram.

The state space process in Figure 5.4 is called a *one-factor model* because at each node in the tree, only one of *two* possibilities can happen (up or down). Each branch also occurs with strictly positive probability. One can conceptualize the tree's being constructed by tossing *one coin* (one-factor).*

If, instead, at each node of the tree there were three branches, each with strictly positive probability, it would be called a *two-factor model* because it would take *two coins* to construct the tree.[†] The first coin would decide between the up branch, on the one hand, and the middle and down branches, on the other hand. The second coin would determine the splitting of the last two branches into middle and down.

Conceptually, the analytics of describing the evolution of the state space process is no more difficult in multiple-factor models than it is in a one-factor model. Multiple-factor models will be discussed in subsequent sections.

This completes the example. □

We next present the abstract representation of this state space tree diagram, see Figure 5.5. At time 0, one of two possible outcomes can

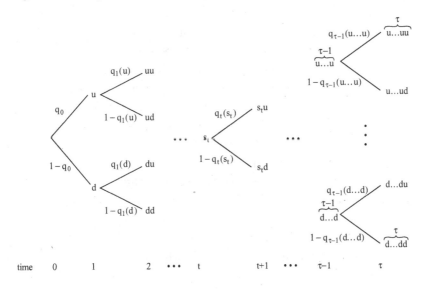

FIGURE 5.5 One-Factor State Space Tree Diagram.

---

* The coin would not be fair, however.
† As in the case of a one-factor model, in general these coins will not be fair.

occur over the next time interval: "up," denoted by u, and "down," denoted by d. The up state occurs with probability $q_0 > 0$, and the down state occurs with probability $1 - q_0 > 0$. Again, at this point, the states up and down have no real economic interpretation. They are just an abstraction formulated to characterize the only uncertainties influencing the term structure of interest rates. At time 1, one of two possible states exists: $\{u, d\}$. We denote the generic state at time 1 by s_1. Thus, $s_1 \in \{u, d\}$.

Over the next time interval, one of two possible outcomes can occur again: up, denoted by u, or down, denoted by d. The resulting state at time 2 is s_1u or s_1d. The up state occurs with probability $q_1(s_1) > 0$, and the down state occurs with probability $1 - q_1(s_1) > 0$. Note that these probabilities sum to one and they can depend on the time 1 and the state s_1.

At time 2, therefore, there are four possible states (s_1u, s_1d for each $s_1 \in \{u, d\}$), namely (uu, ud, du, and dd). The ordering in which the ups and downs occur is important; ud is considered distinct from du.

The process continues in this up-and-down fashion until time π. For an arbitrary time t, there are 2^t possible states. The possible states at time t correspond to all possible t-sequences of u and d.* We let s_t denote a generic state at time t, so $s_t \in$ {all possible t-sequences of u's and d's}.

Over the next time period $[t, t+1]$, one of two possible outcomes can occur, up and down. The resulting state at time $t+1$ is s_tu or s_td. The up outcome occurs with probability $q_t(s_t) > 0$, and the down outcome occurs with probability $1 - q_t(s_t) > 0$. At time $t+1$, therefore, there are 2^{t+1} possible states (s_tu, s_td for each $s_t \in$ {all possible t-sequences of u's and d's}), namely, all possible $t+1$ – sequences of u's and d's. The ordering within the sequence of the u's and d's is important, as distinct orderings are considered different *states*. Note that this specification of a state provides the complete history of the process.

At the last date all uncertainty is resolved, and the state s is some π-sequence of u's and d's. The *state space* consists of all possible τ – sequences of u's and d's, where ordering is important.

As the state space process is constructed, the entire history at any node may be important in determining the probabilities of the next outcome in the tree. This is indicated by making the probabilities at each date t dependent on the state s_t as well as time, i.e., $q_t(s_t)$. Thus, the

---

* There are t empty slots, and each slot can take a u or d. Therefore, there are $2 \cdot 2 \cdots 2$ possibilities, where 2 is multiplied by itself t times. This totals 2^t.

state process is said to be *path dependent*, and the tree is often called *bushy.* Such a tree ensures the most flexibility for modeling term structure evolutions.

5.2.2 The Bond Price Process

The state space process describes the uncertainty underlying and generating the evolution of all the zero-coupon bond prices. The evolution of the zero-coupon bond prices, in turn, determines the evolution of the forward rates and the spot rates. This section describes these stochastic processes. We start via an example.

Example: One-Factor Zero-Coupon Bond Curve Evolution

Figure 5.6 contains an example of a one-factor zero-coupon bond price curve evolution. The zero-coupon bond price evolution is represented as a tree diagram. The underlying state space process is that contained in Figure 5.4. The zero-coupon bond price tree has the same number of branches as the state space tree, and it has the same probabilities. This is by construction.

The initial zero-coupon bond price curve is given at the node at time 0,

$$\begin{bmatrix} P(0,4) \\ P(0,3) \\ P(0,2) \\ P(0,1) \\ P(0,0) \end{bmatrix} = \begin{bmatrix} .923845 \\ .942322 \\ .961169 \\ .980392 \\ 1 \end{bmatrix}.$$

This curve is observed at time 0 (perhaps from a broker's screen or in the financial press). Notice that the last element in this vector is the zero-coupon bond that matures at time 0. This bond pays a dollar at time 0, hence, the "one" in the last element of the vector. We include this price in the vector for consistency across all the vectors in the tree.

After time 0, two possible paths are possible: "up" or "down." With probability ¾ this curve jumps up to:

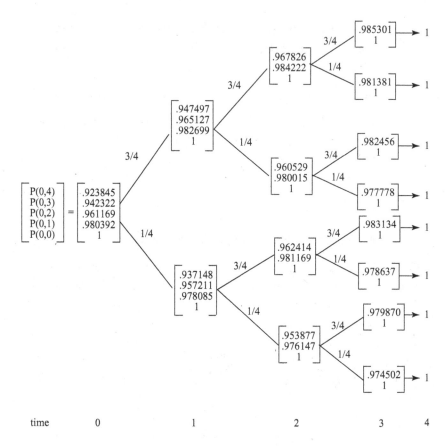

time   0   1   2   3   4

FIGURE 5.6 An Example of a One-Factor Bond Price Curve Evolution. Actual Probabilities Along Each Branch of the Tree.

$$
\begin{bmatrix} P(1,4;u) \\ P(1,3;u) \\ P(1,2;u) \\ P(1,1;u) \end{bmatrix} = \begin{bmatrix} .947497 \\ .965127 \\ .982699 \\ 1 \end{bmatrix}.
$$

In this vector are all the zero-coupon bond prices in the up state. Notice that the notation for the zero-coupon bond price is now augmented to include the state "u." At time 1, the zero-coupon bond that matures at time 1 pays a dollar. This zero-coupon bond is included as the last element in this vector.

With probability ¼ this curve jumps down to:

$$\begin{bmatrix} P(1,4;d) \\ P(1,3;d) \\ P(1,2;d) \\ P(1,1;d) \end{bmatrix} = \begin{bmatrix} .937148 \\ .957211 \\ .978085 \\ 1 \end{bmatrix}.$$

In this vector are the zero-coupon bond prices in the down state. The notation reflects the state "d."

We can now give the "up" and "down" states an economic interpretation. The "up" state refers to the fact that bond prices are greater in state "u" than they are in state "d." This is true for the entire tree.

At time 2, there are four possible zero-coupon bond price curves, depending upon the path through the tree. If at time 1 the "up" path occurred, then if "up" occurs again at time 2, and the zero-coupon bond price vector is:

$$\begin{bmatrix} P(2,4;uu) \\ P(2,3;uu) \\ P(3,3;uu) \end{bmatrix} = \begin{bmatrix} .967826 \\ .984222 \\ 1 \end{bmatrix}.$$

Notice that state "uu" is included in the bond price notation.

The zero-coupon bond price curve evolution given in Figure 5.6 continues in a similar fashion until time 4, when the tree ends.

We see that as each period occurs, the shortest maturity bond matures and then disappears from the tree. Note that between times 3 and 4 only one zero-coupon bond remains in the tree, the 4-period zero-coupon bond. At maturity, it pays off 1 dollar. Hence on the tree, only one arrow with a one appears at time 4. Although there are two states possible at time 4, the payoffs to this bond cannot distinguish them. Hence, the tree really terminates at time 3. For this reason, in all the subsequent examples and theory, the last step in the tree will be a residual, not used for analysis. This completes the example. □

We now present the symbolic representation of a zero-coupon bond price curve evolution. Figure 5.7 depicts this evolution. To understand this

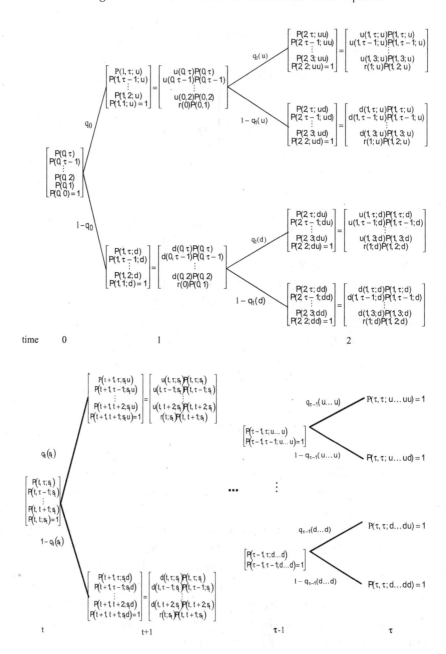

FIGURE 5.7   One-Factor Bond Price Curve Evolution.

figure, we first introduce a new notation for the zero-coupon bond prices in the various states.

Formally, we indicate this state process's influence on the zero-coupon bond prices by expanding the notation, letting $P(t,T;s_t)$ be the T-maturity zero-coupon bond's price at time t under state s_t. Similarly, we expand the notation for yields, forward rates, spot rates, and so on. This is the standard way of denoting random variables given an underlying state space process and probabilities.

We assume that $P(T,T;s_T) = 1$ for all T and s_T. This assumption formalizes the statement that zero-coupon bonds are default-free, i.e., that they are worth a dollar at maturity under all possible states. In addition, we assume that $P(t,T;s_t) > 0$ for all $t \leq T$ and s_t. This ensures that one cannot get something for nothing, i.e., that a sure dollar costs something.

To understand this figure, we first introduce some new notations for the *return* on the zero-coupon bond in the up state and in the down state:

$$u(t,T;s_t) \equiv P(t+1,T;s_tu) / P(t,T;s_t) \quad \text{for} \ t+1 \leq T \qquad (5.1a)$$

and

$$d(t,T;s_t) \equiv P(t+1,T;s_td) / P(t,T;s_t) \quad \text{for} \ t+1 \leq T \qquad (5.1b)$$

where $u(t,T;s_t) > d(t,T;s_t)$ for all $t < T-1$ and s_t.

Expression (5.1a) defines $u(t,T;s_t)$ as the return at time t on the T-maturity zero-coupon bond in the up state. Similarly, expression (5.1b) defines $d(t,T;s_t)$ as the return at time t on the T-maturity bond in the down state. The same symbols "u" and "d" are used for both the states and the returns on the bond. This is done to facilitate understanding by visually linking the state with the changes in the bond price process. This double usage should cause no confusion.

As suggested by the notation, the up and down terminology is given economic meaning by these expressions, for they are seen to describe the relative changes in the magnitudes of the zero-coupon bond prices as* $u(t,T;s_t) > d(t,T;s_t)$. This strict inequality only holds for $t < T-1$ because at time T, its maturity date, the zero-coupon bond pays a sure dollar, i.e., $P(T,T;s_T) = 1$ for all s_T. Therefore, expression (5.1) implies the following:

---

* This restriction is somewhat stronger than that which is actually needed. For the subsequent analysis, we need to only require that $u(t,T;s_t) \neq d(t,T;s_t)$ for all $t < T - 1$ and s_t.

$$u(t,t+1;s_t) = d(t,t+1;s_t) = 1/P(t,t+1;s_t) \equiv r(t;s_t) \quad \text{for all } t \text{ and } s_t. \quad (5.2)$$

Over the last interval in the bond's life, the returns in the up and down states must be identical, nonrandom, and equal to the spot rate.

Expression (5.2) implies an important fact regarding the money market account's value at any time t. Since $B(t) = B(t-1)r(t-1)$, we see that the money market account's value at time t depends only on the previous state s_{t-1} of the process at time $t-1$, and not time t's state. This is because we know the return on the money market account one period before we earn it. Hence, we write $B(t;s_{t-1})$ as the money market account's time t value.

Although Figure 5.7 looks complicated, it is in fact quite simple. The first observation to make is that Figure 5.7 depicts the evolution of the entire zero-coupon bond price curve (a vector). Hence, at each node there is a vector of zero-coupon bond prices, and not just a single point as in Figure 5.5.

At time 0 we start with the initial zero-coupon bond price curve $(P(0,\tau)$, $P(0,\tau-1)$, ..., $P(0,1)$, $P(0,0))$. There are $(\tau+1)$ elements in this vector. For consistency, the price of the bond maturing at time 0 $(P(0,0))$ with a price of unity is included as the last element in this vector.

Over the time interval between 0 and time 1, the zero-coupon bond price curve moves up to $(P(1,\tau;u)$, ..., $P(1,2;u),1)$ with probability $q_0 > 0$ or down to $(P(1,\tau;d)$, ..., $P(1,2;d)$, 1) with probability $1 - q_0 > 0$. These zero-coupon bond price curves, however, consist of only elements, because the 1-period zero-coupon bond at time 0 matures at time 1 and pays a sure dollar. After that date, it no longer trades.

For convenience, these new price vectors are alternatively written in Figure 5.7 as the zero-coupon bond price curve at time 0 multiplied by the return over the period 0 to 1. For the up state the new vector is $(u(0,\tau)P(0,\tau),...,$ $u(0,2)P(0,2), r(0)P(0,1))$ and the down vector is $(d(0,\tau)P(0,\tau), ..., d(0,2)P(0,2),$ $r(0)P(0,1))$. At time 1, therefore, there are two possible zero-bond price curves as determined by the returns.

Between time 1 and time 2, given the state at time 1 (either u or d), the entire curve again shifts up or down. For example, if the state at time 1 is $s_1 = u$, then with probability $q_1(u) > 0$ the curve moves up to $(P(2,\tau;uu), ..., 1)$, or with probability $1 - q_1(u) > 0$ the curve moves down to $(P(2,\tau;ud), ...,1)$. In return form this is $(u(1,\tau;u)P(1,\tau;u), ..., r(1;u)P(1,2;u))$ or $(d(1,\tau;u)P(1,\tau;u), ...,$ $r(1;u)P(1,2;u))$. At time 2 four zero-coupon bond price curves are possible. Each curve now has only 1 elements.

Starting from an arbitrary curve $(P(t,\tau;s_t), ..., P(t,t;s_t) = 1)$ at time t under state s_t, the curve moves up with probability $q_t(s_t) > 0$ to $(P(t+1,\tau;s_t\,u), ..., P(t+1,t+1;s_t u) = 1)$ or down with probability $1 - q_t(s_t) > 0$ to $(P(t+1,\tau;s_t d), ..., P(t+1,t+1;s_t d) = 1)$. In return form these can be written as $(u(t,\tau;s_t)P(t,\tau;s_t), ..., r(t;s_t)P(t,t+1;s_t))$ and $(d(t,\tau;s_t)P(t,\tau;s_t), ..., r(t;s_t)P(t,t+1;s_t))$, respectively.

The evolution of the zero-coupon bond price curve continues in this fashion until time τ. Over the last interval in the model, time $\tau - 1$ to time τ, there is only one zero-coupon bond trading, the bond that matures at time τ. This bond pays a sure dollar at time τ regardless of the state, and after time τ the model ends.

It is important to point out that the traded zero-coupon bonds do not span the relevant uncertainty over this last time interval. For any history $s_{\tau-1}$ at time $\tau - 1$, the state space process tree branches up or down. However, the single traded zero-coupon bond pays a certain amount, a dollar, regardless of the state. Hence, this zero-coupon bond's payoff does not differentiate the up and down states at time τ. This is the only time period in the model that has this property. The last time period in the tree is thus a residual, not really useful for analysis. Indeed, we will later show that in all other time intervals the market is complete with enough zero-coupon bonds trading to span all the relevant states or histories.

We can summarize the evolution of the one-factor zero-coupon bond price curve analytically as in expression (5.3):

$$P(t+1,T;s_{t+1}) = \begin{cases} u(t,T;s_t)P(t,T;s_t) & \text{if } s_{t+1} = s_t u \quad \text{with } q_t(s_t) > 0 \\ d(t,T;s_t)P(t,T;s_t) & \text{if } s_{t+1} = s_t d \quad \text{with } 1 - q_t(s_t) > 0 \end{cases} \tag{5.3}$$

where

$$u(t,T;s_t) > d(t,T;s_t) \quad \text{for } t < T - 1$$

and

$$u(t,t+1;s_t)P(t,t+1;s_t) = d(t,t+1;s_t)P(t,t+1;s_t) = r(t;s_t)P(t,t+1;s_t) = 1.$$

For simplicity of presentation, we have constructed the economy so that at each trading date, a zero-coupon bond matures and it is removed from trading. In addition, no new zero-coupon bonds are issued. It is an easy adjustment to introduce a *newly* issued τ-maturity zero-coupon bond at each trading date. In this situation, the zero-coupon bond price vector would be of constant size containing τ elements at every date. Expression (5.3) is simply expanded to accommodate these newly issued bonds. This extension is not pursued in this text.

5.2.3 The Forward Rate Process

This section describes the evolution of the forward rate curve. This evolution can be derived from the evolution of the zero-coupon bond price curve, given the definition of the forward rate from Chapter 4. As before, we first illustrate the setup with an example.

Example: One-Factor Forward Rate Curve Evolution

Figure 5.8 gives the one-factor forward rate curve evolution implied by the zero-coupon bond price curve evolution example in Figure 5.6. The process starts at time 0 with a flat term structure, with forward rates given by

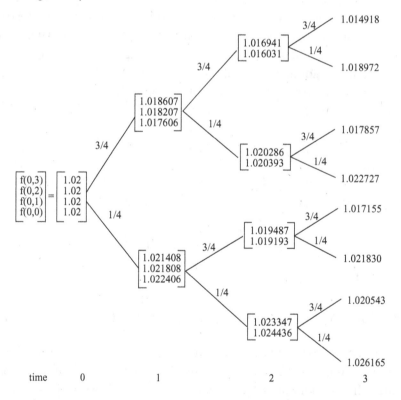

FIGURE 5.8 An Example of a One-Factor Forward Rate Curve. Actual Probabilities Along Each Branch of the Tree.

$$\begin{bmatrix} f(0,3) \\ f(0,2) \\ f(0,1) \\ f(0,0) \end{bmatrix} = \begin{bmatrix} 1.02 \\ 1.02 \\ 1.02 \\ 1.02 \end{bmatrix}.$$

With probability ¾, the forward rate curve shifts up to

$$\begin{bmatrix} f(1,3;u) \\ f(1,2;u) \\ f(1,1;u) \end{bmatrix} = \begin{bmatrix} 1.018607 \\ 1.018207 \\ 1.017606 \end{bmatrix}.$$

The term "up" corresponds to changes in the bond prices. Because rates move inversely to prices, the forward rate curve actually moves down.

With probability ¼, the forward rate curve shifts "down" to

$$\begin{bmatrix} f(1,3;d) \\ f(1,2;d) \\ f(1,1;d) \end{bmatrix} = \begin{bmatrix} 1.021408 \\ 1.021808 \\ 1.022406 \end{bmatrix}.$$

Given the evolution of the forward rate curve, we can deduce the evolution of the bond price curve. For example, at time 1, the up node in the forward rate curve enables us to generate the bond price vector in Figure 5.6 as follows:

$$\begin{bmatrix} P(1,4;u) \\ P(1,3;u) \\ P(1,2;u) \\ P(1,1;u) \end{bmatrix} = \begin{bmatrix} 1/(1.018607)(1.018207)(1.0176066) \\ 1/(1.018207)(1.017606) \\ 1/(1.017606) \\ 1 \end{bmatrix} = \begin{bmatrix} .947497 \\ .965127 \\ .982699 \\ 1 \end{bmatrix}.$$

The remainder of the forward rate process in Figure 5.8 continues in a similar fashion. Note that at time 3, there is only one period left in the example. The only forward rate at time 3, $f(3,3)$, is the spot rate of interest at that date $r(3)$. For example, at time 3 in state uuu, the forward rate is: $f(3,3;uuu) = r(3;uuu) = 1.014918$.

This completes the example. □

We now introduce the symbolic representation of the forward rate curve evolution, as given in Figure 5.9. To understand this figure, we introduce a new notation for the *rate of change* in the forward rate over any interval of time $[t, t+1]$ conditional upon the history at time t, i.e.,

$$\alpha(t,T;s_t) \equiv f(t+1,T;s_t u)/f(t,T;s_t) \quad \text{for } t+1 \le T \le \tau-1 \quad (5.4a)$$

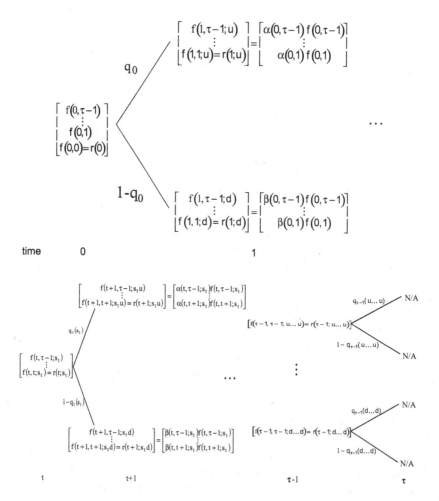

FIGURE 5.9 One-Factor Forward Rate Curve Evolution.

and

$$\beta(t,T;s_t) \equiv f(t+1,T;s_t d) / f(t,T;s_t) \quad \text{for } t+1 \leq T \leq \tau-1. \tag{5.4b}$$

Expression (5.4a) defines the rate of change in the T-maturity forward rate over $[t, t+1]$ to be $\alpha(t,T;s_t)$ in the up state and to be $\beta(t,T;s_t)$ in the down state. These rates of change depend on the history s_t and time.

Like the preceding figures, Figure 5.9 describes the stochastic evolution of an entire vector of forward rates. At time 0 the tree starts with the initial vector of forward rates $(f(0,\tau - 1), \dots, f(0,0) = r(0))$. There are τ elements in this vector, the last element being the spot rate.

This forward rate curve jumps at time 1 to the vector $(f(1,\tau - 1;u), \dots, f(1,1;u))$ in the up state and $(f(1,\tau - 1;d), \dots, f(1,1;d))$ in the down state. These can alternatively be written in return form as $(\alpha(0,\tau - 1)f(0,1), \dots, \alpha(0,1)f(0,1))$ and $(\beta(0,\tau - 1)f(0,1), \dots, \beta(0,1)f(0,1))$, respectively. Each new vector has only $\tau - 1$ elements; its size is reduced by one as the time 0, 1-period zero-coupon bond matures.

Consider an arbitrary time t and state s_t, where the forward rate curve is at $(f(t,\tau - 1;s_t), \dots, f(t,t;s_t))$. It moves with probability $q_t(s_t) > 0$ to $(f(t+1,\tau - 1;s_t u), \dots, f(t+1,t+1;s_t u))$ and with probability $(1 - q_t(s_t)) > 0$ to $(f(t+1,\tau - 1;s_t d), \dots, f(t+1,t+1;s_t d))$. In return form, these can be written as $(\alpha(t,\tau - 1;s_t)f(t,1;s_t), \dots, \alpha(t,t+1;s_t)f(t,t+1;s_t))$ and $(\beta(t,\tau - 1;s_t)f(t,1;s_t), \dots, \beta(t,t+1;s_t)f(t,t+1;s_t))$, respectively. The forward rate curve starts at time t with $\tau - t$ elements and is reduced to $\tau-t-1$ elements at time $t+1$.

We can summarize this evolution for an arbitrary time t as in expression (5.5):

$$f(t+1,T;s_{t+1}) = \begin{cases} \alpha(t,T;s_t)f(t,T;s_t) & \text{if } s_{t+1} = s_t u \quad \text{with } q_t(s_t) > 0 \\ \beta(t,T;s_t)f(t,T;s_t) & \text{if } s_{t+1} = s_t d \quad \text{with } 1-q_t(s_t) > 0 \end{cases} \tag{5.5}$$

where

$$\tau-1 \geq T \geq t+1.$$

Finally, at time $\tau - 1$ only one zero-coupon bond remains in the market and only one forward rate exists, the spot rate. At time τ, when the last zero-coupon bond matures, no additional forward rates can be defined, and the model is terminated.

As seen in the example, the relation between the zero-coupon bond price process and the forward rate process can be easily deduced. First, by the definition of a forward rate,

$$f(t+1,T;s_{t+1}) = P(t+1,T;s_{t+1})/P(t+1,T+1;s_{t+1}). \quad (5.6)$$

Letting $s_{t+1} = s_t u$, we can rewrite expression (5.6) in return form:

$$f(t+1,T;s_t u) = P(t,T;s_t)u(t,T;s_t)/P(t,T+1;s_t)u(t,T+1;s_t). \quad (5.7)$$

Using the definition of a forward rate again yields:

$$f(t+1,T;s_t u) = f(t,T;s_t)u(t,T;s_t)/u(t,T+1;s_t). \quad (5.8a)$$

A similar analysis for $s_{t+1} = s_t d$ yields

$$f(t+1,T;s_t d) = f(t,T;s_t)d(t,T;s_t)/d(t,T+1;s_t). \quad (5.8b)$$

Comparison with expression (5.5) gives the final result:

$$\alpha(t,T;s_t) = u(t,T;s_t)/u(t,T+1;s_t) \quad \text{for} \quad \tau-1 \geq T \geq t+1 \quad (5.9a)$$

and

$$\beta(t,T;s_t) = d(t,T;s_t)/d(t,T+1;s_t) \quad \text{for} \quad \tau-1 \geq T \geq t+1. \quad (5.9b)$$

Expression (5.9) relates the forward rate's rate of change parameters to the zero-coupon bond price process's rate of return parameters in the up (5.9a) and down (5.9b) states, respectively. Expression (5.9) is useful when one parameterizes the bond price process first and then wants to deduce the forward rate process from it.

The parameterization can also work in the reverse direction. Given the forward rate process parameters as in Figure 5.9 one can alternatively deduce the zero-coupon bond price process's rate of return parameters in the up and down states. These relations are given in expression (5.10). The proof follows.

$$u(t,T;s_t) = \frac{r(t;s_t)}{\prod_{j=t+1}^{T-1} \alpha(t,j;s_t)} \quad \text{for} \quad \tau-1 \geq T \geq t+2 \quad (5.10a)$$

and

$$d(t,T;s_t) = \frac{r(t;s_t)}{\prod_{j=t+1}^{T-1} \beta(t,j;s_t)} \quad \text{for} \quad \tau-1 \geq T \geq t+2 \qquad (5.10b)$$

Derivation of Expression (5.10)

From (5.9a) we have that $u(t,t+1;s_t)/\alpha(t,t+1;s_t) = u(t,t+2;s_t)$. But $u(t,t+1;s_t) = r(t;s_t)$, so substitution generates

$$r(t;s_t)/\alpha(t,t+1;s_t) = u(t,t+2;s_t) \qquad (*)$$

Similarly, from (5.9a) we have that

$$u(t,t+2;s_t)/\alpha(t,t+2;s_t) = u(t,t+3;s_t).$$

Substituting in (*) gives

$$r(t;s_t)/\alpha(t,t+1;s_t)\alpha(t,t+2;s_t) = u(t,t+3;s_t).$$

Proceeding inductively gives the result (5.10a). Finally, (5.10b) follows in a similar fashion. This completes the proof. □

Expression (5.10) is useful when one parameterizes the forward rate process first and then wants to deduce the bond price process from it. This is the focal point of the analysis, for example, in the original papers of Heath, Jarrow, and Morton [1, 2, 3]. This perspective is detailed in Chapters 16 and 17 of this text.

In summary, the evolution of the zero-coupon bond price curve can be specified in one of two ways. First, one can directly specify the changes in the zero-coupon bond price curve itself as in Figure 5.7 or expression (5.3). Alternatively, one can specify the changes in the forward rate curve as given in Figure 5.9 or expression (5.5) and then use the relation between the changes in forward rates and zero-coupon bond prices as given in expression (5.10) to deduce the bond price curve evolution. For reasons based on the stability of empirical estimates of the various parameters, the latter approach may often be preferred (see Heath, Jarrow, and Morton [2]).

5.2.4 The Spot Rate Process

The stochastic process for the spot rate can be deduced from the zero-coupon bond price process's evolution in Figure 5.7 or read off the forward rate curve evolution in Figure 5.9. Indeed, as each 1-period zero-coupon bond matures, its return determines the spot rate; or, the shortest maturity forward rate is, by definition, the spot rate. We can apply either of these procedures to the previous example to obtain the spot rate process.

Example: One-Factor Spot Rate Process Evolution

The spot rate process evolution implied by the example in Figure 5.6 is given in Figure 5.10. This curve can be read off Figure 5.8, because the last entry in every forward rate vector is the spot rate of interest.

In this tree diagram, the spot rate of interest at time 0 is 1.02. In the up state it falls to 1.017606 and in the down state it rises to

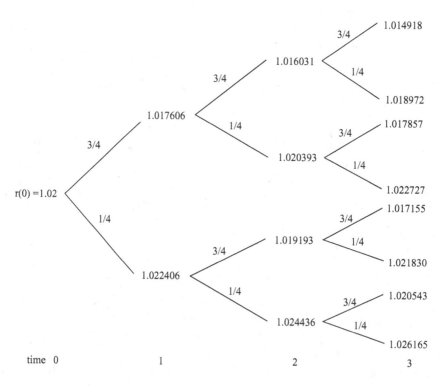

FIGURE 5.10 An Example of a One-Factor Spot Rate Process. Actual Probabilities Along Each Branch of the Tree.

1.022406. Just like with forward rates, the spot rate moves inversely to zero-coupon bond prices. The remainder of the tree evolves in a similar manner.

This completes the example. □

More symbolically, we can represent the evolution for the spot-rate process as in Figure 5.11.

The spot rate curve starts at $r(0)$, and it moves "up" to $r(1;u)$ with probability $q_0>0$ and "down" to $r(1;d)$ with probability $1-q_0>0$. Quotes are placed around "up" and "down" because the spot rate actually moves inversely to the zero-coupon bond price curve movement. So, in fact, u indicates that spot rates move down, and d indicates that spot rates move up. The spot rate process's parameters are deduced from Figure 5.7 (expression 5.3) as $u(1,2;u)$ or $d(1,2;d)$.

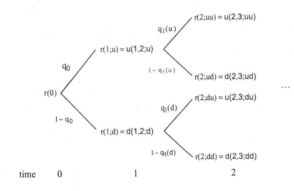

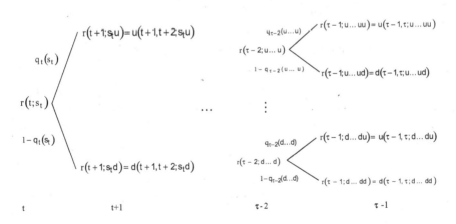

FIGURE 5.11   One-Factor Spot Rate Process.

The spot rate process continues in this branching fashion. At time t, under state s_t, the spot rate $r(t;s_t)$ again moves down to $r(t+1;s_t u) = u(t+1, t+2; s_t u)$ with probability $q_t(s_t) > 0$ and up to $r(t+1;s_t d) = d(t+1, t+2; s_t d)$ with probability $1 - q_t(s_t) > 0$.

The last period's spot rate is determined at time $\tau - 1$: it is $r(\tau - 1; s_{-1})$, and it is either $u(\tau - 1, \tau; s_{\tau-2} u)$ with probability $q_{\tau-2}(s_{\tau-2}) > 0$ or $d(\tau - 1, \tau; s_{\tau-2} d)$ with probability $1 - q_{\tau-2}(s_{\tau-2}) > 0$. We can summarize the spot rate's stochastic process as

$$r(t+1; s_{t+1}) = \begin{cases} u(t+1, t+2; s_t u) & \text{with probability } q_t(s_t) > 0 \\ d(t+1, t+2; s_t d) & \text{with probability } 1 - q_t(s_t) > 0 \end{cases} \quad (5.11)$$

for all s_t and $t + 1 \le T - 1$.

It is in fact possible to go in the opposite direction, but only given additional information. In later sections, using the risk-neutral valuation methodology, we will show that Figure 5.11 is sufficient (given additional information) to deduce the evolution of the zero-coupon bond price curve as given in Figure 5.7.

5.3 THE TWO-FACTOR ECONOMY

Given the analysis for a one-factor economy, the extension to a two-factor economy is straightforward. For brevity, the presentation proceeds using only the analytic representation.

5.3.1 The State Space Process

Between time 0 and time 1, one of three possible outcomes can occur: up, denoted by u; middle, denoted by m; and down, denoted by d, i.e., $s_1 \in \{u, m, d\}$. The probability that $s_1 = u$ is $q_0^u > 0$, the probability that $s_1 = m$ is $q_0^m > 0$, and the probability that $s_1 = d$ is $1 - q_0^u - q_0^m > 0$.

At time $t \in \{1, 2, \dots\}$, the generic initial state is labeled $s_t \in \{$all possible t sequences of u's, m's, and d's$\}$. There exist 3^t possible histories at time t. Over the time interval $[t, t+1]$, the new state s_{t+1} is generated according to expression (5.12):

$$s_{t+1} = \begin{cases} s_t u & \text{with } q_t^u(s_t) > 0 \\ s_t m & \text{with } q_t^m(s_t) > 0 \\ s_t d & \text{with } 1 - q_t^u(s_t) - q_t^m(s_t) > 0. \end{cases} \quad (5.12)$$

The state space at time contains 3 possible histories and is {all possible ordered sequences of u's, m's, and d's}.

5.3.2 The Bond Price Process

The notation for the return on a zero-coupon bond is expanded to include the middle state as in expression (5.13):

$$u(t,T;s_t) \equiv P(t+1,T;s_t u)/P(t,T;s_t) \quad \text{for} \quad t+1 \leq T \qquad (5.13a)$$

$$m(t,T;s_t) \equiv P(t+1,T;s_t m)/P(t,T;s_t) \quad \text{for} \quad t+1 \leq T \qquad (5.13b)$$

and

$$d(t,T;s_t) \equiv P(t+1,T;s_t d)/P(t,T;s_t) \quad \text{for} \quad t+1 \leq T \qquad (5.13c)$$

where $u(t,T;s_t) > m(t,T;s_t) > d(t,T;s_t)$ for $t < T-1$ and s_t, and

$$\begin{bmatrix} 1 & u(t,T;s_t) & u(t,T^*;s_t) \\ 1 & m(t,T;s_t) & m(t,T^*;s_t) \\ 1 & d(t,T;s_t) & d(t,T^*;s_t) \end{bmatrix} \qquad (5.13d)$$

is nonsingular for $T \neq T^*$, $t+1 < min(T,T^*)$ and s_t. By construction,

$$u(t,t+1;s_t) = m(t,t+1;s_t) = d(t,t+1;s_t) = 1/P(t,t+1;s_t) \equiv r(t;s_t). \quad (5.14)$$

Expressions (5.13) and (5.14) are a straightforward extension of the one-factor model with the exception of the nonsingularity condition on the matrix in expression (5.13d). This is included so that different maturity bonds $(T \neq T^*)$ are not identical in their return structure either to each other or to the money market account. (As in the one-factor case, we do not need the restriction that $u(t,T;s_t) > m(t,T;s_t) > d(t,T;s_t)$ for all $t < T-1$ and s_t. It is imposed for clarity of the exposition.)

The evolution of the zero-coupon bond price curve is described by expression (5.15):

$$P(t+1,T;s_{t+1}) = \begin{cases} u(t,T;s_t)P(t,T;s_t) & \text{if } s_{t+1} = s_t u \quad \text{ith } q_t^u(s_t) > 0 \\ m(t,T;s_t)P(t,T;s_t) & \text{if } s_{t+1} = s_t m \quad \text{with } q_t^m(s_t) > 0 \\ d(t,T;s_t)P(t,T;s_t) & \text{if } s_{t+1} = s_t d \quad \text{with } 1 - q_t^u(s_t) - q_t^m(s_t) > 0 \end{cases}$$

$$(5.15)$$

for all $t \leq T-1 \leq \tau-1$ and s_t.

This completes the description of the zero-coupon bond price evolution.

5.3.3 The Forward Rate Process

The new notation for the rate of change of the forward rate is given in expression (5.16):

$$\alpha(t,T;s_t) \equiv f(t+1,T;s_t u) / f(t,T;s_t) \quad \text{for} \quad t+1 \leq T \leq \tau-1 \quad (5.16a)$$

$$\gamma(t,T;s_t) \equiv f(t+1,T;s_t m) / f(t,T;s_t) \quad \text{for} \quad t+1 \leq T \leq \tau-1 \quad (5.16b)$$

$$\beta(t,T;s_t) \equiv f(t+1,T;s_t d) / f(t,T;s_t) \quad \text{for} \quad t+1 \leq T \leq \tau-1. \quad (5.16c)$$

The evolution of the forward rate curve is described by expression (5.17):

$$f(t+1,T;s_{t+1}) = \begin{cases} \alpha(t,T;s_t)f(t,T;s_t) & \text{if } s_{t+1}=s_t u \text{ with } q_t^u(s_t)>0 \\ \gamma(t,T;s_t)f(t,T;s_t) & \text{if } s_{t+1}=s_t m \text{ with } q_t^m(s_t)>0 \\ \beta(t,T;s_t)f(t,T;s_t) & \text{if } s_{t+1}=s_t d \text{ with } 1-q_t^u(s_t)-q_t^m(s_t)>0 \end{cases}$$

$$(5.17)$$

where $\tau-1 \geq T \geq t+1$.

The same derivation that generates expressions (5.9) and (5.10) generates expressions (5.18) and (5.19). We can derive the forward rate process from the bond price process:

$$\alpha(t,T;s_t) = u(t,T;s_t)/u(t,T+1;s_t) \quad \text{for} \quad \tau-1 \geq T \geq t+1 \quad (5.18a)$$

$$\gamma(t,T;s_t) = m(t,T;s_t)/m(t,T+1;s_t) \quad \text{for} \quad \tau-1 \geq T \geq t+1 \quad (5.18b)$$

$$\beta(t,T;s_t) = d(t,T;s_t)/d(t,T+1;s_t) \quad \text{for} \quad \tau-1 \geq T \geq t+1. \quad (5.18c)$$

We can derive the bond price process from the forward rate process:

$$u(t,T;s_t) = \frac{r(t;s_t)}{\displaystyle\prod_{j=t+1}^{T-1} \alpha(t,j;s_t)} \quad \text{for} \quad \tau-1 \geq T \geq t+2 \quad (5.19a)$$

$$m(t,T;s_t) = \frac{r(t;s_t)}{\displaystyle\prod_{j=t+1}^{T-1} \gamma(t,j;s_t)} \quad \text{for} \quad \tau - 1 \geq T \geq t + 2 \qquad (5.19b)$$

and

$$d(t,T;s_t) = \frac{r(t;s_t)}{\displaystyle\prod_{j=t+1}^{T-1} \beta(t,j;s_t)} \quad \text{for} \quad \tau - 1 \geq T \geq t + 2. \qquad (5.19c)$$

This completes the description of the forward rate process evolution.

5.3.4 The Spot Rate Process

The spot rate process evolution is described by expression (5.20):

$$r(t+1;s_{t+1}) = \begin{cases} u(t+1,t+2;s_t u) & \text{with probability } q_t^u(s_t) > 0 \\ m(t+1,t+2;s_t m) & \text{with probability } q_t^m(s_t) > 0 \\ d(t+1,t+2;s_t d) & \text{with probability } 1 - q_t^u(s_t) - q_t^m(s_t) > 0. \end{cases}$$

$$\qquad (5.20)$$

This completes the description of the spot rate process evolution.

5.4 $N \geq 3$-FACTOR ECONOMIES

The extension in Section 5.C from a one-factor economy to a two-factor economy is straightforward. It just corresponds to adding an additional branch on every node in the appropriate tree (or analytic expression). This procedure for extending the economy to three factors, four factors, and so on is similar and is left as an exercise for the reader. This procedure is, in fact, a blueprint for the design of a computer program for generating these evolutions.

5.5 CONSISTENCY WITH EQUILIBRIUM

The preceding material in Chapter 5 exogenously imposes a stochastic structure on the evolution of the zero-coupon bond price curve. The evolution, except for the number of factors, is almost completely unrestricted. But a moment's reflection reveals that this cannot be the case. A T-maturity zero-coupon bond is, of course, a close substitute (for investment purposes) to a $T-1$ or a $T+1$ – maturity zero-coupon bond. Therefore, in an economic *equilibrium*, the returns on these similar maturity zero-coupon bonds cannot be too different. If they were too different, no investor would hold the bond with the smaller return. This difference could not persist in an economic equilibrium.

Furthermore, the entire zero-coupon bond curve is pairwise linked in this manner to adjacent maturity zero-coupon bonds. Consequently, to be consistent with an economic equilibrium, there must be some additional explicit structure required on the parameters of the evolution of the zero-coupon bond price curve. But what are these restrictions? Chapter 7 introduces these restrictions by studying the meaning and existence of *arbitrage opportunities.*

REFERENCES

1. Heath, D., R. Jarrow, and A. Morton, 1990. "Bond Pricing and the Term Structure of Interest Rates: A Discrete Time Approximation." *Journal of Financial and Quantitative Analysis* 25 (4), 419–440.
2. Heath, D., R. Jarrow, and A. Morton, 1990. "Contingent Claim Valuation with a Random Evolution of Interest Rates." *Review of Futures Markets* 9 (1), 54–76.
3. Heath, D., R. Jarrow, and A. Morton, 1992. "Bond Pricing and the Term Structure of Interest Rates: A New Methodology for Contingent Claims Valuation." *Econometrica* 60 (1), 77–105.

CHAPTER 6

The Expectations Hypothesis

6.1 MOTIVATION

This chapter studies the expectations hypothesis. The expectations hypothesis should really be called the expectations *hypotheses* – plural. There are in fact more than one expectations hypothesis, each of which is distinct (see Jarrow [2]). One expectations hypothesis relates to a simple technique for computing present values. This hypothesis is of immense practical importance, and it will be discussed later on in this chapter. The second expectations hypothesis relates to the ability of forward rates to forecast future (realized) spot rates of interest. An implication of this later hypothesis can be illustrated by glancing at Figure 6.1. Figure 6.1 graphs a typical forward rate curve. In Figure 6.1 we are standing at time t, and the forward rates correspond to dates greater than time t. The graph is upward sloping. The second form of the expectations hypothesis implies that an upward sloping forward rate curve, as illustrated, predicts that spot rates will increase as time evolves. This correspondence between the slope of the forward rate curve and the future movement of spot rates is a commonly held belief. Unfortunately, in general, it is not true. In fact, neither of the two expectations hypotheses is empirically valid. We document this statement below.

But then, why study the expectations hypotheses? The answer is surprising. Although the expectations hypothesis is not true in its stated form, and not used again in the text, a modification of it is! A modification

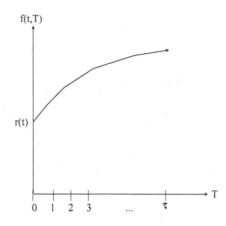

FIGURE 6.1   The Graph of an Upward Sloping Forward Rate Curve.

of the "present value" form of the expectations hypothesis is used in the subsequent chapters to value interest rate derivatives. To understand this application, it is important to first understand the traditional form of the expectations hypothesis. Hence, the motivation for this chapter.

6.2 PRESENT VALUE FORM

This section discusses the present value form of the expectations hypothesis. This version of the expectations hypothesis is related to the expected returns on zero-coupon bonds of different maturities.

To characterize this hypothesis, we first define a *liquidity premium* (sometimes called a *term premium* or *risk premium*), $L_1(t,T)$, as

$$L_1(t,T) \equiv E_t\left(\frac{P(t+1,T)}{P(t,T)}\right) - r(t) \qquad (6.1)$$

for $1 \le t+1 < T \le \tau$.

$E_t(.)$ is the time t expected value using the actual probabilities as given in Figure 5.5 for a one-factor economy or expression (5.12) for a two-factor economy. These are the probabilities that generate the bond prices observed in the markets, reported in the financial press, and viewed on broker screens. These are the probabilities that standard statistical procedures estimate from market prices. These probabilities have various names in the literature; they are called either the *actual* or *statistical* or *empirical* or *true* probabilities.

The *present value form of the expectations hypothesis* is that

$$L_1(t,T) = 0 \text{ or equivalently } E_t\left(\frac{P(t+1,T)}{P(t,T)}\right) = r(t) \qquad (6.2)$$

The present value form of the expectations hypothesis states that the liquidity premium is zero or, equivalently, that the expected return on all zero-coupon bonds is the same and equal to the spot rate of interest.

The motivation for the present value form of the expectations hypothesis can be understood by considering standard investment theory (see Jarrow [3]). The liquidity premium in expression (6.1), $L_1(t,T)$, represents the excess expected return that the T-maturity zero-coupon bond earns above the spot rate.

In the theory of investments, it is normally believed that traders are *risk averse*, and the riskier the investment (financial instrument, e.g., common stock), the higher the expected return must be (in equilibrium) to induce traders to hold it. The risk of an investment is usually measured by its contribution to the variance (standard deviation) of an investor's optimal investment portfolio. In the simplest model of asset pricing, this is called the investment's "beta." In these equilibrium asset pricing models, the liquidity premium is the compensation that risk-averse traders require for bearing risk. As an aside, the empirical literature supports the validity of this theory – riskier assets earn higher expected returns – but the particular form that the asset pricing model assumes is still subject to significant debate (see Jagannathan, Schaumburg, and Zhou [1]).

Also in this investment theory, if investors are *risk-neutral*, then they do not care about risk. By definition, risk-neutral investors only care about expected returns. The higher the expected return on an investment, the more desirable the investment is to the risk-neutral investor. Consequently, risk-neutral investors invest all of their savings into the security with the highest expected return.

In an equilibrium economy consisting of only risk-neutral investors, the excess expected return on all assets, including zero-coupon bonds, must be zero. To see this, consider the contrary. If two assets had different expected returns, then the asset with the highest expected return would be desired by all investors, and the asset with the lowest expected return would be shunned. Supply would not equal demand. Thus, this could not be an equilibrium. The only condition where supply equals demand is when all assets have equal expected returns. In such an economy with a spot rate of interest, in equilibrium, all assets' returns must be equal to it (see expression (6.2)).

This gives the first motivation for the present value form of the expectations hypothesis. It is consistent with economic equilibrium in a *risk-neutral economy*. Unfortunately, few economists believe that the actual economy is populated with risk-neutral investors. Nonetheless, it provides some insight into the conditions under which expression (6.2) would be true.

The second motivation for the present value form of the expectations hypothesis can be obtained by rewriting expression (6.2) in an equivalent form:

$$P(t,T) = \frac{E_t(P(t+1,T))}{r(t)}. \tag{6.3}$$

Expression (6.3) demonstrates that the present value form of the expectations hypothesis is equivalent to the statement that the time t value of a T-maturity zero-coupon bond is its time $t+1$ expected value, discounted at the spot rate. That is, expression (6.3) provides a method for computing the present value of a zero-coupon bond. By a process of iterated substitution, recalling that $P(T,T) = 1$, we can rewrite expression (6.3) as

$$P(t,T) = E_t\left(\frac{1}{r(t)r(t+1)\cdots r(T-1)}\right). \tag{6.4}$$

We see here, more clearly, that this form of the expectations hypothesis gives us a simple technique for computing the present value of a bond. Simply, take the cash flow, discount it to the present using the spot rate of interest, and then take an expectation using the actual probabilities. There is no need to estimate risk premiums or use asset pricing models. If only expression (6.4) was true!

Unfortunately, the empirical evidence is stacked against the validity of expression (6.4). In this form, it does not appear to be true. However, we will return to an expression similar to (6.4), a transformation of expression (6.4), which is true in subsequent chapters. At that time, it will be important to keep in mind that the present value form of the expectations hypothesis is not true (as given by expression (6.4)). Instead, something else is going on. What that something else is awaits a subsequent chapter.

Example: Present Value Form of the Expectations Hypothesis

For the one-factor example given in Figure 6.2, we see that the expected returns on the bonds are given by the following:

$$
\begin{bmatrix}
E(P(1,4)/P(0,4)) \\
E(P(1,3)/P(0,3)) \\
E(P(1,2)/P(0,2)) \\
1/P(0,1)
\end{bmatrix}
=
\begin{bmatrix}
(3/4)1.025601+(1/4)1.014399 \\
(3/4)1.024201+(1/4)1.015800 \\
(3/4)1.022400+(1/4)1.017600 \\
(3/4)1.02+(1/4)1.02
\end{bmatrix}
=
\begin{bmatrix}
1.022801 \\
1.022101 \\
1.021200 \\
1.02
\end{bmatrix}.
$$

where $E(.)$ stands for the expectation based on the actual probabilities $q_t(s_t)$ in Figure 6.2.

We see that the longer-maturity zero-coupon bonds are expected to earn more than the spot rate. This difference is consistent with the existence of a liquidity premium. Alternatively stated, the present value form of the expectations hypothesis does not hold for this example.

In fact, one can also show that the discounted, expected bond prices are not equal to the current zero-coupon bond prices.

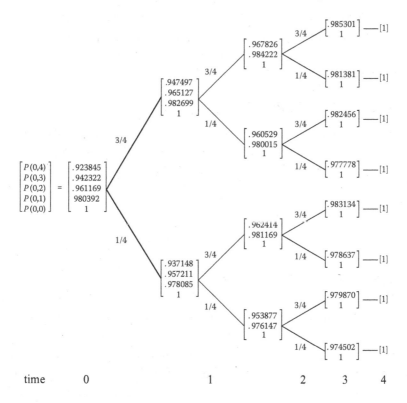

FIGURE 6.2 An Example of a One-Factor Bond Price Curve Evolution. Actual Probabilities Along Each Branch of the Tree.

The discounted expect bond prices are

$$
\begin{bmatrix} E(P(1,4)/r(0)) \\ E(P(1,3)/r(0) \\ E(P(1,2)/r(0)) \\ 1/r(0) \end{bmatrix} = \begin{bmatrix} [(3/4)0.947496+(1/4)0.937148]/1.02 \\ [(3/4)0.965127+(1/4)0.957211]/1.02 \\ [(3/4)0.982699+(1/4)0.978085]/1.02 \\ 1/1.02 \end{bmatrix} = \begin{bmatrix} .926381 \\ .944262 \\ .962299 \\ .980392 \end{bmatrix}.
$$

This does not equal the initial bond price vector $\begin{bmatrix} .923845 \\ .942322 \\ .961169 \\ .980392 \end{bmatrix}$ in Figure 6.2.

In future chapters it is important to remember that in this example, the present value form of the expectations hypothesis does not hold. At that time, a modification of the present value relation will hold, and one must not confuse the two concepts. □

6.3 UNBIASED FORWARD RATE FORM

This section presents the unbiased forward rate form of the expectations hypothesis. This second version of the expectations hypothesis involves the different maturity forward rates.

To characterize this hypothesis, we define a second *liquidity premium*, $L_2(t,T)$, as

$$
L_2(t,T) = f(t,T) - E_t(r(T)) \tag{6.4}
$$

for $0 \le t \le T \le \tau - 1$.

The liquidity premium $L_2(t,T)$ is the difference between the forward rate at time t for date T and the expected spot rate for date T. This liquidity premium has two interpretations.

The first is that it represents the premium required for avoiding the risk of waiting until the future to borrow or lend, versus contracting today. This interpretation can be understood by glancing at Figure 6.3.

Suppose we need to borrow funds at a future date, time T. We are currently standing at time t, with a spot rate of $r(t)$. There are two ways to

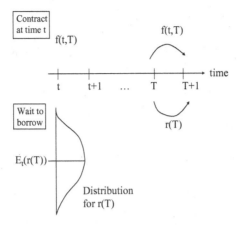

FIGURE 6.3   The Forward Rate f(t, T) versus the Expected Spot Rate $E_t(r(T))$.

borrow in the future. One is to wait until time T to borrow. At that date, we borrow at the spot rate $r(T)$. Viewing this alternative from time t, the future spot rate is random, so there is a probability distribution around the possible values we can borrow at. It could be higher or less than the current rate $r(t)$. Our best guess for the borrowing rate is the mean of this distribution, $E_t(r(T;s_{T-1}))$.

The second alternative is to contract today (time t) to borrow at time T. The rate we can contract for is the forward rate $f(t,T)$. In this alternative, there is no risk associated with the borrowing rate. Expression (6.4) represents the difference between these two alternatives. The difference represents the premium one is willing to pay (in interest) to avoid the risk of waiting to borrow.

The second interpretation of expression (6.4) is more straightforward. It represents the forecasting bias in using the forward rate as an estimate of the expected future spot rate at time T. It is a common belief among traders that the forward rate is a good forecast of the spot rate expected to hold in the future. This belief is formalized as the second version of the expectations hypothesis.

The *unbiased forward rate form of the expectations hypothesis* is that

$$L_2(t,T) = 0 \quad \text{or, equivalently} \quad f(t,T) = E_t(r(T)). \qquad (6.5)$$

This hypothesis states that forward rates are unbiased predictors of future spot rates. Unfortunately, the evidence is inconsistent with this hypothesis; see the last section of this chapter. The satisfaction of this hypothesis depends on the supply and demand of individuals willing to wait versus contracting today, for borrowing and lending in the future. This is a complex decision, involving the availability of funds, risk aversion, and beliefs over the future distribution of the spot rate at time T. There is no reason to believe that equality between rates as in expression (6.5) must hold.

It is also a common belief among traders that the slope of the forward rate curve predicts the future evolution of spot rates. This is a corollary of expression (6.5). For example, if the forward rate slope is positive, traders expect spot rates to increase over time.

Contrary to common belief, the slope of the forward rate curve has no simple relation to the expected increase or decrease in future spot rates. Indeed, let the slope of the forward rate curve be fixed and as given at the node at time 0 in Figure 6.4. At time 0, Figure 6.4 has a flat forward rate curve. Hence, under the common belief, traders would

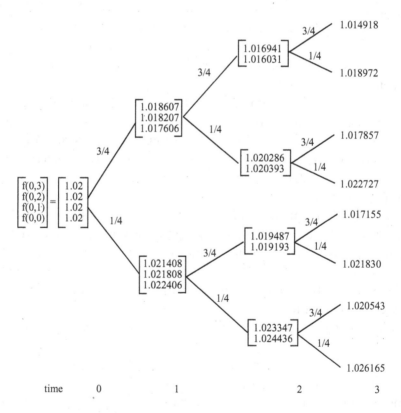

FIGURE 6.4 An Example of a One-Factor Forward Rate Curve. Actual Probabilities Along Each Branch of the Tree.

expect spot rates to stay constant over time. But, the expected spot rate in the future is determined by both the outcomes and the probabilities on the branches of the tree and not by the initial forward rate curve. Keeping the outcomes fixed and altering the probabilities, it is possible to have either increasing or decreasing expected spot rates, even given a flat initial forward rate curve. We clarify these statements through an example.

Example: Unbiased Forward Rates form of the Expectations Hypothesis

Consider the example shown in Figure 6.4. First, we claim that the forward rate process in Figure 6.4 does not satisfy this version of the expectations hypothesis.

Indeed, at time 0, all forward rates $f(0,1)$, $f(0,2)$ and $f(0,3)$ are 1.02. This is a flat term structure.

The expected spot rates are:

$$E(r(1)) = (3/4)1.017606 + (1/4)1.022406 = 1.018806$$

$$E(r(2)) = (3/4)(3/4)1.016031 + (3/4)(1/4)1.020393$$
$$+ (1/4)(3/4)1.019193 + (1/4)(1/4)1.024436$$
$$= 1.017967$$

$$E(r(3)) = (3/4)(3/4)(3/4)1.014918 + (3/4)(3/4)(1/4)1.018972$$
$$+ (3/4)(1/4)(3/4)1.017857 + (3/4)(1/4)(1/4)1.022727$$
$$+ (1/4)(3/4)(3/4)1.017155 + (1/4)(3/4)(1/4)1.021830$$
$$+ (1/4)(1/4)(3/4)1.020543 + (1/4)(1/4)(1/4)1.026165$$
$$= 1.017345$$

This shows that $f(0,T) \neq E(r(T))$ for $T = 1, 2, 3$.

Second, this example shows that the slope of the forward rate curve does not forecast the direction of future spot rates. In this figure, the forward rate curve is flat, but spot rates are expected to decline from time 0 to time 1 to time 2 to time 3 (1.02 to 1.018806 to 1.017967 to 1.017345, respectively).

By changing the probabilities in the tree, we can get the expected spot rate to increase or decrease in any period, leaving the forward rate curve's slope unchanged. This demonstrates the absence of any simple relation between the future course of interest rates and the slope of the forward rate curve. ☐

6.4 RELATION BETWEEN THE TWO VERSIONS OF THE EXPECTATIONS HYPOTHESIS

This section examines the relation between the two versions of the expectations hypothesis. It is more abstract than the previous two sections, and it can be skipped, as it is not used in the remainder of the text.

The two versions of the expectations hypothesis are not equivalent; i.e., $L_1(t,T) = 0$ does not imply, nor is it implied by $L_2(t,T) = 0$. The easiest way to prove this statement is to rewrite expressions (6.2) and (6.4) in an equivalent form. These are given by expressions (6.6) and (6.7):

$$P(t,T) = E_t\left(1\bigg/\prod_{j=t}^{T-1}\left[L_1(j,T)+r(j)\right]\right) \qquad (6.6)$$

$$P(t,T) = 1\bigg/\prod_{j=t}^{T-1} E_t\left[L_2(t,j)+r(j)\right] \qquad (6.7)$$

For general random processes, $L_1(t,T) \equiv 0$ does not imply that $L_2(t,T) \equiv 0$ (and conversely), because the expected value of the inverse of a product does not equal the inverse of the products of the expected values.

Derivation of Expressions (6.6) and (6.7)

From (6.2) we have

$$P(t,T) = E_t\left(P(t+1,T)\big/\left[L_1(t,T)+r(t)\right]\right)$$

This also holds true for time $t+1$. Successive substitution for $P(t,T)$ and the law of iterated expectations, $E_t(E_{t+1}(.)) = E_t(.)$, gives (6.6).

From (6.5), substituting in the definition of $f(t,T) \equiv P(t,T)/P(t,T+1)$ yields

$$P(t,T+1) = P(t,T)\big/\left[L_2(t,T)+E_t(r(T))\right].$$

This holds for maturity T as well. Successive substitution for $P(t,T)$ yields

$$P(t,T+1) = 1 \Big/ \prod_{j=t}^{T} \Big[E_t \big(L_2(t,j) + r(j) \big) \Big].$$

A change of maturity from $T+1$ to T gives (6.7). □

Example: Comparison of the Two Liquidity Premiums

The example of Figures 6.2 and 6.4 illustrates the fact that there is no simple relation between $L_1(t,T)$ and $L_2(t,T)$. This can also be seen by examining Table 6.1, which contains the two liquidity premiums for time 0. The liquidity premiums have different values, and for two cases, they are not even defined for the same maturity. In summary, this table shows that the two liquidity premiums are distinct and not equal. □

TABLE 6.1 Liquidity Premiums $L_1(0,T)$ and $L_2(0,T)$ for the Example of Figure 6.2

T	$L_1(0,T)$	$L_2(0,T)$
0	Not defined	0
1	0	.001194
2	.001200	.002033
3	.002101	.002655
4	.002801	Not defined

6.5 EMPIRICAL ILLUSTRATION

This section provides an illustrative empirical investigation of both forms of the expectations hypothesis. These investigations use weekly price observations of U.S. Treasury bonds from January 2007 to December 2018. From these bond prices, zero-coupon bond prices and (continuously compounded) forward rates are extracted for the maturities: 1 month, 3 months, 6 months, 1 year, 2 years, 3 years, 5 years, 7 years, 10 years, and 20 years. The details of this procedure are described in Chapter 17. For this chapter, one does not need to understand the details of the estimation procedure. Instead, it is sufficient to only understand the reasons for the statistical tests and the summary statistics provided.

The analysis is kept purposely simple in order to facilitate understanding and to illustrate the ease in which one can reject both forms of the expectations hypothesis. More complicated tests of these two hypotheses exist in the literature, and their study is left to independent reading.

6.5.1 Present Value Form

To investigate the present value form of the expectations hypothesis, we computed the excess (continuously compounded) weekly return on the different maturity zero-coupon bonds over the time period, i.e.,

$$\log\left[\frac{P(t+\Delta,t+\bar{T})}{P(t,t+\bar{T})}\right] - r(t)\Delta \tag{6.8}$$

where $\bar{T} \in \{.25,.5,1,2,3,5,7,10,20,30\}$ corresponds to the zero-coupon bond's time-to-maturity and $\Delta = 1/52$ (1 week). The 1-month zero-coupon bond is omitted because its return is the spot rate.

The results are contained in Table 6.2. Note that the mean excess return on all the zero-coupon bonds is strictly positive and increasing in time-to-maturity. The present value form of the expectations hypothesis is rejected

TABLE 6.2   Mean Excess Returns on Zero-coupon Bonds of Different Maturities

Time-to-Maturity	Mean	95% Confidence Interval	
		Lower Bound	Upper Bound
3 Months	.000017384	.0000077797	.0000269883
6 Months	.000061822	.0000400166	.0000836280
1 Year	.000091535	.0000575853	.0001254856
2 Years	.000174464	.0001170825	.0002318464
3 Years	.000303000	.0001621592	.0004438411
5 Years	.000484895	.0002440123	.0007257774
7 Years	.000684481	.0002179723	.0011509887
10 Years	.000791800	.0000942480	.0014893517
20 Years	.000934200	−.0000541616	.0019225622
30 Years	.001252783	−.0007847082	.0032902745

for all zero-coupon bond maturities except for the 20- and 30-year zero-coupon bonds.*

An alternative joint (across all bonds) non-parametric test rejects the present value form of the expectations hypothesis more definitively. If the present value form of the expectations hypothesis were true, then across maturities, we would expect to see a positive difference in the mean excess return half of the time and a negative difference half of the time. The probability of observing ten positive differences in a row across maturities is thus $(1/2)^{10} = .00098$. This is so unlikely to occur by chance that we can easily reject the present value form of the expectations hypothesis.

In fact, as noted previously, the mean excess returns on the zero-coupon bonds increase with time-to-maturity. This could not occur by chance under the present value form of the expectations hypothesis, and it is consistent with the hypothesis that longer maturity zero-coupon bonds are "riskier" in the sense discussed earlier in the chapter.

6.5.2 Unbiased Forward Rate Form

To investigate the unbiased rate form of the expectations hypothesis, we computed the forecasting bias of the various constant maturity (continuously compounded) forward rates over the sample period, i.e.,

$$\tilde{f}(t,T) - r(T) \text{ for } T - t \in \{.083, .25, .5, 1, 2, 3, 5, 7, 10\} \qquad (6.9)$$

where $(T - t)$ corresponds to time-to-maturity and $\tilde{f}(t,T)$ is defined in Chapter 16.

The (continuously compounded) forward rate $\tilde{f}(t,T)$ is observed at time t, and it can be viewed as a forecast of the future time T spot rate $r(T)$. The difference, the forecasting bias, is observed for each week over the sample period. The time-to-maturities 20 and 30 years are excluded from this empirical investigation because we will be studying realized differences across time as in expression (6.9) over our 12-year sample period. For time-to-maturities 20 and 30 years, no realized forward rate forecasting differences exist in our sample.

The mean is calculated and recorded in Table 6.3 along with the 95 percent confidence intervals.

---

* This follows because the 95 percent confidence interval includes the value zero. Hence, the hypothesis that the difference is zero cannot be rejected at the 95 percent level.

TABLE 6.3   Forecasting Bias of Forward Rates of Different Maturities

Time-to-Maturity	Mean	95% Confidence Interval	
		Lower Bound	Upper Bound
1 Month	.00091166	.00072941	.00109391
3 Months	.00322410	.00284610	.00360211
6 Months	.00475482	.00426444	.00524519
1 Year	.00921053	.00851424	.00990681
2 Years	.01581192	.01473416	.01688968
3 Years	.02525952	.02409864	.02642040
5 Years	.03494272	.03362886	.03625658
7 Years	.04076861	.03932894	.04220829
10 Years	.04326958	.04200714	.04453202

Note that the mean forecasting bias is strictly positive and increasing with time-to-maturity. As seen, the unbiased forward rate form of the expectations hypothesis is rejected for all time-to-maturities at the 95 percent confidence interval.

An alternative joint (across all maturities) non-parametric test rejects the unbiased forward rate form of the expectations hypothesis more definitively. If the unbiased forward rate form of the expectations hypothesis were true, then across maturities, we would expect to see a positive difference half of the time and a negative difference half of the time. The probability of observing nine positive differences in a row across maturities is thus $(1/2)^9 = .00195$. This is so unlikely to occur by chance that we can easily reject the unbiased forward rate form of the expectations hypothesis.

In fact, the forward rate forecasting bias for all maturities increases with time-to-maturity. This could not occur by chance under the forecasting bias form of the expectations hypothesis, and it is consistent with the hypothesis that forward rates have an embedded risk premium that increases with time-to-maturity.

REFERENCES

1. Jagannathan, R., E. Schaumburg, and G. Zhou, 2010. "Cross-Sectional Asset Pricing Tests." *Annual Review of Financial Economics* 2, 49–74.
2. Jarrow, R., 1981. "Liquidity Premiums and the Expectations Hypothesis." *Journal of Banking and Finance* 5 (4), 539–546.
3. Jarrow, R., 2018. *Continuous-Time Asset Pricing Theory*, Springer.

CHAPTER 7

Trading Strategies, Arbitrage Opportunities, and Complete Markets

7.1 MOTIVATION

This chapter studies the meaning of trading strategies, arbitrage opportunities, and complete markets. These are the basic concepts needed to develop the pricing and hedging methodology for fixed income securities and interest rate options. An intuitive description of these concepts can be easily obtained by considering an example, say, the pricing of an interest rate cap. An interest rate cap is a type of call option written on an interest rate (see Chapter 2).

As mentioned earlier in Chapter 1, the idea underlying the pricing methodology is to construct a synthetic cap by forming a portfolio in the underlying zero-coupon bonds such that the cash flows and value to this portfolio match the cash flows and value to the traded cap. This portfolio of zero-coupon bonds may be dynamic, involving rebalancing the holdings of the zero-coupon bonds through time. Such a portfolio is called a *trading strategy*.

Then, to avoid riskless profit opportunities – called *arbitrage opportunities* – the cost of constructing this portfolio must equal the market price of the traded cap. The cost of construction is called the fair or *arbitrage-free* price of the cap.

This argument works, however, only if a portfolio of zero-coupon bonds can be found to replicate the cash flows and value of the traded cap. If such a portfolio of zero-coupon bonds cannot be found, then the arbitrage-free price of the cap cannot be determined. If such a portfolio can be found to replicate the cap and, in fact, any arbitrarily selected interest rate option, then the market is said to be *complete*. Complete markets are advantageous for risk management.

This chapter formalizes this intuitive discussion.

7.2 TRADING STRATEGIES

This section formally defines the concept of a trading strategy used throughout the book. Intuitively, a trading strategy is a dynamic investment portfolio involving some or possibly all of the traded zero-coupon bonds. Portfolio rebalancings can occur within the investment horizon, and they are based on the information available at the time that the portfolio is revised. Simply stated, one cannot look into the future to decide which portfolios to hold today.

An example best illustrates how to formally describe a trading strategy.

Example: A Trading Strategy

Figure 7.1 gives a zero-coupon bond price evolution for times 0, 1, and 2. This is the first part of Figure 5.4.

Figure 7.2 repeats Figure 7.1, but this time with a vector of zero-coupon bond and money market account holdings listed underneath the bond price vectors. We will call these the *holdings* vector.

For example, looking at time 0, the holdings vector is:

$$(n_0(0), n_4(0), n_3(0), n_2(0)) = (-1, 0, 2.5, -2).$$

The first vector gives the notation for the holdings in the available securities. The elements in the vector are the number of units of the money market account held at time 0, $n_0(0)$, the number of units of the 4-period zero-coupon bond held at time 0, $n_4(0)$, the number of units of the 3-period zero-coupon bond held at time 0, $n_3(0)$, and the number of units of the 2-period zero-coupon bond held at time 0, $n_2(0)$. The 1-period zero-coupon bond is not included because a position in this zero-coupon bond is implicitly incorporated into the position in the money market account. Note that for simplicity, the

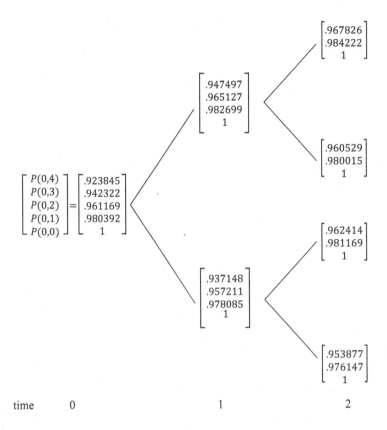

FIGURE 7.1   An Example of a Zero-Coupon Bond Price Curve.

subscript on the holdings of each zero-coupon bond corresponds to the maturity of the zero-coupon bond. The money market account is included as the first element in this vector, and it has a subscript of zero.

This position vector represents selling 1 unit of the money market account, holding 0 units of the 4-period zero-coupon bond, holding 2.5 units of the 3-period zero-coupon bond, and selling 2 units of the 2-period coupon bond.

Similarly, position vectors are given under each price vector possible in the tree at times 1 and 2. As before, these represent the holdings in the money market account and the various zero-coupon bonds.

As depicted on Figure 7.2, this collection of holdings vector represents a *trading strategy*. A trading strategy is thus a complete listing of the holdings of each traded security for each time and state

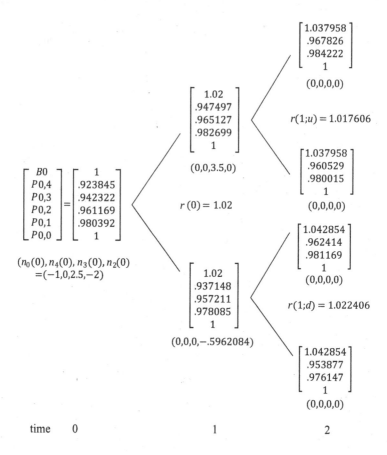

FIGURE 7.2 An Example of a Zero-Coupon Bond Price Curve Evolution and a Trading Strategy $(n_0(t), n_4(t), n_3(t), n_2(t))$.

in the tree. These trading strategies can be time and state dependent. A trading strategy imposes no additional restrictions on these holdings.

Let us examine the trading strategy given in Figure 7.2 in more detail. At time 0, the initial cash flow from forming this trading strategy is:

$$n_o(0)B(0) + n_4(0)P(0,4) + n_3(0)P(0,3) + n_2(0)P(0,2)$$

$$= n_o(0)1 + n_4(0).923845 + n_3(0).942322 + n_2(0).961169$$

$$= +(1)1 + (0).923845 - (2.5).942322 + (2).961169$$

$$= +.566533.$$

This portfolio has a positive cash inflow of .566533 dollars at time 0. A positive *cash flow* implies that the *value* of this position is −.566533 dollars; it is a liability. Other trading strategies may have had an initial cash flow that is negative or zero. When the initial cash inflow is zero, the trading strategy has a special name, a *zero investment trading strategy*.

These holdings are formed at time 0 and held until time 1. At time 1, two outcomes are possible − either up or down occurs. If up occurs, the trading strategy enters time 1 with a value of:

$$n_o(0)B(1)+n_4(0)P(1,4;u)+n_3(0)P(1,3;u)+n_2(0)P(1,2;u)$$

$$= -(1)1.02+(0).947497+(2.5).965127-(2).982699$$

$$= -.5725805.$$

Note that in this valuation, the holdings are the same as those formed at time 0 and only the prices have changed. They represent the "new" prices at time 1 in the up state. As shown, the position has lost value, moving from −.566533 to −.5725805.

Next, at time 1 in the up state, the portfolio is *rebalanced*. The new holdings at time 1 in the up state are 0 units of the money market account, 0 units of the 4-period zero-coupon bond, 3.5 units of the 3-period zero-coupon bond, and 0 units of the 2-period zero-coupon bond. In symbols:

$$(n_o(1;u),n_4(1;u),n_3(1;u),n_2(1;u)) = (0,0,3.5,0)$$

where the notation for the holdings at time 1 includes the state u as the second argument.

In this "new" holdings vector, the positions in the money market account, 4- and 2-period zero-coupon bonds have been liquidated. An additional unit of the 3-period zero-coupon bond has been purchased. The value of this rebalanced position at time 1 in the up state is:

$$n_o(1;u)B(1)+n_4(1;u)P(1,4;u)+n_3(1;u)P(1,3;u)+n_2(1;u)P(1,2;u)$$

$$= (0)1.02+(0).947497+(3.5).965127+(0).982699$$

$$= +3.3779445.$$

This portfolio was rebalanced from $-.5725805$ dollars to $+3.3779445$ dollars. This can occur only if there was *cash input* to the trading strategy at time 1 in state u. The cash inflow to this trading strategy is equal to $(.5725805 + 3.3779445) = 3.950525$ dollars.

If the rebalanced value of this trading strategy at time 1 state u had been equal to the entering value of the portfolio, then the strategy would be called *self-financing*. Self-financing strategies are special and utilized frequently in the text. The rebalancing just discussed is not self-financing.

A self-financing rebalancing is illustrated at time 1 in the down state. The value of the portfolio entering time 1 in state d is given by

$$n_o(0)B(1) + n_4(0)P(1,4;d) + n_3(0)P(1,3;d) + n_2(0)P(1,2;d)$$

$$= -(1)1.02 + (0).937148 + (2.5).957211 - (2).978085$$

$$= -.5831425.$$

The rebalanced portfolio is $(n_0(1; d), n_4(1; d), n_3(1; d), n_2(1; d)) = (0, 0, 0, -.5962084)$. This consists of zero units of the money market account, zero units of the 4- and 3-period zero-coupon bonds, and $-.5962084$ units of the 2-period zero-coupon bond. The value of this rebalanced portfolio is:

$$n_o(1;d)B(1) + n_4(1;d)P(1,4;d) + n_3(1;d)P(1,3;d) + n_2(1;d)P(1,2;d)$$

$$= (0)1.02 + (0).937148 + (0).957211 - (.5962084).978085$$

$$= -.5831425.$$

As these two values are equal, this rebalancing is self-financing.

For this trading strategy, all holdings are liquidated at time 2. This is indicated by the holdings vector at time 2 in state uu, ud, du, and dd having only zero entries, i.e., $(0, 0, 0, 0)$. The liquidated portfolio values at time 2 in the various states are:

$$\left(\text{time 2 state } uu\right) \quad n_o(1;u)B(2;u) + n_4(1;u)P(2,4;uu) + n_3(1;u)P(2,3;uu) + n_2(1;u)P(2,2;uu)$$
$$= (0)1.037958 + (0).967826 + (3.5).984222 + (0)1 = 3.444777$$

$$\left(\text{time 2 state } ud\right) \quad n_o(1;u)B(2;u) + n_4(1;u)P(2,4;ud) + n_3(1;u)P(2,3;ud) + n_2(1;u)P(2,2;ud)$$
$$= (0)1.037958 + (0).960529 + (3.5).980015 + (0)1 = 3.430053$$

$$\left(\text{time 2 state } du\right) \quad n_o(1;d)B(2;d) + n_4(1;d)P(2,4;du) + n_3(1;d)P(2,3;du) + n_2(1;d)P(2,2;du)$$
$$= (0)1.042854 + (0).962414 + (0).981169 - (.5962084)1 = -.5962084$$

$$\left(\text{time 2 state } dd\right) \quad n_o(1;d)B(2;d) + n_4(1;d)P(2,4;dd) + n_3(1;d)P(2,3;dd) + n_2(1;d)P(2,2;dd)$$
$$= (0)1.042854 + (0).953877 + (0).976147 - (.5962084)1 = -.5962084.$$

The money market account's value at time 2 in the uu and ud states is given by $B(1)r(1;u) = 1.02(1.017606) = 1.037958$ and it is given by $B(1)r(1;d) = 1.02(1.022406) = 1.042854$ in the du and dd states. At time 2 in either state uu or ud the trading strategy has a negative $value$ at liquidation and thus a negative $cash\ flow$.

This completes the illustration of a trading strategy. They will be used repeatedly throughout the book. For simplicity in the subsequent formalization of this example, since at the liquidation date, time 2, the holdings vector is identically zero, it will usually be omitted. □

We now formalize this example. First, fix a particular zero-coupon bond with maturity date τ_1, where $0 < \tau_1 \leq \tau$. The simplest $trading\ strategy$ is a pair of security holdings $(n_0(t;s_t),\ n_\tau(t;s_t))$ for all s_t and $t \in \{0,1,\ldots,\min(\tau_1,\tau-1)-1\}$ such that $n_0(t;s_t)$ is the number of units of the money market account held at time t at state s_t, and $n_\tau(t;s_t)$ is the number of shares of the τ_1-maturity zero-coupon bond held at time t at state s_t.

This portfolio is initially formed at time 0, and it is liquidated at the horizon date $\tau^* = \min(\tau_1,\ \tau-1)$. The horizon date is the smaller of the zero-coupon bond's maturity, τ_1, or the last relevant date in the model, $\tau-1$. The last relevant date in the model is time $\tau-1$ because over the last period in the model $[\tau-1,\ \tau]$, the remaining traded τ-maturity zero-coupon bond's payoff is identically one. It does not differ across the nodes in the tree, and therefore, it does not span the relevant uncertainties; see Figure 5.3. Consequently, this period is not useful for analysis, and it is omitted from the trading strategy horizon.

The fact that these time t portfolio holdings $(n_0(t;s_t),\ n_\tau(t;s_t))$ depend only on the past history s_t implies that only currently available information is used in the determination of the time t trading strategy. This is an obvious fact in real life, but an important restriction to impose on the abstract mathematical model.

The initial $value$ of this trading strategy at time 0 is

$$n_0(0)1 + n_{\tau_1}(0)P(0,\tau_1). \qquad (7.1)$$

The initial *cash flow* is minus the value given here in expression (7.1). For example, if the *value* in expression (7.1) is positive, then the initial *cash flow* is negative, because one is "buying" the portfolio. If the *value* in expression (7.1) is negative, then the initial *cash flow* is positive, because one is "shorting" the portfolio. If the initial value of the position is zero, then the trading strategy requires no cash inflow or outflow, and it is said to be a *zero investment trading strategy*.

At an (arbitrary) intermediate time *t*, the trading strategy enters with a value equal to

$$n_0(t-1;s_t)B(t;s_{t-1})+n_{\tau_1}(t-1;s_t)P(t,\tau_1;s_t). \tag{7.2}$$

This represents the holdings at date $t-1$ times the value of the securities at date t in state s_t.

The portfolio is then rebalanced, and the time t value of the trading strategy after rebalancing is

$$n_0(t;s_t)B(t;s_{t-1})+n_{\tau_1}(t;s_t)P(t,\tau_1;s_t). \tag{7.3}$$

This represents the new holdings at date t times the value of the securities at date t in state s_t.

The portfolio value entering time t in expression (7.2) need not equal the portfolio value after rebalancing in expression (7.3). If the portfolio value after rebalancing is larger (smaller) than the value entering, then a positive (negative) cash inflow to the trading strategy has occurred. This cash infusion (receipt) is implicitly determined by the difference between expression (7.2) and (7.3). If the portfolio value after rebalancing equals the portfolio value entering, the rebalancing is said to be *self-financing*. A self-financing rebalancing requires no cash inflow or outflow.

At the liquidation date, $\tau^* = \min(\tau_1, \tau-1)$, the trading strategy's value is

$$n_0(\tau^*-1;s_{\tau^*-1})B(\tau^*;s_{\tau^*-1})+n_{\tau_1}(\tau^*-1;s_{\tau^*-1})P(\tau^*,\tau_1;s_{\tau^*}). \tag{7.4}$$

This is the value of the portfolio entering time τ^*. It equals the shares held at time τ^*-1 times the value of the securities at time τ^*. The portfolio is liquidated at time τ^* (all holdings go to zero). The value of the portfolio at liquidation equals the cash flow. For example, if the value is positive at τ^*, then the cash flow will be positive as well because the assets held are sold.

This trading strategy is said to be *self-financing* if at every intermediate trading date $t \in \{1, 2, \ldots, \tau^* - 1\}$ the value of the portfolio entering time t equals its value after rebalancing, i.e.,

$$n_0(t-1; s_t)B(t; s_{t-1}) + n_{\tau_1}(t-1; s_t)P(t, \tau_1; s_t)$$
$$= n_0(t; s_t)B(t; s_{t-1}) + n_{\tau_1}(t; s_t)P(t, \tau_1; s_t). \tag{7.5}$$

We will be interested only in self-financing trading strategies. For convenience, define the set $\Phi_1 = \{$all self-financing trading strategies involving only the money market account and the τ_1-maturity zero-coupon bond$\}$.

Note that these trading strategies are only defined on the time horizon $\tau^* - 1 = \min(\tau_1, \tau - 1) - 1$ determined by the τ_1-maturity bond. This is because after time τ_1, the τ_1-maturity bond no longer trades.

Similarly, one could investigate a self-financing trading strategy involving three securities: two distinct zero-coupon bonds with maturities τ_1, τ_2 where $0 < \tau_1 < \tau_2 \leq \tau$, and the money market account. This would be represented by a triplet $(n_0(t; s_t), n_{\tau_1}(t; s_t), n_{\tau_2}(t; s_t))$ for all s_t and $t \in \{0, 1, \ldots, \tau_1 - 1\}$ such that $n_0(t; s_t)$ and $n_{\tau_1}(t; s_t)$ are as defined before, and $n_{\tau_2}(t; s_t)$ is the number of shares of the τ_2-maturity zero-coupon bond held at time t under state s_t.

We assume that the following self-financing condition, condition (7.6), is also satisfied:

$$n_0(t-1; s_t)B(t; s_{t-1}) + n_{\tau_1}(t-1; s_t)P(t, \tau_1; s_t) + n_{\tau_2}(t-1; s_t)P(t, \tau_2; s_t)$$
$$= n_0(t; s_t)B(t; s_{t-1}) + n_{\tau_1}(t; s_t)P(t, \tau_1; s_t) + n_{\tau_2}(t; s_t)P(t, \tau_2; s_t) \tag{7.6}$$

for all $t \in \{1, 2, \ldots, \tau_1 - 1\}$. This portfolio is liquidated at time τ_1 with value

$$n_0(\tau_1 - 1; s_{\tau_1 - 1})B(\tau_1; s_{\tau_1 - 1}) + n_{\tau_1}(\tau_1 - 1; s_{\tau_1 - 1})1 + n_{\tau_2}(\tau_1 - 1; s_{\tau_1 - 1})P(\tau_1, \tau_2; s_{\tau_1}). \tag{7.7}$$

The liquidation value equals the holdings at time $\tau_1 - 1$ times the securities' prices at time τ_1. Note that the zero-coupon bond with maturity τ_1 is identically one at time τ_1.

In an analogous manner, we define the set $\Phi_2 = \{$all self-financing trading strategies involving only the money market account, the τ_1-maturity zero-coupon bond, and the τ_2-maturity zero-coupon bond$\}$.

For convenience, we have defined the trading strategy over the time horizon determined by the shortest maturity bond τ_1. We could have, however, equivalently defined it over the time period determined by the longest τ_2-maturity bond, $[0, \tau_2]$. This would have required that we set the position in the τ_1-maturity zero-coupon bond to be identically zero over the later part of this horizon, i.e., $n\tau_1(t;s_t) \equiv 0$ for $t \in [\tau_1, \tau_2]$. This later restriction would be necessary because the τ_1-maturity bond no longer exists after time τ_1. When convenient, we will utilize this alternative definition without additional comment.

The set of self-financing trading strategies Φ_1 is a subset of Φ_2. It is the subset in which the holdings of the τ_2-maturity bond in Φ_2 are identically equal to zero.

In a straightforward fashion, given any $K < \tau$ bonds with maturities $\tau_1 < \tau_2 < \ldots < \tau_K$, we could define analogous sets of self-financing trading strategies Φ_K. The relation $\Phi_1 \subset \Phi_2 \subset \Phi_3 \subset \ldots \subset \Phi_K$ holds.*

One last set of self-financing trading strategies needs to be defined. This is the set simultaneously utilizing all the zero-coupon bonds of different maturities. Formally, consider the self-financing trading strategy $(n_1(t;s_t), n_2(t;s_t), \ldots, n_\tau(t;s_t))$ for all s_t and $t \in \{0, 1, \ldots, \tau - 2\}$ such that

$n_j(t;s_t)$ is the number of units of the jth-maturity zero-coupon bond purchased at time t at state s_t for $j = 1, 2, \ldots, \tau$, and

$n_j(t;s_t) \equiv 0$ for $t \geq j$.

The following self-financing condition (7.8) is satisfied, i.e.,

$$\sum_{j=1}^{\tau} n_j(t-1;s_{t-1})P(t,j;s_t) = \sum_{j=1}^{\tau} n_j(t;s_t)P(t,j;s_t) \tag{7.8}$$

for all s_t and $t \in \{0, 1, 2, \ldots, \tau - 2\}$.

The value of this portfolio at liquidation, time $\tau - 1$, is

$$n_{\tau-1}(\tau-2;s_{\tau-2})1 + n_\tau(\tau-2;s_{\tau-2})P(\tau-1,\tau;s_{\tau-1}). \tag{7.9}$$

The right side consists of only two terms because the only zero-coupon bonds still trading at time $\tau - 1$ are those with maturities $\tau - 1$ and τ. Furthermore, the $\tau - 1$ maturity zero-coupon bond has a value identically equal to one at time $\tau - 1$.

---

* Formally, we are imbedding Φ_1 into Φ_2, Φ_2 into Φ_3, and so on, because they are not strict subsets but only isomorphic to a strict subset.

Define the set $\Phi_\tau = \{$all self-financing trading strategies involving all the different maturity zero-coupon bonds$\}$.

There are a couple of subtle observations to make about this set of trading strategies. First, the horizon for this class of trading strategies is the whole trading interval less one period: $[0, \tau - 1]$. This is because over the last period in the model, the only traded zero-coupon bond is the τ-maturity zero-coupon bond. Its payoff at time τ is identically one. Thus, it cannot span the relevant uncertainties; e.g., see Figure 5.3. This last period in the model is a vestige that has no real economic purpose and is therefore omitted from the trading strategy horizon.

Second, after any zero-coupon bond matures, the holdings in this bond must be set to be identically zero. This is because the bond no longer exists after it matures. This fact is captured by the condition that $n_j(t; s_t) \equiv 0$ for $t \geq j$.

Third, this class of trading strategies takes no explicit position in the money market account, because the money market account can be created within this class of trading strategies. Indeed, the money market account is obtainable (by itself or in conjunction with additional holdings of the zero-coupon bonds) with the following self-financing trading strategy. First, purchase $1/P(0,1)$ units of the 1-period zero-coupon bond at time 0. The cost of this purchase is $(1/P(0,1))P(0,1) = 1$ dollar. This cost equals the money market account's value at time 0, i.e., $B(0) = 1$. This position then pays off $1/P(0,1) = r(0) = B(1)$ dollars at time 1. Next, reinvest this $B(1)$ dollars at time 1 into the zero-coupon bond that matures at time 2 (i.e., purchase $B(1)/P(1,2; s_1)$ of these bonds). The time 2 value of this position is then $B(1)r(1;s_1) = B(2;s_1)$. Continuing in this fashion generates the money market account $B(t;s_{t-1})$ for any t, s_{t-1}.

For this reason, and for consistency of the notation with the previously defined sets Φ_j for $j < \tau$, we can without loss of generality redefine Φ_τ to be the class of all self-financing trading strategies

$$(n_0(t;s_t), n_1(t;s_t), n_2(t;s_t), \ldots, n_\tau(t;s_t)) \quad \text{for all } s_t \text{ and } t\{0,1,\ldots,\tau-2\}$$

where $n_j(t;s_t)$ for $j = 1, \ldots, \tau$ are defined as before, and $n_0(t;s_t)$ is the number of units of the money market account $B(t;s_{t-1})$ held at time t under state s_t.

The self-financing condition holds, i.e.,

$$n_0(t-1;s_{t-1})B(t;s_{t-1}) + \sum_{j=1}^{\tau} n_j(t-1;s_{t-1})P(t,j;s_t)$$

$$(7.10)$$

$$= n_0(t;s_{t-1})B(t;s_{t-1}) + \sum_{j=1}^{\tau} n_j(t;s_t)P(t,j;s_t)$$

for all s_t and $t \in \{0, 1, ..., \tau - 2\}$.

The value of this portfolio at liquidation, time $\tau - 1$, is

$$n_0(\tau-2;s_{\tau-2})B(\tau-1;s_{\tau-2}) + n_{\tau-1}(\tau-2;s_{\tau-2})1$$

$$+ n_\tau(\tau-2;s_{\tau-2})P(\tau-1,\tau;s_{\tau-2}).$$

$$(7.11)$$

This is the liquidation value because at time $\tau - 1$ there are only three securities traded: the money market account, the $\tau - 1$ maturity zero-coupon bond, and the τ-maturity zero-coupon bond. The $\tau - 1$ maturity zero-coupon bond has a price identically equal to one at time $\tau - 1$.

7.3 ARBITRAGE OPPORTUNITIES

This section formally defines the concept of an arbitrage opportunity. Intuitively, an arbitrage opportunity is a zero investment self-financing trading strategy that generates positive cash flows (with positive probability) at no risk of a loss. The notion of an arbitrage opportunity is essential to the subsequent theory. In fact, setting prices such that there are no arbitrage opportunities in a market is the key method for determining fair values for fixed income securities and interest rate options. An example illustrates the notion of an arbitrage opportunity.

Example: An Arbitrage Opportunity

Consider the zero-coupon bond price evolution depicted in Figure 7.3. In this figure, there are two time periods depicted, times 0 and 1. At time 0 there are three zero-coupon bonds trading: $P(0,3)$, $P(0,2)$, and $P(0,1)$. The market prices for the bonds are given. At time 2, the zero-coupon bond prices can move up or down. The time 1 market prices for these zero-coupon bonds are also given. Note that across both paths, the shortest maturity bond, $P(0,1)$, matures and pays off a dollar.

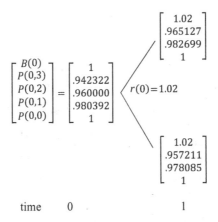

time   0   1

FIGURE 7.3   An Example of a Zero-Coupon Bond Price Curve Evolution with an Arbitrage Opportunity.

There is a mispricing implicit in the time 0 market prices of these zero-coupon bonds. Can you see it?

The 2-period zero-coupon bond is undervalued, and it should sell for .961169 dollars. (This book provides a technique for determining such mispricings.) A clever trader will recognize this mispricing and create a trading strategy based on it, to generate profits at no risk of a loss. The trader should sell .58287 units of the 3-period zero-coupon bond, sell .4119177 units of the money market account, and buy 1 unit of the 2-period zero-coupon bond. The trader should hold this portfolio until time 1, then liquidate his position. This is called a *buy and hold* trading strategy.

The initial cash flow from this position is:

$$+(.58287)P(0,3) - (1)P(0,2) + (.4119177)B(0)$$

$$= +(.58287).942322 - (1).96000 + (.4119177)1 = +.001169.$$

A positive cash flow implies a negative value (a liability). We next determine the time 1 cash flows from liquidating this position. This liquidation requires a determination of the money market account's time 1 value. This is easily computed from the relation $B(1) = B(0)r(0)$ where $B(0) = 1$ and $r(0) = [1/P(0,1)] = [1/.980392] = 1.02$. The money market account has the same time 1 value in both the up- and down states due to the fact that it earns the spot rate of interest over [0, 1], and the spot rate of interest over [0, 1] is known at time 0.

The time 1 cash flow in the up state is:

$$-(.58287)P(1,3;u)+(1)P(1,2;u)-(.4119177)B(1)$$

$$= -(.58287).965127+(1).982699-(.4119177)1.02 = 0.$$

The time 1 cash flow in the down state is:

$$-(.58287)P(1,3;d)+(1)P(1,2;d)-(.4119177)B(1)$$

$$= -(.58287).957211+(1).978085-(.4119177)1.02 = 0.$$

Surprisingly, there is no additional cash flow from liquidating this portfolio at time 1. Thus, this portfolio generates +.001169 dollars at time 0 and has no further liability. It creates cash from nothing! It is a money pump. Obviously, to create more cash, one should increase the size of the holdings in each of the assets proportionately. Why be satisfied with only +.001169 dollars?

The trading strategy employed in this example is an arbitrage opportunity. It generates positive cash flows at time 0 with no further liability at time 1. Arbitrage opportunities are fantastic investments! They increase wealth at no risk of a loss. We would not expect to find many of these arbitrage opportunities in well-functioning markets. The only price for the 2-period zero-coupon bond consistent with no arbitrage is .961169. □

We now formalize the concept of an arbitrage opportunity. An arbitrage opportunity is a particular type of self-financing trading strategy. To be precise, consider the set of self-financing trading strategies Φ_K for some $K \in \{1, 2, \dots, \tau\}$. Recall that these self-financing trading strategies are initiated at time 0 and are liquidated at time τ_1, which is the smallest maturity of the zero-coupon bonds included in the trading strategy.

An *arbitrage opportunity* is a self-financing trading strategy* $(n_0(t;s_t), n\tau_1(t;s_t), \dots, n\tau_K(t;s_t)) \in \Phi_K$ such that either

(Type I)

$$n_0(0)B(0) + \sum_{j=1}^{K} n_{\tau_j}(0)P(0,\tau_J) < 0 \qquad (7.12)$$

---

* A type I arbitrage opportunity can be transformed into a type II arbitrage opportunity by investing the initial cash flow in a zero-coupon bond maturing at time account τ_1.

and

$$n_0(\tau_1 - 1; s_{\tau_1 - 1})B(\tau_1; s_{\tau_1 - 1}) + \sum_{j=1}^{K} n_{\tau_j}(\tau_1 - 1; s_{\tau_1 - 1})P(\tau_1, \tau_j; s_{\tau_1}) \equiv 0 \quad \text{for all } s_{\tau_1},$$

or

(Type II)

$$n_0(0)B(0) + \sum_{j=1}^{K} n_{\tau_j}(0)P(0, \tau_j) = 0 \qquad (7.13)$$

and

$$n_0(\tau_1 - 1; s_{\tau_1 - 1})B(\tau_1; s_{\tau_1 - 1}) + \sum_{j=1}^{K} n_{\tau_j}(\tau_1 - 1; s_{\tau_1 - 1})P(\tau_1, \tau_j; s_{\tau_1}) \geq 0 \quad \text{for all } s_{\tau_1}$$

$$n_0(\tau_1 - 1; s_{\tau_1 - 1})B(\tau_1; s_{\tau_1 - 1}) + \sum_{j=1}^{K} n_{\tau_j}(\tau_1 - 1; s_{\tau_1 - 1})P(\tau_1, \tau_j; s_{\tau_1}) > 0 \quad \text{for some } s_{\tau_1}.$$

There are two types of arbitrage opportunities. A type I arbitrage opportunity, given in expression (7.12), has a negative value at time 0. This corresponds to a positive initial cash flow. Furthermore, at the liquidation date τ_1, there is no further liability. The liquidation value is identically zero. This is the type of arbitrage opportunity that was illustrated in the previous example.

The second type of arbitrage opportunity is given in expression (7.13). It states that the initial value of the portfolio is zero; i.e., it is a zero investment trading strategy. The portfolio is liquidated* at time τ_1. At liquidation, the portfolio's value is nonnegative for all states s_{τ_1} and strictly positive for some states. This type of arbitrage opportunity offers the possibility of turning nothing into something, with no chance of a loss and with a strictly positive probability of a gain. This is a probabilistic money pump!

To illustrate an arbitrage opportunity of type II, suppose the τ_1-maturity bond had a zero price at time 0; i.e., $P(0, \tau_1) = 0$. An arbitrage opportunity of type II in the smallest class of trading strategies Φ_1 would be buying the τ_1-maturity bond and holding it until time τ_1. Indeed, the initial cost is 0.

---

* For $K = \tau$ we make the following identification for the indices: $\tau_1 \equiv 1, \tau_2 \equiv 2, ..., \tau_\tau \equiv \tau$. In this case, the portfolio's liquidation date is $\tau - 1$, not τ_1 as in expressions (7.12) and (7.13).

At maturity, it is worth a sure dollar. Consequently, it satisfies expression (7.13). In well-functioning markets, we would expect $P(0,\tau_1) > 0$ for all τ_1. This is the reason we imposed this strict positivity condition earlier in Chapter 5.

In closing, note that because the trading strategies are subsets of each other, i.e., $\Phi_1 \subset \Phi_2 \subset \ldots \subset \Phi_\tau$, an arbitrage opportunity with respect to a trading strategy set Φ_j is an arbitrage opportunity with respect to a larger trading strategy set Φ_K for $j \leq K \leq \tau$. Thus, if there are no arbitrage opportunities in the larger trading strategy set Φ_K, there are no arbitrage opportunities in the smaller trading strategy set Φ_j for $j \leq K$. The converse is, of course, not true.

The subsequent pricing theory is based on the simple notion that we would not expect to see many arbitrage opportunities in well-functioning markets. Why? Because bright investors would hold these arbitrage opportunities, becoming wealthy in the process. They would desire to hold as many of them as possible. The process of arbitrageurs taking advantage of these arbitrage opportunities would cause equilibrium prices (supply and demand) to change until these arbitrage opportunities are eliminated.

7.4 COMPLETE MARKETS

The concept of a complete market is the last notion that we need to introduce before studying risk-neutral valuation. Roughly, a complete market is one in which any cash flow pattern desired can be obtained via a self-financing trading strategy. This concept is essential in pricing derivative securities. Indeed, if the traded derivative security's cash flows can be duplicated synthetically via a trading strategy, then the initial cost of this trading strategy should be the fair price of the traded derivative security. Otherwise, an arbitrage opportunity can be constructed. The key to this argument is the ability to replicate the traded derivative security's cash flows. This is the concept of a complete market that we next illustrate with an example.

Example: A Complete Market

Figure 7.4 contains an evolution for a term structure of interest rates for two dates (0 and 1) and two zero-coupon bonds ($P(0,1)$, $P(0,2)$). The spot interest rate is included at the intersection of the node on the tree ($r(0) = 1.02$), and the value of the money market account's value is included at the top of each zero-coupon bond price vector.

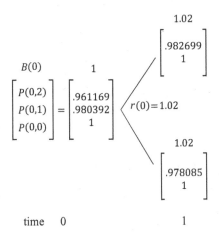

FIGURE 7.4 An Example of a Zero-coupon Bond Price Curve Evolution in a Complete Market.

This term structure evolution implies a complete market. To prove this, suppose at time 1 we desire to construct a portfolio whose value is (x_u, x_d) in the up and down states, respectively, where $x_u \neq x_d$. For example, this cash flow could represent the value of an interest rate cap or a Treasury bond futures option. For convenience, let us call this cash flow the value of an (arbitrary) traded interest rate option.

The idea is to form a trading strategy $(n_0(0), n_2(0))$ in the money market account and 2-period zero-coupon bond at time 0 such that the liquidation value at time 1 matches the cash flow (x_u, x_d).

The initial cost of this trading strategy is

$$n_0(0)B(0) + n_2(0)P(0,2) = n_0(0)1 + n_2(0).961169. \qquad (7.14)$$

The objective is to choose $(n_0(0), n_2(0))$ such that

$$n_0(0)B(1) + n_2(0)P(1,2;u) = x_u$$

and

$$n_0(0)B(1) + n_2(0)P(1,2;d)x_d. \qquad (7.15)$$

Here, the value of the trading strategy entering time 1 equals the desired values of the interest rate option (x_u, x_d).

Substitution of the prices from Figure 7.4 into expression (7.15) yields

$$n_0(0)1.02 + n_2(0).982699 = x_u$$

and

$$n_0(0)1.02 + n_2(0).978085 = x_d. \tag{7.16}$$

The solution is

$$n_0(0) = \frac{x_d(.982699) - x_u(.978085)}{1.02(.005614)} \tag{7.17}$$

and

$$n_2(0) = \frac{x_u - x_d}{.005614}.$$

Expression (7.17) gives the trading strategy that replicates the cash flow (x_u, x_d). This trading strategy is called the *synthetic interest rate option*. The cost of constructing this trading strategy is obtained by substituting expression (7.17) into expression (7.14), i.e.,

$$\frac{x_d(.982699) - x_u(.978085)}{1.02(.005614)}(1) + \frac{x_u - x_d}{.005614}(.961169). \tag{7.18}$$

This cost of constructing the synthetic interest rate option is called the *arbitrage-free* price because if the traded interest rate option had a different market price, then an arbitrage opportunity could be constructed. Indeed, the traded and synthetic interest rate options have identical values at time 1 under all states. Thus, an arbitrage opportunity would be to buy the cheap option and sell the expensive one, pocketing the difference at time 0, and having no future liability. This completes the example. □

To formalize the concept of a complete market, we must first define a simple contingent claim. Simple contingent claims are the building blocks for all possible derivative securities issued against the term structure of interest rates. This includes forward and futures contracts, call and put options both European and American, as well as other interest rate options, exotic or otherwise.

A simple contingent claim has a single maturity date and a value or cash flow on that date. Formally, a *simple contingent claim* with maturity date τ^* for $0 \leq \tau^* \leq \tau - 1$ is defined to be a cash flow $x(\tau^*; s_{\tau^*})$ dependent on time τ^* and state s_{τ^*}.

A market is said to be *complete with respect to the trading strategies* Φ_K for $K \in \{1, 2, 3, \ldots, \tau\}$ if given any simple contingent claim with maturity date* $\tau^* \leq \tau_1 - 1$, there exists a self-financing trading strategy $(n_0(t; s_t): n_{\tau_1}(t; s_t), \ldots, n_{\tau_K}(t; s_t)) \in \Phi_K$ such that at time τ^*,

$$
n_0(\tau^* - 1; s_{\tau^*-1})B(\tau^*; s_{\tau^*-1}) + \sum_{j=1}^{K} n_{\tau_j}(\tau^* - 1; s_{\tau^*-1})P(\tau^*, \tau_j; s_{\tau^*})
$$

$$
= x(\tau^*; s_{\tau^*}) \quad \text{for all } s_{\tau^*}.
$$

(7.19)

The left side of expression (7.19) represents the value of the trading strategy entering time τ^*. The right side of expression (7.19) equals the value of the simple contingent claim. Condition (7.19) guarantees that the trading strategy's value at time τ^* replicates the simple contingent claim's time τ^* cash flow at each state s_{τ^*}. Explicit in expression (7.19) is the condition that only the money market account and the zero-coupon bonds with maturities $\tau_1, \tau_2, \ldots, \tau_K$ are utilized in this trading strategy. We will call the trading strategy in expression (7.19) that replicates the simple contingent claim's time τ^* cash flows a *synthetic simple contingent claim x*.

If the market is complete with respect to the trading strategy set Φ_K, it is also complete with respect to a larger set trading strategy set Φ_j for any $j \geq K$. This follows because the smaller trading strategy set Φ_K is contained in the larger trading strategy set Φ_j for $j \geq K$. So, any portfolio available in the smaller set is available in the larger set as well. This observation will prove useful in subsequent chapters.

We define the *arbitrage-free price of the simple contingent claim with maturity date* τ^* to be the initial cost of constructing the synthetic contingent claim, i.e.,

$$
n_0(0)B(0) + \sum_{j=1}^{K} n_{\tau_j}(0)P(0, \tau_j).
$$

(7.20)

---

* The convention followed for the remainder of the text is that when $K = \tau$, the following identifications hold: (1.) $\tau_1 = 1, \tau_2 = 2, \ldots, \tau_\tau = \tau$. (2.) The trading horizon is $[0, \tau-1]$. (3.) All holdings in bonds that have already matured must be identically zero.

The idea behind this definition is that if the simple contingent claim x is traded, its price would have to be that given by expression (7.20). Otherwise, it would be possible to construct an arbitrage opportunity involving the τ_1-, τ_2-, ... , τ_K- maturity zero-coupon bonds and the traded contingent claim. This logic will be further clarified in subsequent chapters.

We now show that all other derivative securities issued against the term structure of interest rates can be constructed from simple contingent claims. Consider a more *complex contingent claim* than that defined above, namely, one having multiple random cash flows at the different dates $\{0, 1, ... , \tau^*\}$, the cash flows being denoted by $x(t;s_t)$ for each $t \in \{0, 1, ... , \tau^*\}$.

This complex contingent claim, however, is nothing more than a collection of τ^* simple contingent claims each with a different maturity date. So, understanding the simple contingent claims is sufficient for understanding these complex contingent claims.

For example, the self-financing trading strategy that duplicates the complex contingent claim is the sum of the self-financing trading strategies that generate each simple contingent claim composing the complex contingent claim. The holdings in each underlying zero-coupon bond are merely summed to get the aggregate holdings of each zero-coupon bond in the complex contingent claim.

Correspondingly, the arbitrage-free price of this complex contingent claim having multiple cash flows is the sum of the arbitrage-free prices of the simple contingent claims of which it is composed. The motivation for this definition is that if both the simple contingent claims and the complex contingent claim traded simultaneously, this would be the only price consistent with the absence of arbitrage opportunities.

The previous two types of contingent claims are of the *European* type, because the cash flows are independent of any active decision made by the investors holding the claim. Contingent claims in which the cash flows depend on an active decision of the investors holding the claim are of the *American* type.* These contingent claims can also be analyzed using an extension of the above procedure.

Let us denote the cash flows to an American-type complex contingent claim at time $t \in \{0, 1, ... , \tau - 1\}$ given a decision choice $a \in A$ of the investor by $x(t,a;s_t)$ where A represents the set of possible decisions.

---

* A European call option's payoff at maturity is considered to be a *passive* decision, because it is assumed to be exercised if it is in the money.

Given an arbitrary decision $a^* \in A$, the cash flow to the claim is $x(t,a^*;s_t)$ for each $t \in \{0, 1, \ldots, \tau - 1\}$. Now the cash flow no longer depends on the investor making a choice, for it is already determined by a^*. Thus, given $a^* \in A$, the claim is now a European-type claim (sometimes called *pseudo-American*). And, we are back to valuing a complex contingent claim as a portfolio of simple contingent claims. Both the self-financing trading strategy in Φ_K that duplicates the European complex contingent claim under a^* and its arbitrage-free price are determined as discussed previously.

Finally, to get the American contingent claim's synthetic self-financing trading strategy and the arbitrage-free price, we choose that $\tilde{a} \in A$ such that the arbitrage-free price is maximized for the investor holding the claim. The synthetic self-financing trading strategy for $x(t,\tilde{a};s_t)$ duplicates the American claim's cash flows under the optimal decision. The arbitrage-free price of $x(t,\tilde{a};s_t)$ is, thus, the arbitrage-free price for the American contingent claim.

This completes the demonstration that all types of interest rate derivatives can be formulated as portfolios of simple contingent claims. Hence, if one can generate all simple contingent claims synthetically using self-financing trading strategies in the zero-coupon bonds, then all interest rate derivatives can be synthetically constructed, and the market is complete.

CHAPTER 8

Bond Trading Strategies – An Example

8.1 MOTIVATION

This chapter shows how to use the concepts of trading strategies, arbitrage opportunities, and complete markets to investigate mispricings within the yield curve. Taken as given are the stochastic processes for a few zero-coupon bonds (i.e., one bond for the one-factor model, two bonds for the two-factor model, and so forth), the stochastic process for the spot rate, and the assumption that there are no arbitrage opportunities. The purpose is to find the arbitrage-free prices of all the remaining zero-coupon bonds. If the market prices for the remaining zero-coupon bonds differ from these arbitrage-free prices, then arbitrage opportunities have been discovered.

This is the modern analog of arbitraging the yield curve. Prior to the development of these techniques, traders would plot bond yields versus maturities, fitting a smooth curve to the graph (perhaps using spline techniques), and they would buy or sell outliers from the curve. These outliers are analogous to the arbitrage opportunities discovered in the following technique.

This chapter illustrates this bond trading procedure using a 4-period example. The next chapter provides the abstract theory underlying this example. Mastering the example is sufficient for reading the remainder of the book.

This example can be visualized by examining Figure 8.1. Figure 8.1 contains a graph of a zero-coupon bond price curve. Marked with "X's"

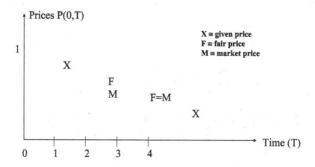

FIGURE 8.1   A Graphical Representation of "Arbitraging the Zero-Coupon Bond Price Curve." X = Given Price; F = Fair Price; M = Market Price.

are the prices of the 1-period and the 4-period zero-coupon bonds. These market prices are taken as given. Presumably, they are the most liquid zero-coupon bonds. Recall that the 1-period zero-coupon bond's price determines the spot rate; thus, the spot rate is given. The purpose of the analysis is to use these prices, plus an exogenous specification of the evolution of the spot rate and 4-period zero-coupon bond, to determine "fair" prices for the remaining zero-coupon bonds. These are the "F's" on the graph. Market quotes are indicated as "M's."

If the market quotes differ from the "F's," then an arbitrage opportunity has been identified. For example, the market quote for the 2-period zero-coupon bond is indicated on the graph as an "M." This market price lies below the "fair" value. Hence, the traded 2-period zero-coupon bond is undervalued. It should be purchased. The arbitrage opportunity is obtained by constructing a synthetic 2-period zero-coupon bond with the fair price "F" and selling it! The mechanics of this procedure are the content of this chapter.

8.2 METHOD 1: SYNTHETIC CONSTRUCTION

There are two equivalent methods for determining the "fair" or arbitrage-free prices of the desired zero-coupon bonds. The first method concentrates on a technique for constructing the zero-coupon bond synthetically, using the money market account and the 4-period zero. The second method uses an elegant theory to derive a present value operator – the technique is called *risk-neutral valuation*.

Both methods are important, and both need to be mastered. This section concentrates on the first method, the next section the second.

To illustrate the procedure for creating synthetic securities, we consider a 4-period example as given in Figure 8.2. Given on this tree diagram are the evolutions for the money market account and the 4-period zero-coupon bond. Also indicated at each node in the tree is the spot rate of interest. The spot rates can be determined from the evolution of the money market account (or conversely). The time 0 prices are the available market quotes.

8.2.1 An Arbitrage-Free Evolution

The first step in the procedure is to investigate whether the exogenous evolution as given in Figure 8.2 is arbitrage-free. The reason is simple. We are taking the evolution in Figure 8.2 as given, and then from it, we will derive arbitrage-free prices for the 2- and 3-period zero-coupon bonds. The procedure will be flawed and nonsensical if the basis for the technique itself contains imbedded arbitrage opportunities. For then, how can we determine arbitrage-free prices from them?

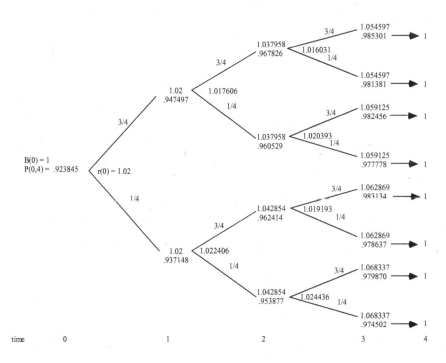

FIGURE 8.2 An Example of a 4-Period Zero-Coupon Bond Price Process. The Value of the Money Market Account and the Spot Rates Are Included on the Tree. Actual Probabilities Along Each Branch of the Tree.

The check to determine if the given evolution is arbitrage-free proceeds node by node. First, consider standing at time 0. There are two securities available for trade at this date, the money market account and the 4-period zero-coupon bond. The money market account is riskless over the next time period earning 1.02. The 4-period zero-coupon bond is risky. The return on the 4-period zero-coupon bond in the up state is ($u(0,4) = .9474$ 97/.923845 = 1.025602) and in the down state it is ($d(0,4) = .937148/.923845$ = 1.014400). In the up state it earns more than the money market account, and in the down state it earns less.

Hence, neither security dominates the other (in terms of returns). There would be no way to form a portfolio of these two securities with zero initial investment that didn't have potential losses at time 1. Furthermore, any portfolio with an initial positive cash flow would have a negative cash flow at time 1 in at least one state.[*] Thus, the fact that neither security dominates the other implies there are no arbitrage opportunities between time 0 and time 1. The converse of this statement is also true. That is, if there are no arbitrage opportunities between these two securities over [0, 1], then neither security should dominate the other in terms of returns.[†]

This condition applies to the first time step. But, if at each and every node in the tree, this condition applies, then the entire tree – the entire evolution – is arbitrage-free. Indeed, then no matter which path in the tree occurred, and at each node, there would be no trading strategy that could create arbitrage profits from that node onward.

Figure 8.3 provides the returns on the money market account and the 4-period zero-coupon bond at each node in the tree. We have argued that the tree is arbitrage-free if and only if

$$u(t,4:s_t) > r(t,s_t) > d(t,4:s_t) \quad \text{for all } t \text{ and } s_t.$$

Inspection of Figure 8.3 shows that this is true for the evolution provided.

8.2.2 Complete Markets

Given the tree is arbitrage-free, we next study the pricing of the 2- and 3-period zero-coupon bonds. This pricing procedure uses the concept of a complete market, introduced in Chapter 7.

---

[*] The reader is encouraged to prove these assertions before reading Chapter 9. These statements exclude arbitrage opportunities of type I and type II.

[†] The proof of this statement is by contradiction; see Chapter 9.

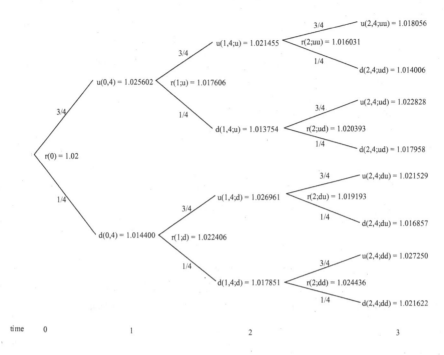

FIGURE 8.3 The Rate of Return Processes for the 4-Period Zero-Coupon Bond and the Money Market Account in Figure 8.2. Actual Probabilities Are Along Each Branch of the Tree.

The trading strategy set considered is $(n_0(t;s_t))$, $(n_4(t;s_t))$ for $t = 0$, 1, 2, and 3. This consists of the number of units of the money market account $(n_0(t;s_t))$ and the 4-period zero-coupon bond $(n_4(t;s_t))$ at time t in state s_t.

From Figure 8.3 we see that in all states and times, $u(t,4) > r(t) > d(t,4)$. Besides implying that the tree is arbitrage-free, this will also be seen to be a sufficient condition for market completeness. The reason for the sufficiency of this condition is clarified by actually constructing synthetic contingent claims for this example.

First, let us construct a *synthetic 2-period zero-coupon bond*. The payoff to this contingent claim is 1 dollar at time 2 across all states (*uu, ud, du, dd*). To construct this contingent claim, we work backward through the tree, starting at time 1. We form a state-dependent portfolio first at time 1, state *u*, and then at time 1, state *d*, to duplicate the 2-period zero-coupon bond's payoff at time 2. Then we move back to time 0 and create a portfolio that duplicates the previous portfolio values needed at time 1, state *u*, and at time 1, state *d*. This procedure will result in a dynamic, self-financing trading strategy that duplicates the payoff to a 2-period zero-coupon bond.

Consider first the node at time 1, state u. The desire is to determine $n_0(1;u)$ and $n_4(1;u)$, the number of units of the money market account and the 4-period zero-coupon bond, such that the portfolio's resulting time 2 value equals 1 dollar in both states, i.e.,

$$n_0(1;u)B(2;u)+n_4(1;u)P(2,4;uu)=n_0(1;u)1.037958+n_4(1;u).967826=1$$

$$(8.1a)$$

and

$$n_0(1;u)B(2;u)+n_4(1;u)P(2,4;ud)=n_0(1;u)1.037958+n_4(1;u).960529=1$$

$$(8.1b)$$

The solution to this system of equations is obtained by subtracting expression (8.1b) from (8.1a) and first solving for $n_4(1;u)$. Then, we substitute the value for $n_4(1;u)$ into (8.1a) and solve for $n_0(1;u)$. The solution is:

$$n_0(1;u)=1/1.037958=.963430 \qquad (8.2a)$$

$$n_4(1;u)=0. \qquad (8.2b)$$

The duplicating portfolio at time 1 under state u is to hold .963430 units of the money market account and zero units of the 4-period zero-coupon bond. This makes sense, because the payoff at time 1 must be certain, and the 4-period zero-coupon bond's payout is risky. So, none of the risky zero-coupon bond should be included.

The time 1 cost at state u of constructing this portfolio is

$$n_0(1;u)B(1)+n_4(1;u)P(1,4;u)=n_0(1;u)1.02+n_4(1;u).947947=.982699 \quad (8.3)$$

To prevent arbitrage, therefore, this *cost of construction* must be the arbitrage-free value of the traded 2-period zero-coupon bond, i.e.,

$$P(1,2;u)=.982699. \qquad (8.4)$$

A similar construction at time 1 under state d necessitates the time 2 payoff to the portfolio to be

$$n_0(1;d)B(2;d)+n_4(1;d)P(2,4;du)=n_0(1;d)1.042854+n_4(1;d).962414=1$$

$$(8.5a)$$

and

$$n_0(1;d)B(2;d)+n_4(1;d)P(2,4;dd) = n_0(1;u)1.042854+n_4(1;d).953877 = 1.$$

(8.5b)

The solution to this system is:

$$n_0(1;d) = 1/1.042854 = .958907$$ (8.6a)

$$n_4(1;d) = 0.$$ (8.6b)

The synthetic 2-period zero-coupon bond at time 1 under state d consists of .958907 units of the money market account and 0 units of the 4-period zero-coupon bond. Its time 1, state d, arbitrage-free value must be

$$P(1,2;d) = n_0(1;d)B(1)+n_4(1;d)P(1,4;d)$$

$$= n_0(1;d)1.02+n_4(1;d).937148$$

$$= .978085.$$

(8.7)

Now, moving backward in the tree to time 0, the values in expressions (8.4) and (8.7) become the new payoffs we need to duplicate at time 1. Let $n_0(0)$ and $n_4(0)$ be the number of units of the money market account and of the 4-period zero-coupon bond, respectively, needed at time 0 such that

$$n_0(0)B(1)+n_4(0)P(1,4;u) = n_0(0)1.02+n_4(0).947497 = .982699 = P(1,2;u)$$

(8.8a)

and

$$n_0(0)B(1)+n_4(0)P(1,4;d) = n_0(0)1.02+n_4(0).937148 = .978085 = P(1,2;d).$$

(8.8b)

A solution to this system exists because

$$u(0,4) = .947497/.923845 = 1.025602 > d(0,4)$$

$$= .937148/.923845 = 1.014399.$$

This is the sufficiency condition mentioned earlier. The solution is:

$$n_0(0) = .549286 \qquad\qquad\qquad (8.9a)$$

$$n_4(0) = .445835. \qquad\qquad\qquad (8.9b)$$

The synthetic 2-period zero-coupon bond consists of .549286 units of the money market account and .445835 units of the 4-period zero-coupon bond. Its time 0 arbitrage-free value is the cost of constructing this portfolio, i.e.,

$$P(0,2) = n_0(0)B(0) + n_4(0)P(0,4) = n_0(0)1 + n_4(0).923845 = .961169. \quad (8.10)$$

These values and portfolio positions are summarized in Figure 8.4.

It is important to note that the value for the 2-period zero-coupon bond does not depend on the actual probabilities for the evolution of

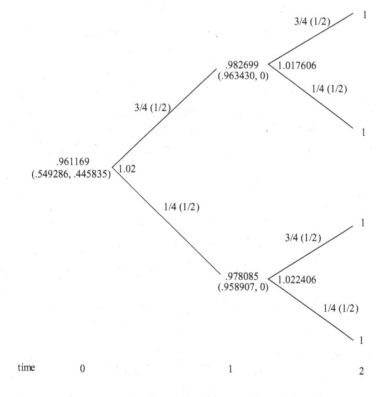

FIGURE 8.4 Arbitrage-Free Values $(P(t,2;s_t))$ and Synthetic Portfolio Positions $(n_0(t;s_t), n_3(t;s_t))$ for the 2-Period Zero-Coupon Bond. The Actual Probabilities Are Along Each Branch of the Tree with the Pseudo Probabilities in Parentheses.

the 4-period zero-coupon bond and the money market account. At first "blush" this appears to be wrong. But, further thought reveals the validity of this insight. Indeed, the method prices by exact replication, in both the up and down states. Regardless of the state which occurs (or its probability of occurrence), the 2-period zero-coupon bond's values are duplicated. Hence, the cost of construction, the arbitrage-free price, is determined independent of the actual probabilities of the future states!

Figure 8.5 presents the arbitrage-free values $P(t,3;s_t)$ and the synthetic portfolio positions $n_0(t;s_t)$ in the money market account and $n_4(t;s_t)$ in the 4-period zero-coupon bond that are needed to construct the 3-period zero-coupon bond. These values and positions are obtained similarly to those in Figure 8.4. The verification of these numbers is left to the reader as an exercise.

This completes the presentation of method 1, valuation by the cost of synthetic construction. This method always works, and it is the more intuitive of the two approaches. The next section discusses method 2, risk-neutral valuation.

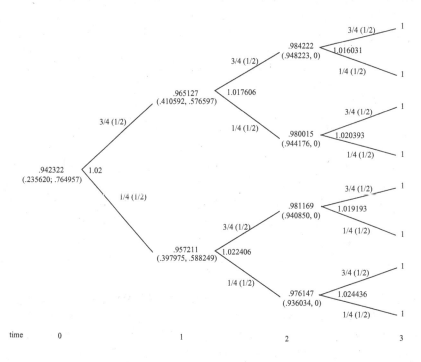

FIGURE 8.5 Arbitrage-Free Values $(P(t,3;s_t))$ and Synthetic Portfolio Positions $(n_0(t;s_t), n_4(t;s_t))$ for the 3-Period Zero-Coupon Bond. The Actual Probabilities Are Along Each Branch of the Tree with the Pseudo Probabilities in Parentheses.

8.3 METHOD 2: RISK-NEUTRAL VALUATION

The second method called risk-neutral valuation is the more useful approach from a computational perspective. This is because the risk-neutral valuation approach relates the derivative security's value to calculating an expectation. Expectations (integrals) are usually easy to compute (if necessary one can use Monte Carlo simulation).

The risk-neutral valuation approach proceeds by determining a present value operator that can be used to value the 2- and 3-period zero-coupon bonds. This present value operator is related to the "present value form of the expectations hypothesis" discussed in Chapter 6. The same example as in Chapter 6 is used below. As noted in Chapter 6, the present value form of the expectations hypothesis does not hold for this example. Something else is going on. This section now explains "the something else."

8.3.1 Risk-Neutral Probabilities

The first step in this method, as with method 1, is to guarantee that the evolution of the money market account and 4-period zero-coupon bond price process is arbitrage-free. We saw before that a necessary and sufficient condition for this is that at every node in the tree, the 4-period bond does not dominate nor is it dominated by the money market.

We wrote this as:

$$u(t,4;s_t) > r(t,s_t) > d(t,4;s_t) \quad \text{for all } t \text{ and } s_t. \tag{8.11}$$

We next investigate an equivalent form of this condition that has an alternative economic interpretation. Simple mathematics allows us to state that expression (8.11) is equivalent to the following condition, *there exists a unique number $\pi(t,s_t)$ strictly between 0 and 1 such that*

$$r(t;s_t) = \pi(t;s_t)u(t,4;s_t) + (1-\pi(t;s_t))d(t,4;s_t) \quad \text{for all } t \text{ and } s_t. \tag{8.12}$$

Expression (8.12) shows that an arbitrage-free evolution is equivalent to the existence of a collection of strictly positive real numbers $\pi(t,s_t) \in (0,1)$ that make a weighted average of the 4-period zero-coupon bond's return in the up state and down state equal to the spot rate of interest.

Expression (8.12), when rearranged, yields a formula for obtaining the $\pi(t,s_t)$'s. This formula is

$$\pi(t, s_t) = [r(t; s_t) - d(t, 4; s_t)] / [u(t, 4; s_t) - d(t, 4; s_t)]. \qquad (8.13)$$

In computing the $\pi(t, s_t)$'s using expression (8.13), if the computed $\pi(t, s_t)$'s are strictly between 0 and 1 for all times t and states s_t, then the tree is arbitrage-free. Otherwise, it is not.

Referring back to Figure 8.3, we can now check to see if the tree is arbitrage-free using this new condition (8.13). We can compute the $\pi(t, s_t)$'s at every node. They are:

$$\pi(0) = \frac{1.02 - 1.014400}{1.025602 - 1.014400} = 0.5$$

$$\pi(1; u) = \frac{1.017606 - 1.013754}{1.021455 - 1.013754} = 0.5$$

$$\pi(1; d) = \frac{1.022406 - 1.017851}{1.026961 - 1.017851} = 0.5$$

$$\pi(2; uu) = \frac{1.016031 - 1.014006}{1.018056 - 1.014006} = 0.5$$

$$\pi(2; ud) = \frac{1.020393 - 1.017958}{1.022828 - 1.017958} = 0.5$$

$$\pi(2; du) = \frac{1.019193 - 1.016857}{1.021529 - 1.016857} = 0.5$$

$$\pi(2; dd) = \frac{1.024436 - 1.021622}{1.027250 - 1.021622} = 0.5$$

The $\pi(t, s_t)$'s are all equal to 0.5. Because this number is strictly between 0 and 1, the evolution in Figure 8.2 is arbitrage-free. The observation that all the $\pi(t, s_t)$'s are equal to 0.5 is not a general result, but just a special case. It would not be satisfied for an arbitrary tree. It was purposely chosen for this example to simplify the subsequent computations.

Notice that each $\pi(t, s_t)$ can be interpreted as a *pseudo probability* of the up state occurring over the time interval $[t, t+1]$. Indeed, just as with the actual probability $q_t(s_t)$, these $\pi(t, s_t)$'s are all strictly between 0 and 1, and sum to 1, i.e., $\pi(t, s_t) + (1 - \pi(t, s_t)) = 1$. We call these $\pi(t, s_t)$'s pseudo probabilities because they are "false" probabilities. They are not the actual probabilities generating the evolution of the money market account and the 4-period zero-coupon bond prices. Nonetheless, these pseudo probabilities have an important usage. To see this usage, we need to transform expression (8.12) once again.

Multiplying the left and right side of expression (8.12) by $P(t,4;s_t)$ and noting that $u(t,4;s_t)P(t,4;s_t) = P(t+1,4;s_tu)$ and $d(t,4;s_t)P(t,4;s_t) = P(t+1,4;s_td)$ we obtain:

$$P(t,4;s_t) = \frac{\left[\pi(t;s_t)P(t+1,4;s_tu)+(1-\pi(t;s_t))P(t+1,4;s_td)\right]}{r(t;s_t)} \quad \text{for all } t \text{ and } s_t.$$

$$(8.14)$$

Expression (8.14) gives a *present value formula*. This present value formula says that to compute the time t value of the 4-period zero-coupon bond (the present value), simply take its expected value at time $t+1$ using the pseudo probabilities, and discount to time t (the present) using the spot rate of interest.

Letting $\tilde{E}_t(\cdot)$ denote taking an expected value using the pseudo probabilities $\pi(t,s_t)$, we can rewrite expression (8.14) as:

$$P(t,4;s_t) = \tilde{E}_t(P(t+1,4;s_{t+1}))/r(t;s_t). \quad (8.15)$$

In this form we see that expression (8.15) is a transformation of the present value form of the expectations hypothesis discussed in Chapter 6. The difference is that the expectations hypothesis holds in terms of the pseudo probabilities and not the actual probabilities! This is the subtlety that is often confused by the uniformed student of this material.

This is an important difference. The transformation to the pseudo probabilities includes the adjustment for the risk of the cash flows. Thus, the risk adjustment takes place in the numerator of expression (8.15). In more traditional present value calculations (say using the capital asset pricing model), the risk adjustment normally is included by increasing the discount factor in the denominator.

In terms of Figure 8.2, it is easy to check that the present value formula expression (8.15) is satisfied:

$$P(2,4;uu) = \tilde{E}_2(P(3,4;s_3))/r(2;uu)$$

$$= \frac{(1/2).985301+(1/2).981381}{1.016031} = .967826$$

$$P(2,4;ud) = \tilde{E}_2(P(3,4;s_3))/r(2;ud)$$

$$= \frac{(1/2).982456+(1/2).977778}{1.020393} = .960529$$

$$P(2,4;du) = \tilde{E}_2(P(3,4;s_3))/r(2;du)$$

$$= \frac{(1/2).983134+(1/2).978637}{1.019193} = .962414$$

$$P(2,4;dd) = \tilde{E}_2(P(3,4;s_3))/r(2;dd)$$

$$= \frac{(1/2).979870+(1/2).974502}{1.024436} = .953877$$

$$P(1,4;u) = \tilde{E}_1(P(2,4;s_2))/r(1;u)$$

$$= \frac{(1/2).967826+(1/2).960529}{1.017606} = .947497$$

$$P(1,4;d) = \tilde{E}_1(P(2,4;s_2))/r(1;d)$$

$$= \frac{(1/2).962414+(1/2).953877}{1.022406} = .937148$$

$$P(0,4) = \tilde{E}_0(P(1,4;s_1))/r(0)$$

$$= \frac{(1/2).947497+(1/2).937148}{1.02} = .923845$$

Thus, having guaranteed that the evolution of the money market account and the 4-period zero-coupon bond prices is arbitrage-free, we now move to valuing the 2- and 3-period zero-coupon bonds.

8.3.2 Risk-Neutral Valuation

It is a surprising result (proven in the next chapter) that the present value formula in expression (8.15) using the pseudo probabilities can be applied to *any random cash flow* whose payoffs depend on the evolution of the money market account and the 4-period zero-coupon bond as in

Figure 8.2. In fact, this implies that this present value formula can be used to price any interest rate derivative.

In general, the present value formula can be written as follows. Let $x(s_{t+1})$ represent a random cash flow received at time $t+1$ in state s_{t+1}. Then, the time t value of this time $t+1$ cash flow is:

$$\text{Present value}_t = \tilde{E}_t(x(s_{t+1}))/r(t;s_t). \tag{8.16}$$

The present value is computed by taking the expected cash flow, using the pseudo probabilities, and discounting at the spot rate of interest.

This present value formula is called *risk-neutral valuation* because it is the value that the random cash flow "x" would have in an economy populated by risk-neutral investors, having the pseudo probabilities as their beliefs (see Chapter 6). Of course, the actual economy is not populated by risk-neutral investors nor are their beliefs given by the pseudo probabilities. Nonetheless, if there are no arbitrage opportunities and the market is complete, then expression (8.16) is valid.

To illustrate this risk-neutral valuation procedure, we consider the 2- and 3-period zero-coupon bonds as examples of random cash flows. To use expression (8.16), we start at time 3 and work backward through the tree to time 0, calculating expected values and discounting according to expression (8.16).

For the 2-period bond, the calculations are as follows:

$$P(1,2;u) = \tilde{E}_1\big(P(2,2;s_2)\big)/r(1;u) = \big[(1/2)1 + (1/2)1\big]/1.017606 = .982699,$$

$$P(1,2;d) = \tilde{E}_1\big(P(2,2;s_2)\big)/r(1;d) = \big[(1/2)1 + (1/2)1\big]/1.022406 = .978085, \text{ and}$$

$$P(0,2) = \tilde{E}_0\big(P(1,2;s_1)\big)/r(0) = \big[(1/2).982699 + (1/2).978085\big]/1.02 = .961169.$$

These values generate the nodes of the tree in Figure 8.4 with less effort than the technique utilized in method 1.

8.3.3 Exploiting an Arbitrage Opportunity

From the above calculations, the price of the 2-period zero-coupon bond at time zero is $P(0,2) = .961169$. Suppose that the actual price quoted in the market is .970000. An arbitrage opportunity exists. To create this arbitrage opportunity, we need to determine the portfolio that generates the synthetic 2-period zero-coupon bond. This can be obtained using the procedure from method 1.

Alternatively, it can be obtained using only expression (8.16) and Figure 8.2. This alternative approach calculates the *delta*. The delta is defined as the change in the value of a derivative security divided by the change in value of the underlying asset. The delta for the 2-period zero-coupon bond at time 1, state u, denoted by $n_4(1;u)$, is:

$$n_4(1;u) = \frac{\Delta P(2,2;s_2)}{\Delta P(2,4;s_2)} = \frac{P(2,2;uu) - P(2,2;ud)}{P(2,4;uu) - P(2,4;ud)} = \frac{1-1}{.967826 - .960529} = 0.$$

This represents the change in the 2-period zero-coupon bond's price relative to that of the 4-period zero-coupon bond. This gives the position in the 4-period zero-coupon bond, i.e., $n_4(1;u)$.

To get the number of units in the money market account, we use:

$$n_0(1;u) = \frac{P(1,2;u) - n_4(1;u)P(1,4;u)}{B(1)} = \frac{.982699 - 0(.947497)}{1.02} = .963430.$$

This last relation is obtained from the cost of construction relation in method 1, expression (8.4). The numerator represents the cost of construction ($P(1,2;u)$) less the dollars invested in the 4-period zero-coupon bond ($n_4(1;u)P(1,4;u)$), that is, the dollars in the money market account. The denominator represents the value of a unit of the money market account. Hence, $n_0(1;u)$ is the number of units in the money market account needed to obtain the synthetic 2-period zero-coupon bond.

Continuing,

$$n_4(1;d) - \frac{\Delta P(2,2;s_2)}{\Delta P(2,4;s_2)} = \frac{P(2,2;du) - P(2,2;dd)}{P(2,4;du) - P(2,4;dd)} = \frac{1-1}{.962414 - .953877} = 0$$

and

$$n_0(1;d) = \left[P(1,2;d) - n_4(1;d)P(1,4;d) \right] / B(1)$$

$$= \left[.978085 - 0(.937148) \right] / 1.02 = .958907.$$

Finally,

$$n_4(0) = \frac{\Delta P(1,2;s_1)}{\Delta P(1,4;s_1)} = \frac{P(1,2;u) - P(1,2;d)}{P(1,4;u) - P(1,4;d)} = \frac{.982699 - .978085}{.947497 - .937148} = .445835$$

and

$$n_0(0) = \left[P(0,2) - n_4(0)P(0,4) \right] / B(0)$$

$$= \left[.961169 - .445835(.923845) \right] / 1 = .549286.$$

To create the arbitrage opportunity, sell the 2-period zero-coupon bond at time 0. This brings in +.97000 dollars in cash. Invest $n_0(0) = .549286$ dollars of this in a money market account and buy $n_4(0) = .445835$ units of $P(0,4) = .923845$, at a total cost of .961169 dollars. This leaves $.97000 - .961169 = .008831$ dollars in excess cash available at time 0.

At time 1, if state u occurs, we rebalance our money market account and 4-period zero-coupon bond holdings to $n_0(1;u)$ and $n_4(1;u)$. No excess cash is generated. This follows from the self-financing condition. At time 1, if state d occurs, we rebalance our money market account and 4-period zero-coupon bond holdings to $n_0(1;d)$ and $n_4(1;d)$. No excess cash is generated. At time 2, our long position in the money market account and 4-period zero-coupon bond exactly offsets our short position in the 2-period zero-coupon bond. This completes the illustration for the 2-period zero-coupon bond.

There is an important observation to make with respect to the arbitrage opportunity outlined above. For the arbitrage opportunity to be executed, in general, one must continue the strategy until the last date relevant for the contingent claim. For this example, it is time 2, the maturity date of the 2-period zero-coupon bond. The arbitrage opportunity was initiated at time 0. The strategy involved rebalancing at time 1, then liquidation at time 2. If instead the trading strategy was liquidated earlier, at time 1, the trading strategy could lose money. Indeed, it is possible for the traded 2-period zero-coupon bond to become even more mispriced at time 1. Since we are short of the traded 2-period zero-coupon bond, we lose money if this occurs. All mispricings must disappear at time 2 when the bond pays the sure dollar. This is the reason we need to continue our strategy until all the derivative's cash flows are paid. This completes the discussion of method 2.

The use of method 2 to obtain the numbers in Figure 8.5 is left to the reader as an exercise. The next chapter formalizes the techniques illustrated in this chapter.

CHAPTER 9

Bond Trading Strategies – The Theory

THIS CHAPTER PRESENTS THE theory underlying the example presented in Chapter 8. The purpose of this chapter, as well as the last,
is to investigate the existence of arbitrage opportunities within the yield
curve. Taken as given are the current market prices and the stochastic
processes for a few zero-coupon bonds and the spot rate of interest. The
theory determines a "fair" or arbitrage-free price for the remaining zero-
coupon bonds. This is the modern analog of arbitraging the yield curve.

We first do the analysis for the one-factor economy, then for the two-
factor economy, and then for the $N \geq 3$-factor economy. The analysis differs between the one- and two-factor cases, but for two or more factors the
analysis is very similar.

9.1 THE ONE-FACTOR ECONOMY

Consider the one-factor economy as described in Chapter 5. We assume
that there are no arbitrage opportunities with respect to the smallest class
of trading strategies Φ_1. In this class of trading strategies, we choose a
particular τ_1-maturity zero-coupon bond, and all trades take place with
respect to it and the money market account.

Taken as given (exogenous) are the stochastic process for the τ_1-maturity
zero-coupon bond, $P(0, \tau_1)$ and $(u(t,\tau_1;s_t), d(t,\tau_1;s_t))$, and the stochastic process for the spot rate, $r(0)$, and $(u(t,t+1;s_t), d(t,t+1;s_t))$. The purpose is to
price all the remaining zero-coupon bonds in the market. Because the

processes for the τ_1-maturity zero-coupon bond and the money market account are exogenously supplied, we need to discuss whether these specifications are consistent with some economic equilibrium. This is also pursued below.

9.1.1 Complete Markets

First, we want to show that the market is complete with respect to the trading strategies Φ_1. To do this, consider any arbitrary simple contingent claim with maturity date $\tau_1 - 1$, denoted by $x(\tau_1 - 1; s_{\tau_1-1})$. In the example of the previous chapter recall that the 2- and 3-period zero-coupon bonds were considered as such simple contingent claims. We need to construct a synthetic contingent claim that replicates the cash flows to x at time $\tau_1 - 1$ under all states s_{τ_1-1}. The easiest way to do this is to first pretend we are at time $\tau_1 - 2$, and form the synthetic contingent claim there to replicate x's cash flows over the last trading period $[\tau_1 - 2, \tau_1 - 1]$.

In this case, we need to choose the number of units of the money market account $n_0(\tau_1 - 2; s_{\tau_1-2})$ and the number of τ_1-maturity zero-coupon bonds purchased $n_{\tau_1}(\tau_1 - 2; s_{\tau_1-2})$ at time $\tau_1 - 2$ such that

$$n_0(\tau_1 - 2; s_{\tau_1-2})B(\tau_1 - 2; s_{\tau_1-3})r(\tau_1 - 2; s_{\tau_1-2})$$

$$+ n_{\tau_1}(\tau_1 - 2; s_{\tau_1-2})P(\tau_1 - 2, \tau_1; s_{\tau_1-2})u(\tau_1 - 2, \tau_1; s_{\tau_1-2}) = x(\tau_1 - 1; s_{\tau_1-2}u)$$

$$(9.1a)$$

and

$$n_0(\tau_1 - 2; s_{\tau_1-2})B(\tau_1 - 2; s_{\tau_1-3})r(\tau_1 - 2; s_{\tau_1-2})$$

$$+ n_{\tau_1}(\tau_1 - 2; s_{\tau_1-2})P(\tau_1 - 2, \tau_1; s_{\tau_1-2})d(\tau_1 - 2, \tau_1; s_{\tau_1-2}) = x(\tau_1 - 1; s_{\tau_1-2}d)$$

$$(9.1b)$$

Expression (9.1a) states that if the τ_1-maturity zero-coupon bond's price goes up, the portfolio's time $\tau_1 - 1$ value matches x. Expression (9.1b) states that if the τ_1-maturity zero-coupon bond's price goes down, the portfolio's time $\tau_1 - 1$ value also matches x. This is a linear system of two equations in two unknowns. The solution can be computed by elementary methods, and it exists because $u(\tau_1 - 2, \tau_1; s_{\tau_1-2}) > d(\tau_1 - 2, \tau_1; s_{\tau_1-2})$. The solution is given by

$$n_0\left(\tau_1-2;s_{\tau_1-2}\right)=\frac{\begin{bmatrix}x\left(\tau_1-1;s_{\tau_1-2}d\right)u\left(\tau_1-2,\tau_1;s_{\tau_1-2}\right)\\-x\left(\tau_1-1;s_{\tau_1-2}u\right)d\left(\tau_1-2,\tau_1;s_{\tau_1-2}\right)\end{bmatrix}}{B\left(\tau_1-2;s_{\tau_1-3}\right)r\left(\tau_1-2;s_{\tau_1-2}\right)\begin{bmatrix}u\left(\tau_1-2,\tau_1;s_{\tau_1-2}\right)\\-d\left(\tau_1-2,\tau_1;s_{\tau_1-2}\right)\end{bmatrix}} \qquad (9.2a)$$

and

$$n_{\tau_1}\left(\tau_1-2;s_{\tau_1-2}\right)=\frac{\left[x\left(\tau_1-1;s_{\tau_1-2}u\right)-x\left(\tau_1-1;s_{\tau_1-2}d\right)\right]}{P\left(\tau_1-2,\tau_1;s_{\tau_1-2}\right)\left[u\left(\tau_1-2,\tau_1;s_{\tau_1-2}\right)-d\left(\tau_1-2,\tau_1;s_{\tau_1-2}\right)\right]}. \qquad (9.2b)$$

As expression (9.2) makes clear, this time τ_1-2 portfolio choice only utilizes information available at time τ_1-2 (because all the terms in expression (9.2) are known at time τ_1-2).

Derivation of Expression (6.12)

First solve expression (9.1a) for $n_0\left(\tau_1-2;s_{\tau_1-2}\right)$:

$$n_0\left(\tau_1-2;s_{\tau_1-2}\right)=\frac{x\left(\tau_1-1;s_{\tau_1-2}u\right)}{B\left(\tau_1-2;s_{\tau_1-3}\right)r\left(\tau_1-2;s_{\tau_1-2}\right)}$$

$$-n_{\tau_1}\left(\tau_1-2;s_{\tau_1-2}\right)\frac{P\left(\tau_1-2,\tau_1;s_{\tau_1-2}\right)}{B\left(\tau_1-2;s_{\tau_1-3}\right)r\left(\tau_1-2;s_{\tau_1-2}\right)}u\left(\tau_1-2,\tau_1;s_{\tau_1-2}\right). \qquad (*)$$

Substitute this expression for $n_0\left(\tau_1-2;s_{\tau_1-2}\right)$ into (9.1b) to obtain:

$$x(\tau_1-1;s_{\tau_1-2}u)-n_{\tau_1}\left(\tau_1-2;s_{\tau_1-2}\right)P(\tau_1-2,\tau_1;s_{\tau_1-2})u(\tau_1-2,\tau_1;s_{\tau_1-2})$$

$$+n_{\tau_1}\left(\tau_1-2;s_{\tau_1-2}\right)P(\tau_1-2,\tau_1;s_{\tau_1-2})d(\tau_1-2,\tau_1;s_{\tau_1-2})=x(\tau_1-1;s_{\tau_1-2}d).$$

Next, solve for $n_{\tau_1}\left(\tau_1-2;s_{\tau_1-2}\right)$, giving expression (9.2b).
Substitution of (9.2b) into (*) gives (9.2a). This completes the proof. □

The *cost of constructing* this portfolio at time τ_1-2 is given by expression (9.3) and is defined to be $x(\tau_1-2;s_{\tau_1-2})$, i.e.,

$$x(\tau_1 - 2; s_{\tau_1-2}) \equiv n_0(\tau_1 - 2; s_{\tau_1-2})B(\tau_1 - 2; s_{\tau_1-3})$$
$$+ n_{\tau_1}(\tau_1 - 2; s_{\tau_1-2})P(\tau_1 - 2, \tau_1; s_{\tau_1-2}).$$

(9.3)

Next, we move back to time $\tau_1 - 3$ and form a portfolio to replicate expression (9.3) at time $\tau_1 - 2$. That is, we need to choose holdings in the money market account, $n_0(\tau_1 - 3; s_{\tau_1-3})$, and holdings in the τ_1-maturity zero-coupon bond, $n_{\tau_1}(\tau_1 - 3; s_{\tau_1-3})$, such that

$$n_0(\tau_1 - 3; s_{\tau_1-3})B(\tau_1 - 3; s_{\tau_1-4})r(\tau_1 - 3; s_{\tau_1-3})$$
$$+ n_{\tau_1}(\tau_1 - 3; s_{\tau_1-3})P(\tau_1 - 3, \tau_1; s_{\tau_1-3})u(\tau_1 - 3, \tau_1; s_{\tau_1-3}) = x(\tau_1 - 2; s_{\tau_1-3}u)$$

(9.4a)

$$n_0(\tau_1 - 3; s_{\tau_1-3})B(\tau_1 - 3; s_{\tau_1-4})r(\tau_1 - 3; s_{\tau_1-3})$$
$$+ n_{\tau_1}(\tau_1 - 3; s_{\tau_1-3})P(\tau_1 - 3, \tau_1; s_{\tau_1-3})d(\tau_1 - 3, \tau_1; s_{\tau_1-3}) = x(\tau_1 - 2; s_{\tau_1-3}d)$$

(9.4b)

Expression (9.4a) duplicates (9.3) in the up outcome and expression (9.4b) duplicates (9.3) in the down outcome. Again, this is a system of two linear equations in two unknowns. A comparison with expression (9.1) reveals that it is the identical system except that all the time and state subscripts are reduced by one. Consequently, the solution is given by expression (9.2) with the corresponding reduction in the time and state subscripts, i.e.,

$$n_0(\tau_1 - 3; s_{\tau_1-3}) = \frac{\begin{bmatrix} x(\tau_1 - 2; s_{\tau_1-3}d)u(\tau_1 - 3, \tau_1; s_{\tau_1-3}) \\ -x(\tau_1 - 2; s_{\tau_1-3}u)d(\tau_1 - 3, \tau_1; s_{\tau_1-3}) \end{bmatrix}}{B(\tau_1 - 3; s_{\tau_1-4})r(\tau_1 - 3; s_{\tau_1-3})\begin{bmatrix} u(\tau_1 - 3, \tau_1; s_{\tau_1-3}) \\ -d(\tau_1 - 3, \tau_1; s_{\tau_1-3}) \end{bmatrix}}$$

(9.5a)

and

$$n_{\tau_1}(\tau_1 - 3; s_{\tau_1-3}) = \frac{\left[x(\tau_1 - 2; s_{\tau_1-3}u) - x(\tau_1 - 2; s_{\tau_1-3}d) \right]}{P(\tau_1 - 3, \tau_1; s_{\tau_1-3})\left[u(\tau_1 - 3, \tau_1; s_{\tau_1-3}) - d(\tau_1 - 3, \tau_1; s_{\tau_1-3}) \right]}.$$

(9.5b)

The cost of constructing this portfolio at time $\tau_1 - 3$, defined to be $x(\tau_1 - 3; s_{\tau_1 - 3})$ is given by:

$$x(\tau_1 - 3; s_{\tau_1 - 3}) \equiv n_0(\tau_1 - 3; s_{\tau_1 - 3})B(\tau_1 - 3; s_{\tau_1 - 4})$$
$$+ n_{\tau_1}(\tau_1 - 3; s_{\tau_1 - 3})P(\tau_1 - 3, \tau_1; s_{\tau_1 - 3}). \tag{9.6}$$

A pattern is emerging. The next step is to construct a portfolio at time $\tau_1 - 4$ to duplicate expression (9.6) at time $\tau_1 - 3$. This portfolio is formulated as before. This process continues backward in time until we reach time 0. At time 0, we choose holdings in the money market account $n_0(0)$ and holdings in the τ_1-maturity zero-coupon bond $n_{\tau_1}(0)$ to duplicate $x(1; s_1)$ at time 1. The solution is given by

$$n_0(0) = \frac{\left[x(1; d)u(0, \tau_1) - x(1; u)d(0, \tau_1)\right]}{r(0)\left[u(0, \tau_1) - d(0, \tau_1)\right]} \tag{9.7a}$$

and

$$n_{\tau_1}(0) = \frac{\left[x(1; u) - x(1; d)\right]}{P(0, \tau_1)\left[u(0, \tau_1) - d(0, \tau_1)\right]}. \tag{9.7b}$$

with an initial cost defined to be $x(0)$, i.e.,

$$x(0) \equiv n_0(0)1 + n_{\tau_1}(0)P(0, \tau_1). \tag{9.8}$$

The trading strategy $(n_0(t; s_t), n_{\tau 1}(t; s_t))$ for all s_t and $0 \le t \le \tau_1 - 2$ constructed above is easily seen to be self-financing. At the start of each period, enough value is generated by the entering portfolio position so that the rebalanced holdings can be obtained without any excess cash flows (either plus or minus). This follows at time $\tau_1 - 2$, for example, from comparing expressions (9.3) and (9.4). Expression (9.3) is the value needed at time $\tau_1 - 2$, and expression (9.4) is the value of the portfolio entering time $\tau_1 - 2$. It also follows at time $\tau_1 - 3$ by similar logic; and so forth.

Thus, this portfolio represents the synthetic contingent claim $x(\tau_1 - 1; s_{\tau_1 - 1})$. Because $x(\tau_1 - 1; s_{\tau_1 - 1})$ is an arbitrary contingent claim, we have just finished our demonstration that the market is complete with respect to the trading strategy set Φ_1. Obviously, therefore, it is complete with respect to the larger sets of trading strategies Φ_K for $K > 1$ as well.

The time 0 *arbitrage-free price* for the contingent claim $x(\tau_1 - 1; s_{\tau_1 - 1})$ is defined to be that value given in expression (9.8). We will return to this observation later on in the section.

9.1.2 Risk-Neutral Probabilities

Arbitrage opportunities with respect to the trading strategies in Φ_1 are assumed to be nonexistent in this economy. This section explores the implication of this assumption for the stochastic processes exogenously imposed on the zero-coupon bond price $P(t, \tau_1; s_t)$ and the spot rate process $r(t; s_t)$. These processes' parameters are not completely arbitrary, as seen below. These restrictions are the ones necessary and sufficient to make these processes consistent with some economic equilibrium.*

Given that the relevant trading strategies in Φ_1 only involve the money market account and the τ_1-maturity zero-coupon bond, no arbitrage implies that neither security dominates the other, i.e.,

$$u(t, \tau_1; s_t) > r(t, s_t) > d(t, \tau_1; s_t) \quad \text{for all } t < \tau_1 - 1 \text{ and } s_t. \tag{9.9}$$

Indeed, suppose expression (9.9) were not true. That is, suppose $r(t, s_t) \geq u(t, \tau_1; s_t)$ for some s_t and $t < \tau_1 - 1$. An arbitrage opportunity of type II with respect to Φ_1 can easily be constructed because the money market account always pays off more than the τ_1-maturity bond at time $t + 1$ given state s_t at time t. The construction is as follows: hold no position until time t. If state s_t does not occur, do nothing. If state s_t occurs, buy $1/B(t; s_t)$ units of the money market account and finance it by selling $1/P(t, \tau_1; s_t)$ units of the τ_1-maturity bond. Liquidate the position at time $t + 1$. This portfolio satisfies the definition of an arbitrage opportunity because it has an initial value of zero. It is self-financing, and it has non-negative payoffs across all states and strictly positive payoffs in one state. Indeed, $r(t, s_t) \geq u(t, \tau_1; s_t) > d(t, \tau_1; s_t)$. This contradiction implies that $u(t, \tau_1; s_t) > r(t, s_t)$ for all s_t and $t < \tau_1 - 1$. The argument that $r(t; s_t) > d(t, \tau_1; s_t)$ for all s_t and $t < \tau_1 - 1$ is similar and is left to the reader as an exercise.

Expression (9.9) implies that there exists a unique, strictly positive number less than one, $\pi(t; s_t)$, such that

---

* This follows when we give this economy the interpretation of being risk-neutral, with equilibrium prices being determined by discounted expected values; see Harrison and Kreps [1] for a formal discussion of these issues.

$$r(t;s_t) = \pi(t;s_t)u(t,\tau_1;s_t) + (1-\pi(t;s_t))d(t,\tau_1;s_t). \qquad (9.10)$$

To interpret the significance of this expression, we need to perform some algebra. First, divide both sides of expression (9.10) by $r(t;s_t)$ to get

$$1 = \frac{\left[\pi(t;s_t)u(t,\tau_1;s_t) + (1-\pi(t;s_t))d(t,\tau_1;s_t)\right]}{r(t;s_t)}. \qquad (9.11)$$

Next, multiply both sides of expression (9.11) by $P(t,\tau_1;s_t)/B(t;s_{t-1})$ and use the facts that

$$B(t+1;s_t) = B(t;s_{t-1})r(t;s_t),$$

$$P(t+1,\tau_1;s_t u) = u(t,\tau_1;s_t)P(t,\tau_1;s_t)$$

and

$$P(t+1,\tau_1;s_t d) = d(t,\tau_1;s_t)P(t,\tau_1;s_t).$$

The resulting expression is

$$\frac{P(t,\tau_1;s_t)}{B(t;s_{t-1})} = \pi(t;s_t)\frac{P(t+1,\tau_1;s_t u)}{B(t+1;s_t)} + (1-\pi(t;s_t))\frac{P(t+1,\tau_1;s_t d)}{B(t+1;s_t)}. \qquad (9.12)$$

This is true for all s_t and $t < \tau_1 - 1$.

To understand expression (9.12), first note that the number $\pi(t;s_t)$ can be interpreted as a probability of the up outcomes occurring at time t. Indeed, it is a strictly positive number less than one. Correspondingly, $1 - \pi(t;s_t)$ can be interpreted as a probability of the down outcomes occurring at time t. Because the true probabilities are $q(t;s_t)$ and $1 - q(t;s_t)$, we call the $\pi(t;s_t)$ *pseudo probabilities*.

With this interpretation, expression (9.12) can be written as

$$\frac{P(t,\tau_1;s_t)}{B(t;s_{t-1})} = \tilde{E}_t\left(\frac{P(t+1,\tau_1;s_{t+1})}{B(t+1;s_t)}\right) \qquad (9.13)$$

where $\tilde{E}_t(\cdot)$ is the time t expected value under the pseudo probabilities $\pi(t;s_t)$.

Expression (9.13) states that the τ_1-maturity zero-coupon bond price at time t, discounted by the money market account's value, is equal to its discounted expected value at time $t+1$. Expression (9.13) also states that the ratio of the τ_1-maturity zero-coupon's bond price to the money market account's value $(P(t,\tau_1)/B(t))$ is a *martingale* under the pseudo probabilities. (A martingale (for a discrete time-discrete state space process) is a stochastic process whose current value equals its expected future value, as in expression (9.13).) This provides us with the second name often used for the pseudo probabilities, *martingale probabilities*.

It is instructive to rewrite expression (9.13) in an alternative form. Multiplying both sides of this expression by $B(t;s_{t-1})$ and recognizing that $B(t;s_{t-1})/B(t+1;s_t) = 1/r(t;s_t)$ gives

$$P(t,\tau_1;s_t) = \frac{\tilde{E}_t\left(P(t+1,\tau_1;s_{t+1})\right)}{r(t;s_t)}. \tag{9.14a}$$

Expression (9.14a) shows that the price of the τ_1-maturity zero-coupon bond at time t is its expected value at time $t+1$ (under the pseudo probabilities) discounted at the spot rate of interest. This form of expression (9.13) is the version most easily programmed on a computer or used in hand calculations.

Expression (9.14a) is the value that the τ_1-maturity zero-coupon bond would have in a *risk-neutral economy* where the actual beliefs of all investors are the pseudo probabilities. This is verified by referring back to Chapter 6 where the present value form of the expectations hypothesis was defined. There, expression (9.14a) appears as the present value form of the expectations hypothesis, but with the actual probabilities used in the expectations operator. Expression (9.14a) uses the pseudo probabilities. Consequently, the pseudo probabilities are also called *risk-neutral probabilities*.

Using the law of iterated expectations, we can rewrite expression (9.14a) only in terms of spot interest rates:

$$P(t,\tau_1;s_t) = \tilde{E}_t\left(\frac{1}{\prod_{j=t}^{\tau_1-1} r(j;s_j)}\right) \tag{9.14b}$$

Expression (9.14b) makes clear the statement that the τ_1-maturity zero-coupon bond's price is its time τ_1 expected value (1 dollar) discounted by the spot interest rates over the intermediate periods.

Derivation of Expression (9.14b)

From expression (9.14a) at time $\tau_1 - 1$, we have

$$P(\tau_1 - 1, \tau_1; s_{\tau_1 - 1}) = \tilde{E}_{\tau_1 - 1}\left(\frac{1}{r(\tau_1 - 1; s_{\tau_1 - 1})}\right). \qquad (*)$$

Continuing backward from (9.14a) at time $\tau_1 - 2$ and using (*), we have

$$P\left(\tau_1 - 2, \tau_1; s_{\tau_1 - 2}\right) = \tilde{E}_{\tau_1 - 2}\left(\tilde{E}_{\tau_1 - 1}\left(\frac{1}{r\left(\tau_1 - 1; s_{\tau_1 - 1}\right)}\right)\frac{1}{r\left(\tau_1 - 2; s_{\tau_1 - 2}\right)}\right).$$

Using the fact that $\tilde{E}_{\tau_1 - 2}(\tilde{E}_{\tau_1 - 1}(\cdot)) = \tilde{E}_{\tau_1 - 2}(\cdot)$, we get

$$P\left(\tau_1 - 2, \tau_1; s_{\tau_1 - 2}\right) = \tilde{E}_{\tau_1 - 2}\left(\frac{1}{r\left(\tau_1 - 1; s_{\tau_1 - 1}\right)r\left(\tau_1 - 2; s_{\tau_1 - 2}\right)}\right).$$

which is (9.14b) at $\tau_1 - 2$. Continuing backward until time t gives the desired result. This completes the proof. □

For future reference, from expression (9.10) we can explicitly write down an expression for the pseudo probabilities in terms of the τ_1-maturity zero-coupon bond's and the spot rate processes' parameters,

$$\pi(t, s_t) = [r(t; s_t) - d(t, \tau_1; s_t)] / [u(t, \tau_1; s_t) - d(t, \tau_1; s_t)] \qquad (9.15a)$$

and

$$1 - \pi(t, s_t) = [u(t, \tau_1; s_t) - r(t; s_t)] / [u(t, \tau_1; s_t) - d(t, \tau_1; s_t)]. \qquad (9.15b)$$

The no-arbitrage condition, expression (9.9), guarantees that each of these probabilities is strictly positive and less than one.

In summary, no arbitrage with respect to the trading strategy set Φ_1 implies the existence of these pseudo probabilities $\pi(t; s_t)$ satisfying expression (9.13). The converse of this statement is also true. The existence of pseudo probabilities $\pi(t; s_t)$ satisfying expression (9.13) implies that there are no arbitrage opportunities with respect to the trading strategy set Φ_1 (the proof of this assertion is contained in the appendix to this chapter). Therefore, the existence of pseudo probabilities $\pi(t;s_t)$ satisfying expression (9.13) is both necessary and sufficient for the absence of arbitrage opportunities with respect to the trading strategy set Φ_1. This is an important observation and is especially useful when constructing stochastic processes for the evolution of the zero-coupon bond price curve.

9.1.3 Risk-Neutral Valuation

Given a simple contingent claim $x(\tau_1 - 1; s_{\tau_1 - 1})$ with maturity date $\tau_1 - 1$, we showed earlier in this chapter that its arbitrage-free price, denoted by $x(0)$, is defined to be

$$x(0) \equiv n_0(0)1 + n_{\tau_1}(0)P(0, \tau_1) \tag{9.16}$$

where $(n_0(t;s_t), n_{\tau_1}(t;s_t)) \in \Phi_1$ is the self-financing trading strategy that creates the synthetic contingent claim.

Using the risk-neutral probabilities, we can rewrite this expression in an alternative but equivalent form. Using a backward inductive argument, it can be shown that expression (9.17a) holds:

$$x(0) = \tilde{E}_0 \left(\frac{x(\tau_1 - 1; s_{\tau_1 - 1})}{B(\tau_1 - 1; s_{\tau_1 - 2})} \right) B(0). \tag{9.17a}$$

This is the *risk-neutral valuation* procedure for calculating current values. The arbitrage-free price of the contingent claim x is seen to be its discounted expected value using the risk-neutral probabilities $\pi(t;s_t)$ as given in expression (9.15). This is the equilibrium value that would obtain in an economy identical to this one except that all investors are risk-neutral and hold the pseudo probabilities as their beliefs (and not $q_t(s_t)$).

Derivation of Expression (9.17a)

Expression (9.3) states that

$$x(\tau_1 - 2; s_{\tau_1 - 2}) \equiv n_0(\tau_1 - 2; s_{\tau_1 - 2})B(\tau_1 - 2; s_{\tau_1 - 3})$$
$$+ n_{\tau_1}(\tau_1 - 2; s_{\tau_1 - 2})P(\tau_1 - 2, \tau_1; s_{\tau_1 - 2}).$$

Substitution of (9.2) into this expression yields

$$x\left(\tau_1-2;s_{\tau_1-2}\right)=\dfrac{\begin{bmatrix}x\left(\tau_1-1;s_{\tau_1-2}d\right)u\left(\tau_1-2,\tau_1;s_{\tau_1-2}\right)\\-x\left(\tau_1-1;s_{\tau_1-2}u\right)d\left(\tau_1-2,\tau_1;s_{\tau_1-2}\right)\end{bmatrix}}{r\left(\tau_1-2;s_{\tau_1-2}\right)\left[u\left(\tau_1-2,\tau_1;s_{\tau_1-2}\right)-d\left(\tau_1-2,\tau_1;s_{\tau_1-2}\right)\right]}$$

$$+\dfrac{\left[x\left(\tau_1-1;s_{\tau_1-2}u\right)-x\left(\tau_1-1;s_{\tau_1-2}d\right)\right]}{\left[u\left(\tau_1-2,\tau_1;s_{\tau_1-2}\right)-d\left(\tau_1-2,\tau_1;s_{\tau_1-2}\right)\right]}.$$

Rearranging terms yields

$$x\left(\tau_1-2;s_{\tau_1-2}\right)=\Bigg\{x\left(\tau_1-1;s_{\tau_1-2}u\right)\dfrac{\left[r\left(\tau_1-2;s_{\tau_1-2}\right)-d\left(\tau_1-2,\tau_1;s_{\tau_1-2}\right)\right]}{\left[u\left(\tau_1-2,\tau_1;s_{\tau_1-2}\right)-d\left(\tau_1-2,\tau_1;s_{\tau_1-2}\right)\right]}$$

$$+x\left(\tau_1-1;s_{\tau_1-2}d\right)\dfrac{\begin{bmatrix}u\left(\tau_1-2,\tau_1;s_{\tau_1-2}\right)\\-r\left(\tau_1-2;s_{\tau_1-2}\right)\end{bmatrix}}{\begin{bmatrix}u\left(\tau_1-2,\tau_1;s_{\tau_1-2}\right)\\-d\left(\tau_1-2,\tau_1;s_{\tau_1-2}\right)\end{bmatrix}}\Bigg\}\dfrac{1}{r\left(\tau_1-2;s_{\tau_1-2}\right)}.$$

Using (9.15) and the definition of $B(t;s_{t-1})$ yields

$$x\left(\tau_1-2;s_{\tau_1-2}\right)=\Big\{x\left(\tau_1-1;s_{\tau_1-2}u\right)\pi\left(\tau_1-2;s_{\tau_1-2}\right)$$

$$+x\left(\tau_1-1;s_{\tau_1-2}d\right)(1-\pi\left(\tau_1-2;s_{\tau_1-2}\right))\Big\}\dfrac{B\left(\tau_1-2;s_{\tau_1-3}\right)}{B\left(\tau_1-1;s_{\tau_1-2}\right)}$$

Using the expectation notation gives

$$x\left(\tau_1-2;s_{\tau_1-2}\right)=\tilde{E}_{\tau_1-2}\left(\dfrac{x\left(\tau_1-1;s_{\tau_1-1}\right)}{B\left(\tau_1-1;s_{\tau_1-2}\right)}\right)B\left(\tau_1-2;s_{\tau_1-3}\right) \qquad (*)$$

This is expression (9.17a) at time τ_1-2.

Next, using the identical algebra as above, but expression (9.16) instead of (9.13), we get

$$x\left(\tau_1-3;s_{\tau_1-3}\right)=\tilde{E}_{\tau_1-3}\left(\dfrac{x\left(\tau_1-2;s_{\tau_1-2}\right)}{B\left(\tau_1-2;s_{\tau_1-3}\right)}\right)B\left(\tau_1-3;s_{\tau_1-4}\right).$$

Substitution of (*) into this expression gives

$$x\left(\tau_1-3;s_{\tau_1-3}\right)=\tilde{E}_{\tau_1-3}\left[\tilde{E}_{\tau_1-2}\left(\frac{x\left(\tau_1-1;s_{\tau_1-1}\right)}{B\left(\tau_1-1;s_{\tau_1-2}\right)}\right)\right]B\left(\tau_1-3;s_{\tau_1-4}\right)$$

because the $B(\tau_1-2;s_{\tau_1-3})$ terms cancel. Using the well-known law of iterated expectations that $\tilde{E}_{\tau_1-3}(\tilde{E}_{\tau_1-2}(\cdot))=\tilde{E}_{\tau_1-3}(\cdot)$ we get

$$x\left(\tau_1-3;s_{\tau_1-3}\right)=\tilde{E}_{\tau_1-3}\left(\frac{x\left(\tau_1-1;s_{\tau_1-1}\right)}{B\left(\tau_1-1;s_{\tau_1-2}\right)}\right)B\left(\tau_1-3;s_{\tau_1-4}\right).$$

This is expression (9.17a) at time τ_1-3. Continuing backward in this fashion until time 0 generates expression (9.17a) as written. This completes the proof. □

For computations, either by hand or on a computer, one normally employs the backward induction procedure underlying the proof of expression (9.17a). For these computations, an equivalent form of expression (9.17a) is needed, i.e.,

$$x(t;s_t)=\left[\pi(t;s_t)x(t+1;s_tu)+(1-\pi(t;s_t))x(t+1;s_td)\right]/r(t;s_t) \qquad (9.17b)$$

for all s_t where $0\le t\le\tau_1-2$.

The proof that this expression is equivalent to (9.17a) is contained in the derivation of expression (9.17a).

From expressions (9.17) and (9.15) it is easily seen that one can determine the arbitrage-free price of any simple contingent claim using only the current price of the τ_1-maturity zero-coupon bond, its stochastic process, the current spot rate, and the current spot rate's stochastic process, i.e., (i) $P(0,\tau_1)$ and $(u(t,\tau_1;s_t), d(t,\tau_1;s_t))$ and (ii) $r(0)$ and $(u(t,t+1;s_t), d(t,t+1;s_t))$. These processes, however, need to satisfy the no-arbitrage restriction given in expression (9.10) that requires the existence of strictly positive pseudo probabilities. We do not need to know the prices of the other zero-coupon bonds nor their stochastic processes. These, in fact, can be determined from the risk-neutral valuation expression (9.17); that analysis is the content of the next section.

It is also important to note that the actual probabilities $q_t(s_t)$ do not enter the valuation formula for the simple contingent claim x. This is seen by examining the risk-neutral valuation expression (9.17), which can be calculated using only the data *(i)* $P(0,\tau_1)$ and $(u(t,\tau_1;s_t), d(t,\tau_1;s_t))$, and *(ii)* $r(0)$ and $(u(t,t+1;s_t), d(t,t+1;s_t))$. This independence of the actual probabilities $q_t(s_t)$ follows because deviations from formula (9.17) can be arbitraged no matter which state occurs. Consequently, the likelihood of a state's occurrence is irrelevant to the procedure. This independence implies that two traders who differ in their beliefs about future probabilities $(q_t(s_t))$, but who agree with the data concerning *(i)* $P(0,\tau_1)$ and $(u(t,\tau_1;s_t), d(t,\tau_1;s_t))$ and *(ii)* $r(0)$ and $(u(t,t+1;s_t), d(t,t+1;s_t))$ will agree upon all the remaining zero-coupon bond prices. Any disagreements in values usually correspond to differences in opinions regarding initial prices or their evolutions.

We need to pause and review the valuation of contingent claims more complex than those previously defined. The simple contingent claim x with maturity $\tau_1 - 1$ defined above consists of a single, random cash flow $x(\tau_1 - 1;s\tau_1 - 1)$ at time $\tau_1 - 1$. Consider a complex contingent claim having multiple random cash flows at the different dates $\{0, 1, \dots, \tau^*\}$ over its life, each denoted by $x(t;s_t)$ for $t \in \{0, 1, \dots, \tau^*\}$. This claim is nothing more than a collection of τ^* separate simple contingent claims each with a different maturity date.

The *arbitrage-free value* of this complex contingent claim is defined to be the sum of the arbitrage-free values of the simple contingent claims into which it can be decomposed; therefore, using expression (9.17) gives

$$\sum_{j=0}^{\tau^*} \tilde{E}_0 \left(\frac{x(j;s_j)}{B(j;s_{j-1})} \right) B(0). \tag{9.18}$$

Finally, consider an American-type contingent claim with cash flows $x(t,a;s_t)$ at time $t \in \{0, 1, \dots, \tau^*\}$ given decision choice $a \in A$. Now, if we arbitrarily fix a decision $a^* \in A$, the cash flows to this claim for each t are $x(t,a^*;s_t)$. Using expression (9.18), its arbitrage-free price at time 0 is then

$$\sum_{j=0}^{\tau^*} \tilde{E}_0 \left(\frac{x(j,a^*;s_j)}{B(j;s_{j-1})} \right) B(0) \tag{9.19}$$

But a^* might not be the best action for the person making the decision. This person, if holding the American contingent claim, would obviously want to select $a^* \in A$ to maximize expression (9.19). This maximum value is defined to be the *arbitrage-free price of the American contingent claim*.

This is the standard approach used in the option pricing literature for pricing American-type contingent claims. For more discussion, we refer the reader to Jarrow and Chatterjea [2]. We will see this procedure in Chapter 12.*

9.1.4 Bond Trading Strategies

The purpose of this chapter is to develop a procedure for valuing the entire zero-coupon bond price curve given *(i)* the price of one zero-coupon bond $P(0, \tau_1)$ and its stochastic process $(u(t,\tau_1;s_t), d(t,\tau_1;s_t))$, *(ii)* the current spot rate $r(0)$ and its stochastic process $(u(t,t+1;s_t), d(t,t+1;s_t))$, and *(iii)* the assumption of no-arbitrage opportunities in the trading strategy set Φ_1. We are now in a position to do just this.

All the distinct zero-coupon bonds maturing at times $T \leq \tau - 1$ can be viewed as contingent claims (of the simple, European type) and valued using either of the risk-neutral valuation expressions (9.8) or (9.14). Formally, however, the procedure was only derived for contingent claims maturing at or prior to time $\tau_1 - 1$. To value the zero-coupon bonds with maturities greater than $\tau_1 - 1$, we need to specify their state-contingent payoffs at time $\tau_1 - 1$. Alternatively, we could define $\tau_1 \equiv \tau$, so that no bonds mature after time τ_1.

In this context the arbitrage-free price of the T-maturity discount bonds with $T < \tau_1$ is given by

$$P(t,T;s_t) = \tilde{E}_t \left(\frac{1}{\prod_{j=t}^{T-1} r(j;s_j)} \right) \tag{9.20a}$$

or, equivalently,

$$P(t,T;s_t) = \frac{\tilde{E}_t\left(P(t+1,T;s_{t+1})\right)}{r(t;s_t)}. \tag{9.20b}$$

---

* The standard approach for solving $\displaystyle \max_{a^* \in A} \sum_{j=0}^{\tau^*} \tilde{E}_0 \left(\frac{x\left(j, a^*; s_j\right)}{B\left(j; s_{j-1}\right)} \right) B(0)$ is by using stochastic dynamic programming. This is the approach taken to solve this problem in Chapter 12.

The procedure given in expressions (9.2) and (9.5)–(9.8) enables us to construct a synthetic T-maturity zero-coupon bond using a self-financing trading strategy involving the τ_1-maturity bond and the money market account. The initial cost of this synthetic T-maturity zero-coupon bond gives its arbitrage-free price. This price does not need to equal the traded price. If the traded price of a T-maturity bond differs from the arbitrage-free price given in expression (9.20), an arbitrage opportunity exists. Indeed, suppose the traded price exceeds the price in expression (9.20). Then, sell the traded T-maturity zero-coupon bond short, and create the T-maturity zero-coupon bond synthetically. The combined position yields an arbitrage opportunity. It is important to emphasize that expression (9.20) is independent of the actual probabilities $q_t(s_t)$.

9.2 THE TWO-FACTOR ECONOMY

Having illustrated the procedure for pricing the zero-coupon bond price curve in the one-factor economy, we can proceed more swiftly in the two-factor case. Consider the two-factor economy described in Chapter 5. We assume that there are no arbitrage opportunities with respect to the trading strategy set Φ_2. In this class of trading strategies, we fix two zero-coupon bonds with maturities τ_1 and τ_2, and all trades take place with respect to these bonds and the money market account.

Taken as given (exogenous) are the stochastic processes for the τ_1- and τ_2-maturity zero-coupon bonds, i.e., $\{P(0,\tau_1), (u(t,\tau_1;s_t), m(t,\tau_1;s_t), d(t,\tau_1;s_t))\}$ and $\{P(0,\tau_2), (u(t,\tau_2;s_t), m(t,\tau_2;s_t), d(t,\tau_2;s_2))\}$; and the stochastic process for the spot rate $r(0), (u(t,t+1;s_t), m(t,t+1,s_t), d(t,t+1,s_t))$. The purpose is to price all the remaining zero-coupon bonds in the market. Because the τ_1- and τ_2-maturity zero-coupon bonds and the money market account processes are exogenously supplied, we need to discuss the restrictions implied by there being no-arbitrage opportunities in the trading strategy set Φ_2. We proceed as we did for the one-factor economy.

9.2.1 Complete Markets

This economy is complete with respect to the trading strategy set Φ_2. To prove this, consider an arbitrary simple contingent claim with maturity date τ_1-1, $x(\tau_1-1;s_{\tau_1-1})$. To duplicate x's cash flows over $[\tau_1-2, \tau_1-1]$, we need to choose $n_0(\tau_1-2;s_{\tau_1-2}), n_{\tau_1}(\tau_1-2;s_{\tau_1-2}), n_{\tau_2}(\tau_1-2;s_{\tau_1-2})$ at time τ_1-2 such that the portfolio's payoff at time τ_1-1 duplicates the contingent claim's payoffs. For simplicity of notation, we will denote

$$n_1(\tau_1 - 2; s_{\tau_1 - 2}) \equiv n_{\tau_1}(\tau_1 - 2; s_{\tau_1 - 2}) \quad \text{and} \quad n_2(\tau_1 - 2; s_{\tau_1 - 2}) \equiv n_{\tau_2}(\tau_1 - 2; s_{\tau_1 - 2}).$$

The payoff matching conditions are:

$$n_0(\tau_1 - 2; s_{\tau_1 - 2})B(\tau_1 - 2; s_{\tau_1 - 3})r(\tau_1 - 2; s_{\tau_1 - 2})$$

$$+ n_1(\tau_1 - 2; s_{\tau_1 - 2})P(\tau_1 - 2, \tau_1; s_{\tau_1 - 2})u(\tau_1 - 2, \tau_1; s_{\tau_1 - 2})$$

$$+ n_2(\tau_1 - 2; s_{\tau_1 - 2})P(\tau_1 - 2, \tau_2; s_{\tau_1 - 2})u(\tau_1 - 2, \tau_2; s_{\tau_1 - 2}) = x(\tau_1 - 1; s_{\tau_1 - 2}u)$$

$$(9.21a)$$

$$n_0(\tau_1 - 2; s_{\tau_1 - 2})B(\tau_1 - 2; s_{\tau_1 - 3})r(\tau_1 - 2; s_{\tau_1 - 2})$$

$$+ n_1(\tau_1 - 2; s_{\tau_1 - 2})P(\tau_1 - 2, \tau_1; s_{\tau_1 - 2})m(\tau_1 - 2, \tau_1; s_{\tau_1 - 2})$$

$$+ n_2(\tau_1 - 2; s_{\tau_1 - 2})P(\tau_1 - 2, \tau_2; s_{\tau_1 - 2})m(\tau_1 - 2, \tau_2; s_{\tau_1 - 2}) = x(\tau_1 - 1; s_{\tau_1 - 2}m)$$

$$(9.21b)$$

and

$$n_0(\tau_1 - 2; s_{\tau_1 - 2})B(\tau_1 - 2; s_{\tau_1 - 3})r(\tau_1 - 2; s_{\tau_1 - 2})$$

$$+ n_1(\tau_1 - 2; s_{\tau_1 - 2})P(\tau_1 - 2, \tau_1; s_{\tau_1 - 2})d(\tau_1 - 2, \tau_1; s_{\tau_1 - 2})$$

$$+ n_2(\tau_1 - 2; s_{\tau_1 - 2})P(\tau_1 - 2, \tau_2; s_{\tau_1 - 2})d(\tau_1 - 2, \tau_2; s_{\tau_1 - 2}) = x(\tau_1 - 1; s_{\tau_1 - 2}d)$$

$$(9.21c)$$

This is a linear system of three equations in three unknowns. Because of the fact that the matrix

$$\begin{bmatrix} 1 & u(t,T;s_t) & u(t,T^*;s_t) \\ 1 & m(t,T;s_t) & m(t,T^*;s_t) \\ 1 & d(t,T;s_t) & d(t,T^*;s_t) \end{bmatrix}$$

is nonsingular for $T \neq T^*$, $t + 1 < min(T,T^*)$ and s_t, this system has a unique solution given by

$$n_2\left(\tau_1-2;s_{\tau_1-2}\right)=\frac{\left(x\left(\tau_1-1;s_{\tau_1-2}m\right)-x\left(\tau_1-1;s_{\tau_1-2}d\right)\right)}{\times\left(u\left(\tau_1-2,\tau_1;s_{\tau_1-2}\right)-m\left(\tau_1-2,\tau_1;s_{\tau_1-2}\right)\right)}\Bigg/\Psi$$

$$-\frac{\left(x\left(\tau_1-1;s_{\tau_1-2}u\right)-x\left(\tau_1-1;s_{\tau_1-2}m\right)\right)}{\times\left(m\left(\tau_1-2,\tau_1;s_{\tau_1-2}\right)-d\left(\tau_1-2,\tau_1;s_{\tau_1-2}\right)\right)}\Bigg/\Psi$$

(9.22a)

$$n_1\left(\tau_1-2;s_{\tau_1-2}\right)=\left[\frac{x\left(\tau_1-1;s_{\tau_1-2}u\right)-x\left(\tau_1-1;s_{\tau_1-2}m\right)}{u\left(\tau_1-2,\tau_1;s_{\tau_1-2}\right)-m\left(\tau_1-2,\tau_1;s_{\tau_1-2}\right)}\right]\frac{1}{P\left(\tau_1-2,\tau_1;s_{\tau_1-2}\right)}$$

$$-\frac{n_2\left(\tau_1-2;s_{\tau_1-2}\right)P\left(\tau_1-2,\tau_2;s_{\tau_1-2}\right)}{P\left(\tau_1-2,\tau_1;s_{\tau_1-2}\right)}$$

$$\times\left[\frac{u\left(\tau_1-2,\tau_2;s_{\tau_1-2}\right)-m\left(\tau_1-2,\tau_2;s_{\tau_1-2}\right)}{u\left(\tau_1-2,\tau_1;s_{\tau_1-2}\right)-m\left(\tau_1-2,\tau_1;s_{\tau_1-2}\right)}\right]$$

(9.22b)

and

$$n_0\left(\tau_1-2;s_{\tau_1-2}\right)=\frac{\left[x\left(\tau_1-1;s_{\tau_1-2}u\right)-n_1\left(\tau_1-2;s_{\tau_1-2}\right)P\left(\tau_1-2,\tau_1;s_{\tau_1-2}\right)u\left(\tau_1-2,\tau_1;s_{\tau_1-2}\right)\right.}{B\left(\tau_1-2;s_{\tau_1-3}\right)r\left(\tau_1-2;s_{\tau_1-2}\right)}$$

$$\frac{\left.-\,n_2\left(\tau_1-2;s_{\tau_1-2}\right)P\left(\tau_1-2,\tau_2;s_{\tau_1-2}\right)u\left(\tau_1-2,\tau_2;s_{\tau_1-2}\right)\right]}{B\left(\tau_1-2;s_{\tau_1-3}\right)r\left(\tau_1-2;s_{\tau_1-2}\right)}$$

(9.22c)

where

$$\Psi\equiv P\left(\tau_1-2,\tau_2;s_{\tau_1-2}\right)\left[\left(u\left(\tau_1-2,\tau_1;s_{\tau_1-2}\right)-m\left(\tau_1-2,\tau_1;s_{\tau_1-2}\right)\right)\right.$$

$$\times\left(m\left(\tau_1-2,\tau_2;s_{\tau_1-2}\right)-d\left(\tau_1-2,\tau_2;s_{\tau_1-2}\right)\right)$$

$$-\left(u\left(\tau_1-2,\tau_2;s_{\tau_1-2}\right)-m\left(\tau_1-2,\tau_2;s_{\tau_1-2}\right)\right)$$

$$\left.\times\left(m\left(\tau_1-2,\tau_1;s_{\tau_1-2}\right)-d\left(\tau_1-2,\tau_1;s_{\tau_1-2}\right)\right)\right].$$

Derivation of Expression (9.22)

To simplify the notation, define

$$n_j \equiv n_j(\tau_1 - 2; s_{\tau_1-2}) \qquad \text{for } j = 0,1,2$$
$$B \equiv B(\tau_1 - 2; s_{\tau_1-3})$$
$$P_j \equiv P(\tau_1 - 2, \tau_j; s_{\tau_1-2}) \qquad \text{for } j = 1,2$$
$$r \equiv r(\tau_1 - 2; s_{\tau_1-2})$$
$$u_j \equiv u(\tau_1 - 2, \tau_j; s_{\tau_1-2}) \qquad \text{for } j = 1,2$$
$$m_j \equiv m(\tau_1 - 2, \tau_j; s_{\tau_1-2}) \qquad \text{for } j = 1,2$$
$$d_j \equiv d(\tau_1 - 2, \tau_j; s_{\tau_1-2}) \qquad \text{for } j = 1,2$$
$$x_u \equiv x(\tau_1 - 1; s_{\tau_1-2}u)$$
$$x_m \equiv x(\tau_1 - 1; s_{\tau_1-2}m)$$
$$x_d \equiv x(\tau_1 - 1; s_{\tau_1-2}d)$$

System (9.21) rewritten using this simplified notation is

$$n_0 Br + n_1 P_1 u_1 + n_2 P_2 u_2 = x_u$$

$$n_0 Br + n_1 P_1 m_1 + n_2 P_2 m_2 = x_m$$

$$n_0 Br + n_1 P_1 d_1 + n_2 P_2 d_2 = x_d$$

We derive the equations in reverse order. First, to obtain (9.22c), take the first equation, divide by Br, and rearrange terms to isolate n_0 on the left side.

Next, subtract the second equation from the first, and the third equation from the second, to get the new system:

$$n_1 P_1(u_1 - m_1) + n_2 P_2(u_2 - m_2) = x_u - x_m$$

$$n_1 P_1(u_1 - d_1) + n_2 P_2(u_2 - d_2) = x_u - x_d.$$

These are two equations in two unknowns. To obtain expression (9.22b), take the first equation, divide by $u_1 - m_1$, and rearrange terms. This gives

$$n_1 P_1 = \frac{x_u - x_m}{(u_1 - m_1)} - n_2 P_2 \frac{(u_2 - m_2)}{(u_1 - m_1)}.$$

Substitution of this expression into the second equation above gives

$$\frac{x_u - x_m}{(u_1 - m_1)}(m_1 - d_1) - n_2 P_2 \frac{(u_2 - m_2)}{(u_1 - m_1)}(m_1 - d_1) + n_2 P_2 (m_2 - d_2) = x_m - x_d.$$

Algebra yields

$$n_2 P_2 ((u_1 - m_1)(m_1 - d_2) - (u_2 - m_2)(m_1 - d_1))$$
$$= (x_m - x_d)(u_1 - m_1) - (x_u - x_m)(m_1 - d_1).$$

Dividing by $P_2((u_1 - m_1)(m_2 - d_2) - (u_2 - m_2)(m_1 - d_1))$ gives the result (9.22a). This is nonzero since

$$\begin{bmatrix} 1 & u_1 & u_2 \\ 1 & m_1 & m_2 \\ 1 & d_1 & d_2 \end{bmatrix}$$

is nonsingular, so its determinant is nonzero. This completes the proof. $\square$

The cost of constructing this portfolio at time $\tau_1 - 2$, denoted by $x(\tau_1 - 2; s_{\tau_1 - 2})$, is

$$x(\tau_1 - 2; s_{\tau_1 - 2}) \equiv n_0(\tau_1 - 2; s_{\tau_1 - 2})B(\tau_1 - 2; s_{\tau_1 - 3})$$
$$+ n_1(\tau_1 - 2; s_{\tau_1 - 2})P(\tau_1 - 2, \tau_1; s_{\tau_1 - 2}) \qquad (9.23)$$
$$+ n_2(\tau_1 - 2; s_{\tau_1 - 2})P(\tau_1 - 2, \tau_2; s_{\tau_1 - 2}).$$

Repeating the above procedure inductively backward gives for an arbitrary time t

$$n_2(t; s_t) = \frac{\begin{bmatrix} (x(t+1, s_t m) - x(t+1; s_t d))(u(t, \tau_1; s_t) - m(t, \tau_1; s_t)) \\ -(x(t+1, s_t u) - x(t+1; s_t m))(m(t, \tau_1; s_t) - d(t, \tau_1; s_t)) \end{bmatrix}}{D} \qquad (9.24a)$$

where

$$D \equiv P(t,\tau_2;s_t) \begin{bmatrix} \big(u(t,\tau_1;s_t)-m(t,\tau_1;s_t)\big)\big(m(t,\tau_2;s_t)-d(t,\tau_2;s_t)\big) \\ -\big(u(t,\tau_2;s_t)-m(t,\tau_2;s_t)\big)\big(m(t,\tau_1;s_t)-d(t,\tau_1;s_t)\big) \end{bmatrix}$$

$$n_1(t;s_t) = \left[\frac{x(t+1,s_tu)-x(t+1;s_tm)}{u(t,\tau_1;s_t)-m(t,\tau_1;s_t)} \right] \frac{1}{P(t,\tau_1;s_t)}$$

$$- \frac{n_2(t;s_t)P(t,\tau_2;s_t)}{P(t,\tau_1;s_t)} \left[\frac{u(t,\tau_2;s_t)-m(t,\tau_2;s_t)}{u(t,\tau_1;s_t)-m(t,\tau_1;s_t)} \right]$$

(9.24b)

and

$$n_0(t;s_t) = \frac{1}{B(t;s_{t-1})r(t;s_t)} \left[\begin{array}{l} x(t+1,s_tu)-n_1(t;s_t)P(t,\tau_1;s_t)u(t,\tau_1;s_t)x(t+1;s_tm) \\ -n_2(t;s_t)P(t,\tau_2;s_t)u(t,\tau_2;s_t) \end{array} \right].$$

(9.24c)

The cost of constructing this portfolio at time t is the arbitrage-free price, denoted by $x(t;s_t)$, and it is given by

$$x(t;s_t) \equiv n_0(t;s_t)B(t;s_{t-1})+n_1(t;s_t)P(t,\tau_1;s_t)+n_2(t;s_t)P(t,\tau_2;s_t). \quad (9.25)$$

This construction illustrates that the market is complete by showing that it is possible to generate an arbitrary simple contingent claim at time t using only the τ_1- and τ_2-maturity zero-coupon bonds and the money market account.

9.2.2 Risk-Neutral Probabilities

Arbitrage opportunities with respect to the trading strategy set Φ_2 are assumed to be nonexistent in this economy. This section analyzes the implication of this assumption for the stochastic processes exogenously imposed on the zero-coupon bond prices $P(t,\tau_1;s_t)$ and $P(t,\tau_2;s_t)$ and the spot rate process $r(t;s_t)$. This analysis is somewhat different from and more complicated than the analysis for the one-factor economy.

Given that the relevant trading strategies in Φ_2 only involve the money market account, the τ_1-maturity zero-coupon bond, and the τ_2-maturity zero-coupon bond, no arbitrage implies that no security dominates any of the others, i.e.,

$$u(t,\tau_1;s_t) > r(t;s_t) > d(t,\tau_1;s_t) \quad \text{for all } s_t \text{ and } t < \tau_1 - 1 \quad (9.26a)$$

$$u(t,\tau_2;s_t) > r(t;s_t) > d(t,\tau_2;s_t) \quad \text{for all } s_t \text{ and } t < \tau_1 - 1 \quad (9.26b)$$

It is not the case that

$$u(t,\tau_i;s_t) \ge u(t,\tau_j;s_t), m(t,\tau_i;s_t) \ge m(t,\tau_j;s_t), \quad (9.26c)$$

and

$$d(t,\tau_i;s_t) \ge d(t,\tau_j;s_t) \quad \text{for } i \ne j; i,j \in \{1,2\} \text{ for all } s_t \text{ and } t < \tau_1 - 1$$

with one inequality strict for some s_t and $t < \tau_1 - 1$.

Condition (9.26a) states that the τ_1-maturity bond neither dominates nor is dominated by the money market account. Condition (9.26b) states that the τ_2-maturity bond neither dominates nor is dominated by the money market account. Finally, condition (9.26c) states that neither the τ_1-maturity or the τ_2-maturity zero-coupon bond dominates the other.

The proofs of these conditions are straightforward. If one is violated, an arbitrage opportunity in the trading strategy set Φ_2 is available. For example, suppose condition (9.26c) is violated; i.e., suppose the return on the τ_1-maturity zero-coupon bond dominates the τ_2-maturity zero-coupon bond at some time and state. The arbitrage opportunity in the trading strategy set Φ_2 is created as follows: Do nothing until the first time the return over the next period on the τ_1-maturity bond dominates that on the τ_2-maturity bond (i.e., the first time expression (9.26c) is violated). Then buy one τ_1-maturity bond and finance this position by shorting $P(t,\tau_1)/P(t,\tau_2)$ units of the τ_2-maturity bond. This initial position has zero value. Liquidate this portfolio at the end of the next time period. The value of the τ_1-maturity bond is always enough to cover the short position in the τ_2-maturity bond, generating an arbitrage opportunity. The proofs of conditions (9.26a–b) are similar.

These expressions, along with the fact that the matrix

$$\begin{bmatrix} 1 & u(t,T;s_t) & u(t,T^*;s_t) \\ 1 & m(t,T;s_t) & m(t,T^*;s_t) \\ 1 & d(t,T;s_t) & d(t,T^*;s_t) \end{bmatrix},$$

is nonsingular for $T \neq T^*$, $t+1 < \min(T,T^*)$ and s_t, imply that there exists a unique pair of pseudo probabilities $[\pi^u(t;s_t), \pi^m(t;s_t)]$ whose sum is strictly between 0 and 1 such that the following hold:[*]

$$r(t;s_t) = \pi^u(t;s_t)u(t,\tau_1;s_t) + \pi^m(t;s_t)m(t,\tau_1;s_t)$$
$$+(1-\pi^u(t;s_t) - \pi^m(t;s_t))d(t,\tau_1;s_t) \tag{9.27a}$$

and

$$r(t;s_t) = \pi^u(t;s_t)u(t,\tau_2;s_t) + \pi^m(t;s_t)m(t,\tau_2;s_t)$$
$$+(1-\pi^u(t;s_t) - \pi^m(t;s_t))d(t,\tau_2;s_t). \tag{9.27b}$$

for all s_t and $t < \tau_1 - 1$.

To interpret expression (9.27), we perform some algebra. Dividing expression (9.27) by $r(t;s_t)$ and multiplying it by $P(t,\tau_1;s_t)/B(t;s_{t-1})$ for (9.27a) and by $P(t,\tau_2;s_t)/B(t;s_{t-1})$ for (9.27b) gives

$$\frac{P(t,\tau_i;s_t)}{B(t;s_{t-1})} = \pi^u(t;s_t)\frac{P(t+1,\tau_i;s_tu)}{B(t+1;s_t)} + \pi^m(t;s_t)\frac{P(t+1,\tau_i;s_tm)}{B(t+1;s_t)}$$

$$+(1-\pi^u(t;s_t) - \pi^m(t;s_t))\frac{P(t+1,\tau_1;s_td)}{B(t+1;s_t)} \quad \text{for } i = 1,2 \tag{9.28}$$

Using the expectations operator, we can rewrite this as

$$\frac{P(t,\tau_i;s_t)}{B(t;s_{t-1})} = \tilde{E}_t\left(\frac{P(t+1,\tau_i;s_{t+1})}{B(t+1;s_t)}\right) \quad \text{for } i = 1,2 \tag{9.29}$$

where $\tilde{E}_t(\cdot)$ is the time t expectation using the pseudo probabilities $\pi^u(t;s_t)$ and $\pi^m(t;s_t)$.

This shows that the discounted τ_1-maturity zero-coupon bond price is a martingale under the pseudo probabilities. For the τ_1-maturity bond this specializes to

---

[*] This is true since $\begin{bmatrix} 1 & u(t,\tau_1;s_t) & u(t,\tau_2;s_t) \\ 1 & m(t,\tau_1;s_t) & m(t,\tau_2;s_t) \\ 1 & d(t,\tau_1;s_t) & d(t,\tau_2;s_t) \end{bmatrix}$ being nonsingular implies that

$\begin{bmatrix} u(t,\tau_1;s_t) - d(t,\tau_1;s_t) & m(t,\tau_1;s_t) - d(t,\tau_1;s_t) \\ u(t,\tau_2;s_t) - d(t,\tau_2;s_t) & m(t,\tau_2;s_t) - d(t,\tau_2;s_t) \end{bmatrix}$ is nonsingular. To see this, just calculate the

determinant of each matrix.

$$P(t,\tau_1;s_t) = \tilde{E}_t \left(\frac{1}{\displaystyle\prod_{j=t}^{\tau_1-1} r(j;s_j)} \right). \tag{9.30}$$

The proof of expression (9.30) is identical to that which generates expression (9.14) from expression (9.13) and is therefore left to the reader.

For future reference, we write explicit expressions for the pseudo probabilities.

$$\pi^u(t;s_t) = \frac{\begin{bmatrix} \big(r(t;s_t)-d(t,\tau_2;s_t)\big)\big(m(t,\tau_1;s_t)-d(t,\tau_1;s_t)\big) \\ -\big(r(t;s_t)-d(t,\tau_1;s_t)\big)\big(m(t,\tau_2;s_t)-d(t,\tau_2;s_t)\big) \end{bmatrix}}{\begin{bmatrix} \big(u(t,\tau_2;s_t)-d(t,\tau_2;s_t)\big)\big(m(t,\tau_1;s_t)-d(t,\tau_1;s_t)\big) \\ -\big(u(t,\tau_1;s_t)-d(t,\tau_1;s_t)\big)\big(m(t,\tau_2;s_t)-d(t,\tau_2;s_t)\big) \end{bmatrix}} \tag{9.31a}$$

$$\pi^m(t;s_t) = \frac{\begin{bmatrix} \big(u(t,\tau_2;s_t)-d(t,\tau_2;s_t)\big)\big(r(t;s_t)-d(t,\tau_1;s_t)\big) \\ -\big(u(t,\tau_1;s_t)-d(t,\tau_1;s_t)\big)\big(r(t;s_t)-d(t,\tau_2;s_t)\big) \end{bmatrix}}{\begin{bmatrix} \big(u(t,\tau_2;s_t)-d(t,\tau_2;s_t)\big)\big(m(t,\tau_1;s_t)-d(t,\tau_1;s_t)\big) \\ -\big(u(t,\tau_1;s_t)-d(t,\tau_1;s_t)\big)\big(m(t,\tau_2;s_t)-d(t,\tau_2;s_t)\big) \end{bmatrix}} \tag{9.31b}$$

and

$$1-\pi^u(t;s_t)-\pi^m(t;s_t)$$

$$= \frac{\begin{bmatrix} \big(u(t,\tau_1;s_t)-r(t;s_t)\big)\big(u(t,\tau_2;s_t)-m(t,\tau_2;s_t)\big) \\ -\big(u(t,\tau_2;s_t)-r(t;s_t)\big)\big(u(t,\tau_1;s_t)-m(t,\tau_1;s_t)\big) \end{bmatrix}}{\begin{bmatrix} \big(u(t,\tau_2;s_t)-d(t,\tau_2;s_t)\big)\big(m(t,\tau_1;s_t)-d(t,\tau_1;s_t)\big) \\ -\big(u(t,\tau_1;s_t)-d(t,\tau_1;s_t)\big)\big(m(t,\tau_2;s_t)-d(t,\tau_2;s_t)\big) \end{bmatrix}} \tag{9.31c}$$

Derivation of Expression (9.31)

We use the abbreviated notation from the proof of expression (9.22) along with

$$\pi^u(t;s_t) \equiv \pi^u$$

$$\pi^m(t;s_t) \equiv \pi^m$$

The system of linear equations (9.27) can be written as:

$$r = \pi^u u_1 + \pi^m m_1 + (1 - \pi^u - \pi^m) d_1$$

$$r = \pi^u u_2 + \pi^m m_2 + (1 - \pi^u - \pi^m) d_2.$$

Rewritten:

$$r - d_1 = \pi^u (u_1 - d_1) + \pi^m (m_1 - d_1)$$

$$r - d_2 = \pi^u (u_2 - d_2) + \pi^m (m_2 - d_2).$$

Solving the first equation for π^m gives

$$\pi^m = \frac{(r - d_1)}{(m_1 - d_1)} - \pi^u \frac{(u_1 - d_1)}{(u_1 - d_1)} \qquad (*)$$

Substitution of π^m into the second equation aboveand solving for π^u gives expression (9.31a), i.e.,

$$\pi^u = \frac{(r - d_2)(m_1 - d_1) - (r - d_1)(m_2 - d_2)}{(u_2 - d_2)(m_1 - d_1) - (u_1 - d_1)(m_2 - d_2)}$$

Substitution of π^u into (*) and some algebra yield (9.31b), i.e.,

$$\pi^m = \frac{(u_2 - d_2)(r - d_1) - (u_1 - d_1)(r - d_2)}{(u_2 - d_2)(m_1 - d_1) - (u_1 - d_1)(m_2 - d_2)}$$

Finally, substitution of π^u and π^m into $1 - \pi^u - \pi^m$ yields (9.31c), i.e.,

$$1 - \pi^u - \pi^m = \frac{(u_1 - r)(u_2 - m_2) - (u_2 - r)(u_1 - m_1)}{(u_2 - d_2)(m_1 - d_1) - (u_1 - d_1)(m_2 - d_2)}$$

This completes the proof. □

In summary, no arbitrage with respect to the trading strategy set Φ_2 implies the existence of pseudo probabilities $\pi^u(t;s_t)$, $\pi^m(t;s_t)$ satisfying expression (9.29). The converse of this statement is also true. Similar to the one-factor case, the existence of pseudo probabilities $\pi^u(t;s_t)$, $\pi^m(t;s_t)$ satisfying expression (9.29) implies that there are no arbitrage opportunities with respect to the trading strategy set Φ_2. (The proof is identical to that presented in the appendix to this chapter for the one-factor case and is therefore omitted.) This is an important observation, useful for constructing arbitrage-free evolutions of the zero-coupon bond price curve.

9.2.3 Risk-Neutral Valuation

Given a simple contingent claim $x(\tau_1 - 1; s_{\tau_1-1})$ with maturity date $\tau_1 - 1$, its arbitrage-free price, denoted by $x(0)$, is

$$x(0) \equiv n_0(0)1 + n_1(0)P(0,\tau_1) + n_2(0)P(0,\tau_2) \tag{9.32}$$

where $(n_0(t;s_t), n_1(t;s_t), n_2(t;s_t)) \in \Phi_2$ is the self-financing trading strategy that creates the synthetic contingent claim. Such a self-financing trading strategy was constructed in expression (9.24).

Using the risk-neutral probabilities as given in expression (9.21), it can now be shown that expression (9.33) holds:

$$x(0) = \tilde{E}_0 \left(\frac{x(\tau_1 - 1_1; s_{\tau_1-1})}{B(\tau_1 - 1; s_{\tau_1-2})} \right) B(0). \tag{9.33}$$

This is an alternate but equivalent expression for the value of the contingent claim x. The time 0 value of the contingent claim is seen to be its discounted expected value using the pseudo probabilities. This is the *risk-neutral valuation* formula for calculating arbitrage-free values.

Derivation of Expression (9.33)

This derivation uses matrix algebra and is easily generalized to $N \geq 3$ factors. Expression (9.21) can be written (using the notation from the derivation of expression (9.22)) as

$$\begin{bmatrix} 1 & u_1 & u_2 \\ 1 & m_1 & m_2 \\ 1 & d_1 & d_2 \end{bmatrix} \begin{bmatrix} n_0 B \\ n_1 P_1 \\ n_2 P_2 \end{bmatrix} = \begin{bmatrix} x_u \\ x_m \\ x_d \end{bmatrix}$$

Similarly, expression (9.27) can be written as

$$\begin{bmatrix} \pi^u & \pi^m & \pi^d \end{bmatrix} \begin{bmatrix} 1 & u_1 & u_2 \\ 1 & m_1 & m_2 \\ 1 & d_1 & d_2 \end{bmatrix} = \begin{bmatrix} r & r & r \end{bmatrix}$$

where $\pi^d \equiv 1 - \pi^u - \pi^m$. Now, let

$$U \equiv \begin{bmatrix} 1 & u_1 & u_2 \\ 1 & m_1 & m_2 \\ 1 & d_1 & d_2 \end{bmatrix}$$

Then

$$\begin{bmatrix} \pi^u & \pi^m & \pi^d \end{bmatrix} = \begin{bmatrix} r & r & r \end{bmatrix} U^{-1}$$

So

$$\begin{bmatrix} \pi^u & \pi^m & \pi^d \end{bmatrix} \begin{bmatrix} x_u \\ x_m \\ x_d \end{bmatrix} = \begin{bmatrix} r & r & r \end{bmatrix} U^{-1} U \begin{bmatrix} n_0 B \\ n_1 P_1 \\ n_2 P_2 \end{bmatrix} = \begin{bmatrix} n_0 B + n_1 P_1 + n_2 P_2 \end{bmatrix} r$$

That is, $(\pi^u x^u + \pi^m x^m + \pi^d x^d)/r = n_0 B + n_1 P_1 + n_2 P_2$.

This gives expression (9.33) for time $\tau_1 - 2$. Using the backward-inductive procedure and the law of iterated expectations as in the derivation of expression (9.17) gives the result. This completes the proof. $\square$

Note that the arbitrage-free price of the contingent claim x can be determined using only knowledge of $\{r(0),\ u(t,t+1;s_t),\ m(t,t+1;s_t),\ d(t,t+1;s_t)\}$, $\{P(0,\ \tau_1),\ (u(t,\tau_1;s_t),\ m(t,\tau_1;s_t),\ d(t,\tau_1;s_t))\}$, and $\{P(0,\tau_2),\ (u(t,\tau_2;s_t),\ m(t,\tau_2;s_t),\ d(t,\tau_2;s_t))\}$. No information regarding the stochastic processes of the other zero-coupon bonds is needed. The extension of expression (9.33) to multiple cash flows and American-type features is identical to that extension in the one-factor economy, and is therefore omitted.

9.2.4 Bond Trading Strategies

The purpose of this section is to price the entire zero-coupon bond price curve given *(i)* the prices of two zero-coupon bonds $P(0,\tau_1)$ and $P(0,\tau_2)$ and their stochastic processes $\{(u(t,\tau_1;s_t), m(t,\tau_1;s_t)), d(t,\tau_1);s_t), (u(t,\tau_2;s_t), m(t,\tau_2;s_t), d(t,\tau_2;s_t))\}$, *(ii)* the current spot rate $r(0)$ and its stochastic process $(u(t,t+1;s_t), m(t,t+1;s_t), d(t,t+1;s_t))$, and *(iii)* the assumption of no-arbitrage opportunities in the trading strategy set Φ_2.

The distinct maturity discount bonds $T \le \tau_1 - 1$ can all be viewed as contingent claims (of the simple, European type) and valued using the risk-neutral valuation expressions (9.32) or (9.33). In this context, the value of a T-maturity zero-coupon bond is

$$P(t,T;s_t) = \tilde{E}_t \left(\frac{1}{\displaystyle\prod_{j=t}^{T-1} r(j;s_j)} \right). \tag{9.34}$$

If the price of the traded T-maturity bond differs from its arbitrage-free price in expression (9.34), an arbitrage opportunity exists.

9.3 MULTIPLE FACTOR ECONOMIES

The extension of the economy in the previous section to $N \ge 3$ factors is straightforward. As long as $N \le \tau_1 - 1$, the markets are complete with respect to the trading strategy set Φ_N. For each additional factor added, however, one additional zero-coupon bond is needed to form the appropriate trading strategy. The risk-neutral valuation expressions (9.29), (9.30), (9.33), and (9.34) generalize to the $N > 3$-factor economy with little effort.

APPENDIX

This appendix provides the proof that the existence of pseudo probabilities $\pi(t;s_t)$ satisfying expression (9.13) implies that there are no arbitrage opportunities with respect to the trading strategy set Φ_1.

First, note that $B(t;s_{t-1}) > 0$ for all t, s_{t-1} because

$$r(t;s_t) = \frac{1}{P(t,t+1;s_t)} > 0$$

for all t, s_t.

Suppose there exists $\pi(t;s_t)$ between 0 and 1 satisfying expression (9.13). Suppose also there exists an arbitrage opportunity of type II denoted by $(n_0(t;s_t), n_{\tau_1}(t;s_t))$ in Φ_1 with liquidation date $\tau^* < \tau_1$.*

We search for a contradiction.

From the fact that an arbitrage opportunity of type II has nonnegative payoffs in all states, strictly positive payoffs in some states, and from the fact that $B(t;s_{t-1}) > 0$ for all t, s_{t-1}, we obtain

$$B\left(\tau^*-1;s_{\tau_1-2}\right)\tilde{E}_{\tau^*-1}\left(\left[n_0\left(\tau^*-1;s_{\tau^*-1}\right)B\left(\tau^*;s_{\tau^*-1}\right)\right.\right.$$

$$\left.\left.+n_{\tau_1}\left(\tau^*-1;s_{\tau^*-1}\right)P\left(\tau^*,\tau_1;s_{\tau_1}\right)\right]\big/B\left(\tau^*;s_{\tau^*-1}\right)\right) > 0.$$

But by the martingale condition, expression (9.13), this expression equals

$$n_0\left(\tau^*-1;s_{\tau^*-1}\right)B\left(\tau^*-1;s_{\tau^*-2}\right)+n_{\tau_1}\left(\tau^*-1;s_{\tau^*-1}\right)P\left(\tau-1^*,\tau_1;s_{\tau_1-1}\right) > 0.$$

Using the self-financing condition (7.8) of Chapter 7 gives

$$n_0\left(\tau^*-2;s_{\tau^*-2}\right)B\left(\tau^*-1;s_{\tau^*-2}\right)+n_{\tau_1}\left(\tau^*-2;s_{\tau^*-2}\right)P\left(\tau-1^*,\tau_1;s_{\tau_1-1}\right) > 0.$$

Repeating this procedure again by applying $B(\tau_1-2;s_{\tau_1-3})\tilde{E}_{\tau_1-2}(\cdot)/B(\tau_1-1;s_{\tau_1-2})$ to the above inequality yields

$$n_0\left(\tau^*-3;s_{\tau^*-3}\right)B\left(\tau^*-2;s_{\tau^*-3}\right)+n_{\tau_1}\left(\tau^*-3;s_{\tau^*-3}\right)P\left(\tau-2^*,\tau_1;s_{\tau_1-2}\right) > 0.$$

Continuing backward to time 0 yields

$$n_0(0)B(0)+n_{\tau_1}(0)P(0,\tau_1) > 0.$$

This contradicts the fact that an arbitrage opportunity of type II has zero net investment. Therefore, there exist no arbitrage opportunities with respect to Φ_1. This completes the proof.

---

* Since an arbitrage opportunity of type I can be transferred into an arbitrage opportunity of type II, we need only consider arbitrage opportunities of type II in this proof.

REFERENCES

1. Harrison, J. M., and D. M. Kreps, 1970. "Martingales and Arbitrage in Multiperiod Securities Markets." *Journal of Economic Theory* 20, 381–408.
2. Jarrow, R., and A. Chatterjea, 2019. *An Introduction to Derivative Securities, Financial Markets, and Risk Management*, 2nd edition, World Scientific Press.

CHAPTER 10

Contingent Claims Valuation – Theory

10.1 MOTIVATION

The previous two chapters showed how to determine and take advantage of arbitrage opportunities within the yield curve. That procedure takes as given knowledge of the spot rate (and its evolution) and the prices (and evolution) of only one bond in a one-factor economy, or two bonds in a two-factor economy, and so forth. This information is then used to compute arbitrage-free prices for the remaining zero-coupon bonds. If the market prices differ from these arbitrage-free values for any of the remaining zero-coupon bonds, then arbitrage opportunities exist. The analysis of Chapters 8 and 9 provides the insights necessary to construct trading strategies for arbitraging these opportunities.

The purpose of this chapter is different. Its purpose is to price interest rate options given the market prices of the *entire* zero-coupon bond price curve and its stochastic evolution. We do not ascertain whether these zero-coupon bond prices are correct relative to each other, as in arbitraging the yield curve. These initial zero-coupon bond prices are assumed to be fair. Rather, the desire is to price options relative to this zero-coupon bond price curve. We therefore need to impose conditions on the zero-coupon bond price curve's evolution, given the current market prices, so that it is arbitrage-free. Options are priced relative to this evolution. This is the motivation for the original development of the HJM model, and it explains the primary use of these techniques in the investment community.

To some extent, the purpose of Chapters 8 and 9 was to provide the foundation for the material in this chapter.

This change in perspective is dictated by the circumstances surrounding the application of arbitrage pricing methods to valuing interest rate options. If the initial zero-coupon bond price curve is not taken as given, and the techniques of Chapters 8 and 9 are applied, then mispricings observed in interest rate options could be due to either mispriced zero-coupon bonds or mispriced options. If the mispricing is due to the underlying zero-coupon bonds, then using options to take advantage of this situation is an indirect and highly levered strategy. These strategies are particularly sensitive to model risk. A better approach is to trade these zero-coupon bond mispricings directly, using the bond trading strategies of Chapters 8 and 9. But, if the mispricings in interest rate options are due to the options themselves, then the techniques of this chapter apply. The techniques of this chapter therefore seek mispricings of interest rate options relative to the market prices of the entire zero-coupon bond price curve (not just the prices of a few bonds). Then, one is guaranteed that the arbitrage opportunities discovered are due to the options themselves.

This chapter emphasizes the theory underlying this methodology, which is decomposed into the one-factor case, the two-factor case, and the multiple-factor case. The details of specific applications are postponed to subsequent chapters. For example, Chapter 11 studies coupon bonds. Chapter 12 considers both European and American options on bonds. Chapter 13 studies forward and futures contracts. Chapter 14 studies swaps, caps, floors, and swaptions. Finally, interest rate exotics are the topic of Chapter 15.

10.2 THE ONE-FACTOR ECONOMY

The key insight to be understood in this chapter is how to determine when a given zero-coupon bond price curve evolution is arbitrage-free. This is a nontrivial exercise, and it is the primary contribution of the HJM model to the academic literature.

To facilitate understanding, we illustrate the key concepts through an example.

Example: An Arbitrage-Free Zero-Coupon Bond Price Curve Evolution

Consider the zero-coupon bond price curve evolution given in Figure 10.1. In practice the initial zero-coupon bond price curve is

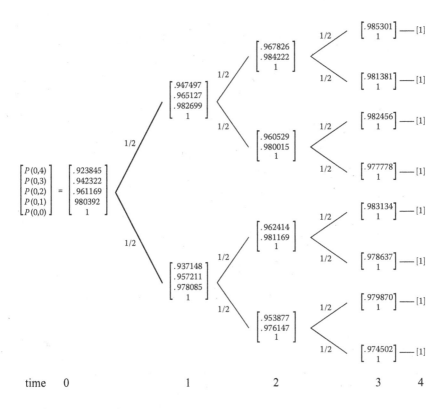

FIGURE 10.1 An Example of a One-Factor Bond Price Curve Evolution. Pseudo Probabilities Are Along Each Branch of the Tree.

based on available market quotes. This initial curve is assumed to evolve for four time periods, as given in the tree. At time 3, there is only one bond remaining, and it matures at time 4.

From this evolution we can compute the spot rate of interest. Recall that at each node in the tree the spot rate is the return on the shortest maturity bond over the next period. From the spot rates, we can compute the value of the money market account. The money market account starts at a dollar and grows by compounding at the spot rate. These computations are left as an exercise for the reader, and the resulting values are recorded in Figure 10.2. Figure 10.2 is identical to Figure 10.1 with the exception that it includes both the value of the money market account at the top of each vector of zero-coupon bond prices and the spot rate of interest to the right of each node on the tree. These additional values will prove useful in the subsequent computations.

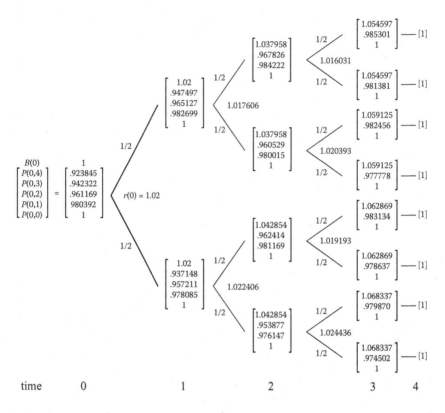

FIGURE 10.2 An Example of a One-Factor Bond Price Curve Evolution. The Money Market Account Values and Spot Rates Are Included on the Tree. Pseudo Probabilities Are Along Each Branch of the Tree.

The purpose of this example is to determine whether this evolution is arbitrage-free. With this goal in mind, note that embedded within this tree is a simpler tree that provides just the evolution of the 4-period zero-coupon bond and the money market account. From the analysis in Chapter 8 we know that this reduced tree is arbitrage-free if there exist a *unique* set of pseudo probabilities $\pi(t,4; s_t)$'s such that

$$r(t; s_t) = \pi(t, 4; s_t)u(t, 4; s_t) + (1 - \pi(t, 4; s_t))d(t, 4; s_t) \quad \text{for all } t \text{ and } s_t.$$

$$(10.1)$$

In this condition we have included a four in the notation for the pseudo probabilities to remind us that it is based on the 4-period bond/money market account pair. Also, it is important to emphasize the fact that these pseudo probabilities are unique.

So, as a beginning, we know that the entire tree cannot be arbitrage-free unless this reduced tree is arbitrage-free. So, we know that condition (10.1) must hold. Condition (10.1) is thus a necessary condition of an arbitrage-free evolution. Condition (10.1) can be checked by rewriting it to see if the $\pi(t,4; s_t)$'s computed according to

$$\pi(t, 4; s_t) = [r(t; s_t) - d(t, 4; s_t)] / [u(t, 4; s_t) - d(t, 4; s_t)] \qquad (10.2)$$

are strictly between 0 and 1. We now check this computation. First, one must determine the returns on the bonds at each node in the tree $(u(t, 4; s_t), d(t, 4; s_t))$. This calculation is left as an exercise for the reader. Then, given these returns, one uses (10.2) to compute the pseudo probabilities.

$$\pi(0, 4) = \frac{1.02 - 1.014400}{1.025602 - 1.014400} = 0.5$$

$$\pi(1, 4; u) = \frac{1.017606 - 1.013754}{1.021455 - 1.013754} = 0.5$$

$$\pi(1, 4; d) = \frac{1.022406 - 1.017851}{1.026961 - 1.017851} = 0.5$$

$$\pi(2, 4; uu) = \frac{1.016031 - 1.014006}{1.018056 - 1.014006} = 0.5$$

$$\pi(2, 4; ud) = \frac{1.020393 - 1.017958}{1.022828 - 1.017958} = 0.5$$

$$\pi(2, 4; du) = \frac{1.019193 - 1.016857}{1.021529 - 1.016857} = 0.5$$

$$\pi(2, 4; dd) = \frac{1.024436 - 1.021622}{1.027250 - 1.021622} = 0.5.$$

As all of these pseudo probabilities are strictly between 0 and 1, we know that the reduced tree involving the 4-period zero-coupon bond and the money market account is arbitrage-free.

Just as an aside, let us recall two facts that we will use momentarily. First, given the reduced tree is arbitrage-free, we have that

$$P(t,4;s_t) = \frac{\left[\pi(t,4;s_t)P(t+1,4;s_tu) + (1-\pi(t,4;s_t))P(t+1,4;s_td)\right]}{r(t;s_t)}$$

for all t and s_t.

(10.3)

This is just the present value computation for the 4-period zero-coupon bond.

Second, by the risk-neutral valuation procedure, we know that the price of the 2- and 3-period zero-coupon bond can be written as:

$$P(t,2;s_t) = \frac{\left[\pi(t,4;s_t)P(t+1,2;s_tu) + (1-\pi(t,4;s_t))P(t+1,2;s_td)\right]}{r(t;s_t)}$$

for all t and s_t

(10.4a)

and

$$P(t,3;s_t) = \frac{\left[\pi(t,4;s_t)P(t+1,3;s_tu) + (1-\pi(t,4;s_t))P(t+1,3;s_td)\right]}{r(t;s_t)}$$

for all t and s_t.

(10.4b)

The reader should check to see that expression (10.4) does indeed hold for this tree. One important observation needs to be made about expression (10.4). In the computation, the pseudo probabilities as determined from the 4-period zero-coupon bond/money market account pair. Expression (10.4) just emphasizes the fact that the 4-period zero-coupon bond/money market account reduced tree uniquely determines the prices of the 2- and 3-period zero-coupon bonds. These are the prices that the other bonds must satisfy in the enlarged tree for it to be arbitrage-free relative to the 4-period bond and the money market account.

Prior to checking the satisfaction of this condition, let us also investigate the two other reduced trees possible. One involves the 3-period zero-coupon bond and the money market account pair. The second involves the 2-period zero-coupon bond and the money market account pair. As with the 4-period zero-coupon bond and the money market account pair, each reduced tree must be arbitrage-free.

Let us first consider the reduced tree involving the 3-period zero-coupon bond and the money market account pair. This reduced tree is arbitrage-free if there exist a *unique* set of pseudo probabilities $\pi(t,3;s_t)$'s such that

$$r(t;s_t) = \pi(t,3;s_t)u(t,3;s_t) + (1 - \pi(t,3;s_t))d(t,3;s_t) \quad \text{for all } t \text{ and } s_t.$$

(10.5)

The check is to compute

$$\pi(t,3;s_t) = [r(t;s_t) - d(t,3;s_t)] / [u(t,3;s_t) - d(t,3;s_t)] \quad (10.6)$$

and see if the $\pi(t,3;s_t)$'s are strictly between 0 and 1.

A sample calculation gives

$$d(1,3;u) = \frac{P(2,3;ud)}{P(1,3;u)} = \frac{.980015}{.965127} = 1.015426$$

$$u(1,3;u) = \frac{P(2,3;uu)}{P(1,3;u)} = \frac{.984222}{.965127} = 1.019785$$

and

$$\pi(1,3;u) = \frac{1.017606 - 1.015426}{1.019785 - 1.015426} = .5.$$

The two remaining calculations show that $\pi(0,3) = \pi(1,3;d) = .5$. These pseudo probabilities are all between 0 and 1 as well, so this reduced tree is arbitrage-free.

Using the present value formula, we get that

$$P(t,3;s_t) = \frac{[\pi(t,3;s_t)P(t+1,3;s_tu) + (1 - \pi(t,3;s_t))P(t+1,3;s_td)]}{r(t;s_t)}$$

for all t and s_t.

(10.7)

Using risk-neutral valuation, we get that

$$P(t,4;s_t) = \frac{\left[\pi(t,3;s_t)P(t+1,4;s_tu)+(1-\pi(t,3;s_t))P(t+1,4;s_td)\right]}{r(t;s_t)}$$

for all t and s_t

(10.8a)

and

$$P(t,2;s_t) = \frac{\left[\pi(t,3;s_t)P(t+1,2;s_tu)+(1-\pi(t,3;s_t))P(t+1,2;s_td)\right]}{r(t;s_t)}$$

for all t and s_t.

(10.8b)

The reader should check to see that these conditions are satisfied by Figure 10.2. Note that in expressions (10.7) and (10.8) the πs have a "3" in the argument, reflecting the fact that these expressions were obtained from the 3-period bond/money market account reduced tree.

Now, we return to the satisfaction of condition (10.4) by the prices of the 3-period zero-coupon bond. Expression (10.4) is identical to expression (10.7) with the exception that the first uses $\pi(t,4;s_t)$ and the second uses $\pi(t,3;s_t)$. The only way this can happen, given the uniqueness of the πs in each of them, is if

$$\pi(t,3;s_t) = \pi(t,4;s_t).$$

So, this is another necessary condition for the (enlarged) arbitrage-free tree.

A similar argument for the reduced tree involving the 2-period zero-coupon bond and the money market account pair shows that this reduced tree is arbitrage-free, and that the only way this can happen is if

$$\pi(t,2;s_t) = \pi(t,4;s_t).$$

As these provide all the possible zero-coupon bond and money market account pairs, we see that a necessary condition for the tree to be arbitrage-free is that all the pseudo probabilities are strictly between 0 and 1 and that at each node in the tree, i.e., for a given t and s_t,

$$\pi(t,2;s_t) = \pi(t,3;s_t) = \pi(t,4;s_t). \tag{10.9}$$

It turns out that these conditions are also sufficient (the proof is in the next section).

In summary, then, to show that Figure 10.1 is arbitrage-free, one needs to first compute the pseudo probabilities for each zero-coupon bond/money market account pair just as in Chapter 8, and show that these pseudo probabilities are strictly between 0 and 1. Second, one needs to show that at each node in the tree, all the pseudo probabilities are equal. Notice that the values of the pseudo probabilities at a particular time t and state s_t can differ from those at a different time and state. The condition only requires equality of the pseudo probabilities at a particular time and state s_t pair, not throughout the entire tree.*

There is an economic interpretation of expression (10.9) that aids our understanding as to why it must be sufficient. This interpretation is based on recognizing two facts. The first is that expression (10.9) is equivalent to

$$1 - \pi(t,2;s_t) = 1 - \pi(t,3;s_t) = 1 - \pi(t,4;s_t) \quad \text{for all } t \text{ and } s_t.$$

The second is that the quantity

$$1 - \pi(t,T;s_t) = \frac{u(t,T;s_t) - r(t;s_t)}{u(t,T;s_t) - d(t,T;s_t)} \quad \text{for } T = 2,3,4$$

is a measure of the excess return per unit of risk for holding the T-maturity zero-coupon bond. Indeed, the numerator is the potential extra return that one earns on the T-maturity bond above the spot rate over the next time period. This is the extra compensation for the risk involved in holding the T-period zero-coupon bond. The denominator is a measure of the spread in the returns on the T-maturity bond possible – a measure of risk. Hence, expression

---

* For this particular example, they are all equal to (1/2) at every node and state. This is a special case that need not be true in general.

(10.9) states that *a zero-coupon bond price curve evolution is arbitrage-free if and only if all bonds earn the same excess return per unit of risk.*

This makes intuitive sense, because if it were not true, then some zero-coupon bonds would provide better risk/reward opportunities than others. No one would desire to hold the bond with the lowest risk/reward tradeoff. In fact, as shown above, this would imply the existence of an arbitrage opportunity. This discussion completes the example. □

We now formalize this example. Consider the one-factor economy as described in Chapter 5. We assume that there are no arbitrage opportunities with respect to the largest class of self-financing trading strategies possible Φ_τ. This class of trading strategies uses all the available zero-coupon bonds and the money market account (see Chapter 7).

Taken as given (exogenous) are the current prices and the stochastic processes for all the zero-coupon bonds, $\{P(0,T)$ for all $0 \leq T \leq \tau$ and $(u(t,T;s_t), d(t,T;s_t))$ for all $0 \leq t \leq T \leq \tau$ and $s_t\}$. Because these processes are exogenously specified, we need to investigate the restrictions that must be imposed so that they are arbitrage-free. To do this, we need to recall some facts proven in previous chapters.

10.2.1 Complete Markets

The one-factor economy is complete with respect to the trading strategy set Φ_τ. This follows directly from the observation in Chapter 7 that the one-factor economy is complete with respect to the trading strategy set Φ_1. Indeed, if only one zero-coupon bond and the money market account are needed to synthetically construct any contingent claim using Φ_1, then adding more bonds to the trading strategy set (as in Φ_τ) does not alter this conclusion. The advantage of using the larger trading strategy set Φ_τ is that having all zero-coupon bonds trade simultaneously allows the possibility of constructing more complicated portfolios involving multiple zero-coupon bonds.

10.2.2 Risk-Neutral Probabilities

Arbitrage opportunities with respect to the trading strategy set Φ_τ are assumed to be nonexistent in this economy. This section analyzes the restrictions that this assumption imposes upon the stochastic processes for the evolution of the *entire* zero-coupon bond price curve. Recall that the assumption of no arbitrage opportunities with respect to the trading strategy set Φ_τ is more restrictive than the assumption of no-arbitrage opportunities with respect to the trading strategy set Φ_1. Indeed, the set of self-financing trading strategies with respect to the larger trading strategy set Φ_τ allows the use of *all* the zero-coupon bonds rather than just the *one* zero-coupon bond with maturity τ_1. Consequently, the no-arbitrage restrictions proven earlier with respect to the trading strategy set Φ_1 still apply with respect to the trading strategy set Φ_τ.

Recalling that analysis from Chapter 9, no arbitrage with respect to the trading strategy set Φ_1 (for a given τ_1-maturity zero-coupon bond) implies that there exists a *unique* pseudo probability $\pi(t,\tau_1;s_t)$ such that $P(t,\tau_1;s_t)/B(t;s_{t-1})$ is a martingale, i.e.,

$$\frac{P(t,\tau_1;s_t)}{B(t;s_{t-1})} = \pi(t,\tau_1;s_t)\frac{P(t+1,\tau_1;s_t u)}{B(t+1;s_t)} + \left(1-\pi(t,\tau_1;s_t)\right)\frac{P(t+1,\tau_1;s_t d)}{B(t+1;s_t)}.$$

(10.10a)

The form of expression (10.10a) most useful for calculations is

$$P(t,\tau_1;s_t) = \left[\pi(t,\tau_1;s_t)P(t+1,\tau_1;s_t u) + (1-\pi(t,\tau_1;s_t))P(t+1,\tau_1;s_t d)\right]/r(t;s_t).$$

(10.10b)

The pseudo probabilities are determined by:

$$\pi(t,\tau_1;s_t) = [r(t;s_t)-d(t,\tau_1;s_t)]/[u(t,\tau_1;s_t)-d(t,\tau_1;s_t)] \qquad (10.11a)$$

and

$$1-\pi(t,\tau_1;s_t) = [u(t,\tau_1;s_t)-r(t,\tau_1;s_t)]/[u(t,\tau_1;s_t)-d(t,\tau_1;s_t)]. \qquad (10.11b)$$

In expression (10.11) we have made explicit in the notation the dependence of the pseudo probabilities upon the τ_1-maturity zero-coupon bond.

Because all the zero-coupon bonds are available for trade in the set Φ_τ, expression (10.11) must simultaneously hold for every possible τ_1-maturity zero-coupon bond selected, i.e., for all τ_1 such that $0 \le \tau_1 \le \tau$. This argument could not be made if we only used the smaller set of trading strategies Φ_1.

Therefore, there exists a unique and potentially different pseudo probability for each zero-coupon bond that makes that particular zero-coupon bond a martingale (after division by the money market account's value). This condition ensures that there are no arbitrage opportunities between any single zero-coupon bond and the money market account, i.e., that neither of these securities dominates the other. This says nothing, however, about the absence of arbitrage opportunities across zero-coupon bonds of different maturities. These are the restrictions we now seek.

We know that since the market is complete with respect to the trading strategy set Φ_1, it is also complete with respect to the larger trading strategy set Φ_τ. Thus, given the τ-maturity bond and the money market account, we can create a synthetic T-maturity zero-coupon bond of any maturity. No arbitrage with respect to Φ_τ should imply that the traded T-maturity bond equals its arbitrage-free price* as defined in Chapter 7.

To see this argument, let $(n_0(t;s_t),\ n_\tau(t;s_t)\colon\ 0 \le t \le \tau_1 - 1)$ be the self-financing trading strategy in the money market account and τ-maturity zero-coupon bond that duplicates the τ_1- maturity zero-coupon bond. The portfolio holdings are given (see Chapter 7) by

$$n_0(t;s_t) = \frac{x(t+1;s_t d)u(t,\tau;s_t) - x(t+1;s_t u)d(t,\tau;s_t)}{B(t;s_t)r(t;s_t)[u(t,\tau;s_t) - d(t,\tau;s_t)]} \qquad (10.12a)$$

$$n_\tau(t;s_t) = \frac{x(t+1;s_t u) - x(t+1;s_t d)}{P(t,\tau;s_t)[u(t,\tau;s_t) - d(t,\tau;s_t)]} \qquad (10.12b)$$

where

$$x(\tau_1;s_t) \equiv P(\tau_1,\tau_1;s_{\tau_1}) = 1 \ \text{ for all } s_{\tau_1} \qquad (10.12c)$$

and

$$x(t;s_t) = n_0(t;s_t)B(t;s_{t-1}) + n_\tau(t;s_t)P(t,\tau;s_t) \quad \text{ for } \ 0 \le t \le \tau_1 - 1. \ (10.13)$$

Given that there are no arbitrage opportunities with respect to Φ_τ, it must be the case that the market price of the τ_1- maturity zero-coupon bond equals the cost of the synthetic construction, i.e.,

---

* Recall that in Chapter 7 trading strategies with respect to Φ_1 only involved the τ_1-maturity bond and the money market account. No other zero-coupon bond could be traded. Hence, we *defined* the arbitrage-free price to be expression (7.20). This section, however, uses the assumption of no arbitrage opportunities with respect to Φ_τ to *prove* that expression (7.20) holds.

$$P(t,\tau_1;s_t) = n_0(t;s_t)B(t;s_{t-1}) + n_\tau(t;s_t)P(t,\tau;s_t). \tag{10.14}$$

To prove expression (10.14), we use proof by contradiction. Proof by contradiction works by supposing the contrary to what you want to prove, and then showing that this supposition leads to a contradiction. Therefore, to prove expression (10.14), let us suppose the contrary: that is, suppose

$$P(t,\tau_1;s_t) > n_0(t;s_t)B(t;s_{t-1}) + n_\tau(t;s_t)P(t,\tau;s_t) \quad \text{for some } t \text{ and } s_t.$$

An arbitrage opportunity with respect to Φ_τ can now be constructed. The arbitrage opportunity is as follows. At time 0, start monitoring prices to determine when the violation of expression (10.14) occurs. When it occurs, sell the τ_1-maturity bond, buy $n_\tau(t;s_t)$ units of the τ-maturity bond, and buy $n_0(t;s_t)$ units of the money market account. This generates a strictly positive cash inflow of $[P(t,\tau_1;s_t) - n_0(t;s_t)B(t;s_{t-1}) - n_\tau(t;s_t)P(t,\tau;s_t)]$ dollars. Invest this additional inflow in a τ_1-maturity bond and hold the bond until its maturity. With this cash inflow $[P(t,\tau_1;s_t) - n_0(t;s_t)B(t;s_{t-1}) - n_\tau(t;s_t)P(t,\tau;s_t)]/P(t,\tau_1;s_t) > 0$ units of the τ_1-maturity bond can be purchased. In addition, hold the portfolio that generated this time t cash inflow, expression (10.12), rebalancing $n_0(t;s_t)$ and $n_\tau(t;s_t)$ as required until time τ_1, when it is also liquidated. Recall that this rebalancing is self-financing. By construction, this strategy has zero cash flows at time t and a strictly positive cash flow of $[P(t,\tau_1;s_t) - n_0(t;s_t)B(t;s_{t-1}) - n_\tau(t;s_t)P(t,\tau;s_t)]/P(t,\tau_1;s_t) > 0$ dollars at time τ_1.

Hence, an arbitrage opportunity has been constructed. But this contradicts the assumption of no arbitrage opportunities! Therefore,

$$P(t,\tau_1;s_t) \le n_0(t;s_t)B(t;s_{t-1}) + n_\tau(t;s_t)P(t,\tau;s_t) \quad \text{must be true.}$$

The reverse inequality is proven in an equivalent fashion* and is left to the reader as an exercise.

From Chapter 9, the risk-neutral valuation method, we know that we can rewrite expression (10.14) as

$$\frac{P(t,\tau_1;s_t)}{B(T;s_t)} = \left[\pi(t,\tau;s_t)\frac{P(t+1,\tau_1;s_t u)}{B(t+1;s_t)} + (1-\pi(t,\tau;s_t))\frac{P(t+1,\tau_1;s_t d)}{B(t+1;s_t)} \right]$$

$$\tag{10.15a}$$

---

* This same argument could not have been made if we had used only the smaller set of trading strategies Φ_1. The set Φ_1 does not allow holdings in more than one zero-coupon bond.

or equivalently,

$$P(t,\tau_1;s_t) = \left[\pi(t,\tau;s_t)P(t+1,\tau_1;s_tu) + (1-\pi(t,\tau;s_t))P(t+1,\tau_1;s_td)\right]/r(t;s_t)$$

(10.15b)

where $\pi(t,\tau;s_t)$ are the *unique* pseudo probabilities associated with the τ-maturity bond. Expression (10.15) must hold for all $0 \le t < \tau_1 < \tau$ and s_t. But expression (10.10) is also valid. The uniqueness of the pseudo probabilities $\pi(t,\tau_1;s_t)$ in that expression implies that

$$\pi(t,\tau;s_t) = \pi(t,\tau_1;s_t) \text{ for all } \tau_1, t, s_t \text{ such that } 0 \le t \le \tau_1 - 1 \le \tau - 1. \quad (10.16)$$

In words, the pseudo probabilities must be independent of the particular τ_1-maturity bond selected. Consequently, under the assumption of no-arbitrage opportunities with respect to Φ_τ, we can go back to our earlier notation $\pi(t;s_t)$ for the pseudo probabilities.

Expression (10.16) can be given an economic interpretation. From expression (10.11b), we see that $1 - \pi(t,\tau;s_t)$ is a measure of the excess return per unit of risk provided on the τ-maturity zero-coupon bond. First, the numerator $(u(t,\tau;s_t) - r(t;s_t))$ is the excess return provided by the τ-maturity bond above the spot rate in the good state of nature. This is the compensation for bearing the risk of the τ-maturity zero-coupon bond. Second, the denominator $(u(t,\tau;s_t) - d(t,\tau;s_t))$ is a measure of risk since it measures the spread in returns possible from holding the τ-maturity bond.* The ratio, therefore, is a measure of excess return, per unit of risk, for the τ-maturity bond. Another, more analytic justification for this economic interpretation can be found in the appendix to this chapter.

The economic interpretation of condition (10.16) is that, to prevent arbitrage, all zero-coupon bonds must have the same excess return per unit of risk. Otherwise, those zero-coupon bonds with higher excess returns, per unit of risk, are good buys. Those zero-coupon bonds with lower excess returns, per unit of risk, are good sells. Combined, these differences imply that there can be no equilibrium and that these strategies should generate arbitrage profits.

---

* Simple algebra shows that $E_t(P(t+1,\pi;s_{t+1})/P(t,\pi;s_t)) = q_t(s_t)u(t,\pi;s_t) + (1-q_t(s_t))d(t,\pi;s_t)$ and $\text{Var}_t(P(t+1,\pi;s_{t+1})/P(t,\pi;s_t)) = q_t(s_t)(1-q_t(s_t))[u(t,\pi;s_t) - d(t,\pi;s_t)]^2$. Thus, the spread in returns possible from holding the τ-maturity bond is proportional to the bond return's standard derivation. This is another justification for this spread as a risk measure.

Note that from expression (10.11), expression (10.16) implies that there are cross-restrictions on the parameters in the bond processes $(u(t,T;s_t)$, $d(t,T;s_t))$ across different maturities T. The stochastic processes chosen for the evolution of the zero-coupon bond price curve cannot be arbitrary, given an initial zero-coupon bond price curve. Rather, the processes must be restricted so that expression (10.16) holds. Otherwise, they would not be arbitrage-free. It is this additional restriction that differentiates the bond-trading strategy techniques of Chapters 8 and 9 from the option pricing techniques of Chapter 10. This restriction is the contribution of the HJM model.

In summary, no arbitrage with respect to the trading strategy set Φ_τ for the one-factor economy implies the existence of unique pseudo probabilities $\pi(t,s_t)$ that are independent of any particular zero-coupon bond and are such that $P(t,T;s_t)/B(t,s_{t-1})$ is a martingale for all T-maturity zero-coupon bonds.

The converse of this statement is also true. The existence of pseudo probabilities $\pi(t;s_t)$ that are independent of any particular zero-coupon bond selected and are such that $P(t,T;s_t)/B(t;s_{t-1})$ is a martingale for all T-maturity bonds implies that there are no arbitrage opportunities with respect to the trading strategy set Φ_τ. The proof of this assertion is identical to that contained in the appendix to Chapter 9 and is therefore omitted.

10.2.3 Risk-Neutral Valuation

The pseudo probabilities given in expression (10.16) are used as in Chapter 9 to price all interest rate-dependent contingent claims. The difference between this calculation and that given in Chapter 9 is that the pseudo probabilities in Chapter 9 depended upon the τ_1-maturity bond's specification and could differ from those given by the τ-maturity bond. Now, however, these probabilities must be identical (as given in expression (10.16)). Given this specification, contingent claims are priced and hedged in an identical fashion as in Chapter 9. This procedure is illustrated in subsequent chapters.

10.3 THE TWO-FACTOR ECONOMY

The case of the two-factor economy is almost identical to that of the one-factor economy, so the description of the methodology will be brief. Consider the two-factor economy as described in Chapter 5. We assume that there are no arbitrage opportunities with respect to the trading strategy set Φ_τ. Taken as given (exogenous) is the stochastic process for all the zero-coupon bonds, $\{P(0,T)$ for all $0 \leq T \leq \tau$ and $(u(t,T;s_t), m(t,T;s_t), d(t,T;s_t))$ for all $0 \leq t \leq T \leq \tau$ and $s_t\}$. The purpose is to price interest rate-dependent contingent claims.

10.3.1 Complete Markets

Given that the two-factor economy is complete with respect to the trading strategy set Φ_2 (see Chapter 9), the two-factor economy is complete with respect to the larger trading strategy set Φ_τ. Indeed, the trading strategy set Φ_τ allows the self-financing trading strategies to utilize more zero-coupon bonds, so if the market is complete with respect to the smaller trading strategy set Φ_2, it is complete with respect to the larger trading strategy set Φ_τ as well.

10.3.2 Risk-Neutral Probabilities

The assumption that there are no arbitrage opportunities with respect to the larger trading strategy set Φ_τ implies that there are no arbitrage opportunities with respect to the smaller trading strategy set Φ_2 for all possible pairs of τ_1-maturity and τ_2-maturity zero-coupon bonds selected.

From Chapter 9, this implies that there exist *unique* pseudo probabilities $\pi^u(t,\tau_1,\tau_2;s_t)$, $\pi^m(t,\tau_1,\tau_2;s_t)$ such that

$$\frac{P(t,\tau_i;s_t)}{B(t;s_{t-1})} = \begin{bmatrix} \pi^u(t,\tau_1,\tau_2;s_t)\dfrac{P(t+1,\tau_i;s_t u)}{B(t+1;s_t)} + \pi^m(t,\tau_1,\tau_2;s_t)\dfrac{P(t+1,\tau_i;s_t m)}{B(t+1;s_t)} \\ \\ +(1-\pi^u(t,\tau_1,\tau_2;s_t)-\pi^m(t,\tau_1,\tau_2;s_t))\dfrac{P(t+1,\tau_i;s_t m)}{B(t+1;s_t)} \end{bmatrix}$$

(10.17)

for $i = 1, 2$ where the dependence on the τ_1- and τ_2-maturity zero-coupon bonds selected is indicated in the notation for the pseudo probabilities. Expression (10.17) guarantees that none of the money market account, the τ_1-maturity zero-coupon bond, nor the τ_2-maturity zero-coupon bond dominates either of the others. But no arbitrage with respect to the larger trading strategy set Φ_τ is stronger.

Because the market is complete with respect to the smaller trading strategy set Φ_2, given the $\tau-1$- and τ-maturity zero-coupon bonds and the money market account, there exists another (different) self-financing trading strategy that duplicates both the τ_1-maturity zero-coupon bond and the τ_2-maturity zero-coupon bond. No arbitrage opportunities with respect to the larger trading strategy set Φ_τ imply that the initial cost of these synthetic zero-coupon bonds equals their market prices. The proof is identical to the one used to prove expression (10.15). This, in turn, implies by risk-neutral valuation that

$$\frac{P(t,\tau_i;s_t)}{B(t;s_{t-1})} = \left[\begin{array}{l} \pi^u(t,\tau-1,\tau;s_t)\dfrac{P(t+1,\tau_i;s_t u)}{B(t+1;s_t)} + \pi^m(t,\tau-1,\tau;s_t)\dfrac{P(t+1,\tau_i;s_t m)}{B(t+1;s_t)} \\[3mm] +(1-\pi^u(t,\tau-1,\tau;s_t)-\pi^m(t,\tau-1,\tau;s_t))\dfrac{P(t+1,\tau_i;s_t m)}{B(t+1;s_t)} \end{array} \right]$$

(10.18)

for $i = 1$, 2. The distinction between this system and expression (10.17) is the replacement of the pseudo probabilities $\pi^u(t,\tau_1,\tau_2;s_t)$ and $\pi^m(t,\tau_1,\tau_2;s_t)$ by $\pi^u(t,\tau-1,\tau;s_t)$ and $\pi^m(t, \tau-1,\tau;s_t)$.

By the uniqueness of the pseudo probabilities $\pi^u(t,\tau_1,\tau_2;s_t)$ and $\pi^m(t,\tau_1,\tau_2;s_t)$ in expression (10.17), we get

$$\pi^u(t,\tau-1,\tau;s_t) = \pi^u(t,\tau_1,\tau_2;s_t)$$
$$\pi^m(t,\tau-1,\tau;s_t) = \pi^m(t,\tau_1,\tau_2;s_t)$$

(10.19)

for all s_t and $0 \le t < \tau_1 - 1 < \tau_2 - 1 \le \tau - 1$. That is, the pseudo probabilities must be independent of the particular pair of zero-coupon bonds with maturities τ_1 and τ_2 selected.

In summary, no arbitrage with respect to the larger trading strategy set Φ_τ for the two-factor economy implies the existence of unique pseudo probabilities $\pi^u(t;s_t)$, $\pi^m(t;s_t)$, independent of any pair of zero-coupon bonds selected, such that $P(t,T;s_t)/B(t;s_{t-1})$ is a martingale for all T maturity zero coupon bonds.

The converse of this statement is also true. The existence of pseudo probabilities $\pi^u(t;s_t)$, $\pi^m(t;s_t)$ independent of any pair of zero-coupon bonds selected, such that $P(t,T;s_t)/B(t;s_{t-1})$ is a martingale for all T-maturity zero-coupon bonds, implies that there are no arbitrage opportunities with respect to the trading strategy set Φ_τ. The proof of this statement mimics the proof contained in the appendix to Chapter 9 and is left as an exercise for the reader.

10.3.3 Risk-Neutral Valuation

The pseudo probabilities given in expression (10.19) are used as in Chapter 9, risk-neutral valuation, to price interest rate-dependent contingent claims. The difference between this calculation and that given in Chapter 9 is that the pseudo probabilities in Chapter 9 depended on the τ_1- and τ_2-maturity bonds selected, whereas these do not. The valuation procedure is otherwise identical, and it is illustrated in subsequent chapters.

10.4 MULTIPLE FACTOR ECONOMIES

The extension of the last section to $N \geq 3$ factors is straightforward. Because of the results in Chapter 9, the markets are complete with respect to the trading strategy set Φ_τ. Furthermore, the same analysis as that given above guarantees that no arbitrage with respect to the trading strategy set Φ_τ implies the existence of unique pseudo probabilities, independent of any N-collection of different-maturity zero-coupon bonds selected, such that $P(t,T;s_t)/B(t;s_{t-1})$ is a martingale for all T. Risk-neutral valuation proceeds in the usual fashion.

APPENDIX

This appendix shows that $\pi(t,T;s_t)$ is a measure of the risk premium for a T-maturity, zero-coupon bond.

The standard measure for the risk premium on the T-maturity zero-coupon bond is given by

$$\phi(t,T;s_t) = \frac{E_t(P(t+1,T;s_t)/P(t,T;s_t)) - r(t;s_t)}{\sqrt{\mathrm{Var}_t\,(P(t+1,T;s_t)/P(t,T;s_t))}}.$$

Simple algebra shows that

$$E_t(P(t+1,T;s_t)/P(t,T;s_t)) = q_t(s_t)u(t,T;s_t) + (1-q_t(s_t))d(t,T;s_t)$$

and

$$\mathrm{Var}_t(P(t+1,T;s_t)/P(t,T;s_t)) = E([P(t+1,T;s_t)/P(t,T;s_t)]^2)$$

$$-[E(P(t+1,T;s_t)/P(t,T;s_t))]^2$$

$$= q_t(s_t)(1-q_t(s_t))[u(t,T;s_t)-d(t,T;s_t)]^2.$$

Thus,

$$\phi(t,T;s_t) = \frac{q_t(s_t)u(t,T;s_t)+(1-q_t(s_t))d(t,T;s_t)-r(t;s_t)}{\sqrt{q_t(s_t)(1-q_t(s_t))}\,[u(t,T;s_t)-d(t,T;s_t)]}$$

$$= \frac{\sqrt{q_t(s_t)}}{\sqrt{1-q_t(s_t)}} + \left(\frac{d(t,T;s_t)-r(t;s_t)}{u(t,T;s_t)-d(t,T;s_t)}\right)\left(\frac{1}{\sqrt{q_t(s_t)(1-q_t(s_t))}}\right)$$

$$= \frac{\sqrt{q_t(s_t)}}{\sqrt{1-q_t(s_t)}} - \frac{\pi(t,T;s_t)}{\sqrt{q_t(s_t)(1-q_t(s_t))}}.$$

This last expression shows that $\pi(t,T;s_t)$ is linearly related to the risk premium, given that $q_t(s_t)$ is fixed.

Note that the equality of $\pi(t,T;s_t)$ across all T holds if and only if $\tau\phi(t,T;s_t)$ is equal across all T. This completes the derivation.

III

Applications

CHAPTER 11

Coupon Bonds

T HE CHAPTER STUDIES COUPON bonds from the perspective of the arbitrage-free pricing methodology. This is in contrast to the classical approach to fixed income analysis or coupon bond pricing that was presented in Chapter 3. The differences between the two approaches are numerous. First, the arbitrage-free pricing methodology can be used to risk manage a portfolio of bonds given an arbitrary evolution for the term structure of interest rates. The classical approach cannot. The classical approach can only hedge parallel shifts in the term structure of interest rates. Second, the arbitrage-free pricing approach can be used to price interest rate derivatives in a manner consistent with that used to price coupon bonds. The classical approach cannot. Third, the arbitrage-free pricing approach can be extended to handle foreign currency risk and credit risk. The classical approach cannot.

This chapter contains three insights. First, without specifying an evolution for the term structure of interest rates, we show that a coupon bond is equivalent to a portfolio of zero-coupon bonds. This insight provides a robust method for pricing. Second, specifying a particular evolution for the term structure of interest rates allows for a more refined hedging analysis. Here, it is shown that in a one-factor model a coupon bond can be hedged with one zero-coupon bond, in a two-factor model a coupon bond can be hedged with two zero-coupon bonds, and so forth. Examples are provided to illustrate this second insight. Third, the one-factor model hedge is compared to the classical duration hedge, further illustrating the errors inherent in the classical technique.

11.1 THE COUPON BOND AS A PORTFOLIO OF ZERO-COUPON BONDS

A coupon-bearing bond is a loan for a fixed amount of dollars (e.g., $10,000), called the principal or face value. The loan extends for a fixed time period, called the life or maturity of the bond (usually 5, 10, 20, or 30 years). Over its life, the issuer is required to make periodic interest payments (usually semi-annually) on the loan's principal. These interest payments are called coupons. Hence, the term a coupon bond.

A callable coupon bond is a coupon bond that can be repurchased by the issuer, at predetermined times and prices, prior to its maturity. This repurchase provision is labeled a "call" provision. For example, a typical callable coupon bond includes a provision where the issuer can repurchase the bond at face value (the predetermined price), anytime during the last 5 years of its life (the predetermined times).

This section studies the arbitrage-free pricing of noncallable coupon bonds. Callable coupon bonds are studied in the next chapter. The valuation method of this section is independent of the particular evolution of the term structure of interest rates selected; in particular, it does not depend on the number or specification of the factors in the economy, one, two, or three factors.

Formally, we define a *coupon bond* with principle L, coupons C, and maturity T to be a financial security that is entitled to receive coupon payments of C dollars at times $1, \ldots, T$ with a principal repayment of L at time T. The times $1, 2, \ldots, T$ represent the payment dates on the loan. The coupon rate on the bond is $c = 1 + C/L$. This cash flow pattern is illustrated in Table 11.1.

The coupon bond's cash flows can be obtained from a portfolio of zero-coupon bonds. The duplicating portfolio consists of C zero-coupon bonds maturing at times $1, \ldots, T-1$ and $C+L$ zero-coupon bonds maturity at time T.

TABLE 11.1 The Cash Flows to a Typical Coupon Bond with Price $\mathcal{B}(0)$, Principal L, Coupon C, and Maturity T

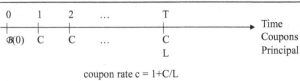

0	1	2	...	T	
					Time
$\mathcal{B}(0)$	C	C	...	C	Coupons
				L	Principal

coupon rate c = 1+C/L

Let the market price of the coupon bond be denoted by $\mathcal{B}(t)$. This represents a trader's quote.

The cost of constructing this portfolio of zero-coupon bonds is:

$$\sum_{i=t+1}^{T} CP(t,i) + LP(t,T).$$

In constructing this portfolio, it is assumed that the construction occurs after the coupon payment has been paid at time t (i.e., it represents the ex-coupon value at time t).

Thus, the arbitrage-free price of the coupon bond is:

$$\mathcal{B}(t) = \sum_{i=t+1}^{T} CP(t,i) + LP(t,T). \tag{11.1}$$

Indeed, if the market price of the coupon differed from expression (11.1), then an arbitrage opportunity with respect to the trading strategy set can be constructed. For example, if $\mathcal{B}(t)$ exceeded the right side of expression (11.1), one could sell the coupon bond short at time t and buy C zero-coupon bonds maturing at time i for each $i = t+1, \dots, T-1$ and $(C+L)$ zero-coupon bonds maturing at time T. Holding this portfolio until time T generates an arbitrage opportunity.

Let's check the conditions. At time t, this portfolio has a positive cash inflow that equals the difference between the left and right side of expression (11.1). At each future date, when the short position requires a cash outflow, the portfolio of zero-coupon bonds provides the cash inflow to exactly satisfy it. Hence, there is no cash flow after time t on this portfolio. The conditions for an arbitrage opportunity are all satisfied.

Note that the arbitrage-free price for the coupon bond can be computed without any knowledge of the evolution of the term structure of interest rates. It depends solely on the initial zero-coupon bond price curve. We now illustrate this computation with an example.

Example: Coupon Bond Calculation

To illustrate this calculation, we use the initial zero-coupon bond price curve as given in Table 11.2. Four zero-coupon bonds are traded, the longest maturity bond being the 4-period zero-coupon bond.

TABLE 11.2   An Example of an Initial
Zero-Coupon Bond Price Curve

$P(0,4) = .923845$
$P(0,3) = .942322$
$P(0,2) = .961169$
$P(0,1) = .980392$

TABLE 11.3   An Example of the Cash Flows to a Coupon Bond

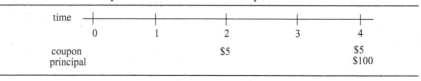

time	0	1	2	3	4
coupon			$5		$5
principal					$100

Consider a coupon bond whose cash flows are as given in Table 11.3. This coupon bond pays 5 dollars at times 2 and 4, with a principal repayment of 100 dollars at time 4. We interpret the coupon payments as being paid semi-annually, hence, the spacing between payments.

Using expression (11.1), the value of this coupon bond at time 0, ex-coupon, is:

$$\mathcal{B}(0) = 5P(0,2) + 5P(0,4) + 100P(0,4)$$

$$= 5[0.961169] + 105[0.923845] = 101.8096.$$

If the market price for the coupon bond differed, an arbitrage opportunity would exist. For example, if the market price of this coupon bond were 102.000, then an arbitrage opportunity is represented by: *(i)* shorting and holding until maturity the coupon bond, *(ii)* buying and holding until maturity five units of the 2-period zero-coupon bond, and *(iii)* buying and holding until maturity 105 units of the 4-period zero-coupon bond. The initial position brings in 102–101.8096 > 0 dollars. Subsequently, the cash flows to the short coupon bond are satisfied by the cash flows from the zero-coupon bond portfolio, leaving no further obligation. This completes the example. ☐

The above arbitrage-free pricing technique does not depend on a particular evolution for the term structure of interest rates. This makes the approach quite useful for pricing. This procedure is less worthwhile, however, for hedging. In this approach, a synthetic coupon bond is constructed via a buy and hold strategy involving a portfolio of zero-coupon bonds. Zero-coupon bonds are needed for each date on which a cash payment to the coupon bond is made. For example, given a 20-year bond with semi-annual coupon payments, 40 zero-coupon bonds are required. This requirement has two practical problems. One, the particular zero-coupon bonds that match the coupon payment dates most likely do not trade, making the replication impossible. Two, if they all trade, the initial transaction costs will be quite large. There is an alternative approach, however. Given an explicit representation of the evolution for the term structure of interest rates fewer zero-coupon bonds can be used to construct a synthetic coupon bond. We illustrate this approach in the next section.

11.2 THE COUPON BOND AS A DYNAMIC TRADING STRATEGY

This section shows how to use the HJM model to synthetically construct a coupon bond using fewer zero-coupon bonds than the number of payment dates. This approach is dependent, however, on a particular evolution for the term structure of interest rates. Fortunately, this evolution can be selected by the user to match the best prediction for the term structure's future evolution.

The theory underlying this technique was presented in Chapter 10. It can be best illustrated with an example.

Example: Synthetic Coupon Bond
Construction in a One-Factor Model

Consider the evolution for the zero-coupon bond price curve as given in Figure 11.1. This is the same evolution as given in Chapter 10, Figure 10.2. The first step in applying this technology is to check to see if the given evolution is arbitrage-free. Fortunately, this determination was already performed in Chapter 10. There it was shown that this evolution is arbitrage-free because the pseudo probabilities calculated for each maturity zero-coupon bond/money market account pair at each node in the tree are strictly between 0 and 1, and equal to each other (i.e., 1/2). These pseudo probabilities appear along the branches of the tree.

To illustrate the synthetic construction of a coupon bond using only one zero-coupon bond, we consider the coupon bond in Table 11.3.

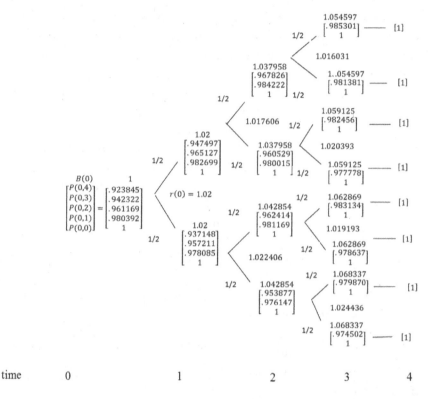

time   0   1   2   3   4

FIGURE 11.1   An Example of a One-Factor Bond Price Curve Evolution. Pseudo Probabilities Are Along Each Branch of the Tree.

We use method 2, risk-neutral valuation, which was illustrated in Chapter 8 and documented in Chapter 9. The calculations are as follows.

Step 1: Risk-Neutral Valuation

Step 1 requires that we determine the arbitrage-free price of the coupon bond at each node in the tree. The risk-neutral valuation procedure is employed. This approach proceeds by backward induction.

At time 4, we know the cash flows to the coupon bond. They are 105 dollars, across all states.

Moving back to time 3, at state uuu, we compute the price of the coupon bond as:

$$B(3;uuu) = \frac{(1/2)B(4;uuuu)+(1/2)B(4;uuud)}{r(3;uuu)}$$

$$= \frac{(1/2)105+(1/2)105}{1.0149182} = 103.4566.$$

This is the discounted expected payoff on the bond. The calculations for the coupon bond's price at time 3 for the remaining states are similar. The details are left to the reader. The resulting values are given in Figure 11.2.

Next, at time 2, we repeat the previous procedure. For example, at time 2 state uu, the computation is:

$$\mathcal{B}(2;uu) = \frac{(1/2)\mathcal{B}(3;uuu) + (1/2)\mathcal{B}(3;uud)}{r(2;uu)}$$

$$= \frac{(1/2)103.4566 + (1/2)103.0450}{1.016031} = 101.6218.$$

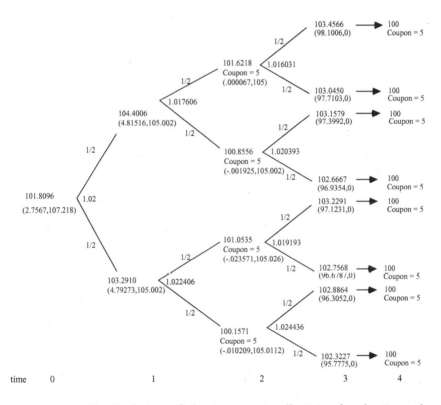

FIGURE 11.2 The Evolution of the Coupon Bond's Price for the Example in Table 11.3. The Coupon Payment at Each Date Is Indicated by the Nodes. The Synthetic Coupon-Bond Portfolio $(n_0(t;s_t),\ n_4(t;s_t))$ in the Money Market Account and 4-Period Zero-Coupon Bond Are Given under Each Node. Pseudo Probabilities Along the Branches of the Tree.

Again, as the remaining time 2 calculations are similar, they are left to the reader as an exercise.

Stepping back to time 1, state u, we repeat the previous procedure, but this time with a twist. When computing the discounted expected value, we need to include both the future value and *future cash flow* in the numerator. The calculation is:

$$\mathcal{B}(1;u) = \frac{(1/2)[\mathcal{B}(2;uu)+C]+(1/2)[\mathcal{B}(2;ud)+C]}{r(1;u)}$$

$$= \frac{(1/2)[101.6218+5]+(1/2)[100.8556+5]}{1.017606} = 104.4006.$$

The remaining time 1 calculation is similar, and recorded on the tree. Finally, at time 0, the procedure yields:

$$\mathcal{B}(0) = \frac{(1/2)\mathcal{B}(1;u)+(1/2)\mathcal{B}(1;d)}{r(0)}$$

$$= \frac{(1/2)104.4006+(1/2)103.2910}{1.02} = 101.8096.$$

There are a number of important observations to make about this procedure.

First, the time 0 price of the coupon bond is identical to that obtained in the previous section based on the portfolio of zero-coupon bonds. This is not by chance, but by construction. Recall that the HJM approach takes the initial zero-coupon bond price curve as given, and then determines an arbitrage-free evolution consistent with it. Interest rate derivatives (the coupon bond) are then priced off this curve. As the interest rate derivative's price is arbitrage-free and it can be computed using only the initial zero-curve, the risk-neutral valuation procedure will generate the same arbitrage-free value!

Second, it is instructive to note that at time 2, the coupon bond's price given is ex-coupon. At time 1 in state u, the coupon bond's price is 104.4006. If the state moves down, the coupon bond's ex-coupon price is 100.8556. There is a drop in the market price! This does not mean, however, that the bond holder has lost money on the position. The coupon bond's net value at time 1, including the coupon, still exceeds the time 1 price, which it should (100.8556 + 5 = 105.8556).

Step 2: Delta Hedging

The next step is to construct the synthetic coupon bond using a dynamic trading strategy in the 4-period zero-coupon bond $(n_4(t;s_t))$ and the money market account $(n_0(t;s_t))$. We use the delta approach that was described in Chapter 8. This approach proceeds by backward induction, and it uses the prices of the coupon bond generated in step 1 above.

First, at time 3 there is only one zero-coupon bond trading, the 4-period zero. It is used to construct the money market account. Thus we can invest in either the 4-period zero or the money market account (since both are identical). We choose, arbitrarily, the money market account:

Time 3, state uuu:

$$n_4(3;uuu) = 0$$

$$n_0(3;uuu) = \frac{\mathcal{B}(3;uuu) - n_4(3;uuu)P(3,4;uuu)}{B(3;uu)} = \frac{103.4566}{1.054597} = 98.1006$$

Time 3, state uud:

$$n_4(3;uud) = 0$$

$$n_0(3;uud) = \frac{\mathcal{B}(3;uud) - n_4(3;uud)P(3,4;uud)}{B(3;uu)} = \frac{103.0450}{1.054697} = 97.7103$$

As the remaining states at time 3 calculations are similar, they are left to the reader and only recorded on the tree.

Moving back to time 2, we apply the delta construction again. Recall that the delta construction computes the change in the price and cash flow of the "hedged" instrument divided by the change in price and cash flow of the "hedging" instrument. The computation is:

Time 2, state uu

$$n_4(2;uu) = \frac{\mathcal{B}(3;uuu) - \mathcal{B}(3;uud)}{P(3,4;uuu) - P(3,4;uud)} = \frac{103.4566 - 103.0450}{.985301 - .981381} = 105$$

$$n_0(2;uu) = \frac{\mathcal{B}(2;uu) - n_4(2;uu)P(2,4;uu)}{B(2;u)} = \frac{101.6218 - (105)0.967826}{1.037958}$$

$$= .000067$$

Time 2, state ud:

$$n_4(2;ud) = \frac{\mathcal{B}(3;udu) - \mathcal{B}(3;udd)}{P(3,4;udu) - P(3,4;udd)} = \frac{103.1579 - 102.6667}{.982456 - .977778} = 105.002$$

$$n_0(2;ud) = \frac{\mathcal{B}(2;ud) - n_4(2;ud)P(2,4;ud)}{B(2;u)}$$

$$= \frac{100.8556 - (105.002)0.960529}{1.037958} = -.001925$$

The remaining time 2 calculations are left for the reader.

Moving back to time 1, we get

Time 1, state u:

$$n_4(1;u) = \frac{[\mathcal{B}(2;uu) + C] - [\mathcal{B}(2;ud) + C]}{P(2,4;uu) - P(2,4;ud)}$$

$$= \frac{[101.6218 + 5] - [100.8556 + 5]}{.967826 - .960529} = 105.002$$

$$n_0(1;u) = \frac{\mathcal{B}(1;u) - n_4(1;u)P(1,4;u)}{B(1)}$$

$$= \frac{104.4006 - (105.002)0.947497}{1.02} = 4.81516$$

Note that we need to replicate the coupon bond's price plus coupon at time 2 across the up and down states. The numerator in $n_4(1;u)$ reflects this joint cash flow.

Time 1, state d:

$$n_4(1;d) = \frac{[\mathcal{B}(2;du) + C] - [\mathcal{B}(2;dd) + C]}{P(2,4;du) - P(2,4;dd)}$$

$$= \frac{[101.0535 + 5] - [100.1571 + 5]}{.962414 - .953877} = 105.002$$

$$n_0(1;u) = \frac{\mathcal{B}(1;d) - n_4(1;d)P(1,4;d)}{B(1)}$$

$$= \frac{103.291 - (105.002)0.937148}{1.02} = 4.79273.$$

Finally, at time 0:

$$n_4(0) = \frac{\mathcal{B}(1;u) - \mathcal{B}(1;d)}{P(1,4;u) - P(1,4;d)} = \frac{104.4006 - 103.291}{.947497 - .937148} = 107.218$$

$$n_0(0) = \frac{\mathcal{B}(0) - n_4(0)P(0,4)}{\mathcal{B}(0)} = \frac{101.8096 - (107.218)0.923845}{1} = 2.75670.$$

The cost of constructing this synthetic coupon bond at time 0 is

$$n_0(0) + n_4(0)P(0,4) = 2.7567 + 107.218(.923845) = 101.8096.$$

This is the arbitrage-free price for the coupon bond at time 0. It matches the previous calculations of its value, as it should!

This synthetic construction is more complicated than the buy and hold strategy discussed previously. This synthetic construction is dynamic, and it involves rebalancing the portfolio across time. The rebalancing, however, is self-financing at time 1. For example, at time 0 we hold $n_0(0) = 2.7567$ units of the money market account and $n_4(0) = 107.218$ shares of the 4-period zero. If at time 1 we move to the up state, we enter time 1, state u with

$$n_0(0)B(1;u) + n_4(0)P(1;u) = 2.7567(1.02) + 107.218(0.947497)$$

$$= 104.4006 \text{ dollars.}$$

We rebalance to $n_0(1;u) = 4.81516$ units of the money market account and $n_4(1;u) = 105.002$ shares of the 4-period zero-coupon bond at a cost of

$$n_0(1;u)B(1;u) + n_1(1;u)P(1,4;u) = 4.81516(1.02) + 105.002(0.947497)$$

$$= 104.4006 \text{ dollars.}$$

This is exactly the value of our portfolio before rebalancing. Thus, the trading strategy is self-financing at time 1. The rebalancing is not self-financing at time 2. At time 2, the synthetic coupon bond must generate a cash flow of 5 dollars to match the traded coupon bond. The strategy represented in Table 11.3 does this. Indeed, the cash flow entering time 2 in state uu is:

$$n_0(1;u)B(2;u) + n_1(1;u)P(2,4;uu) = 4.81516(1.037958) + 105.002(0.967826)$$

$$= 106.6218 \text{ dollars.}$$

The cost of constructing the portfolio leaving time 2 in state uu is:

$$n_0(2;uu)B(2;u)+n_1(2;uu)P(2,4;uu)=.000067(1.037958)+105(0.967826)$$

$$=101.6218.$$

The difference is an implicit cash outflow of 5 dollars.

Lastly, the time 3 trading strategy rebalancing is self-financing as there are no cash flows to the coupon bond at this time. The verification of this statement is left to the reader. This completes the example. □

11.3 COMPARISON OF HJM HEDGING VERSUS DURATION HEDGING

This section compares HJM hedging versus the classical duration hedging of Chapter 3. Via a simple example, we illustrate the error inherent in duration hedging for a term structure evolution that does not have a parallel shift. As seen in Chapter 1, Figure 1.2, historical term structure evolutions almost never experience parallel shifts, so this is a common problem with duration hedging.

To illustrate these facts, consider the term structure evolution given in Figure 11.1.

Example: Error in Modified Duration Hedging Zero-Coupon Bonds

Consider holding a unit position in coupon bond "a" and desiring to hedge this position with a coupon bond "b." Let the two bonds (a, b) considered be zero-coupon bonds of various maturities, i.e.,

$$\mathcal{B}_a(0)=P(0,2)=.961169$$

and

$$\mathcal{B}_b(0)=P(0,4)=.923845.$$

The details of the example are illustrated in Figure 11.3. We explain the content of this diagram below.

The HJM Hedge

From Chapter 8, we know that a hedged portfolio involving these two zero-coupon bonds can be obtained only by creating the "a" bond synthetically

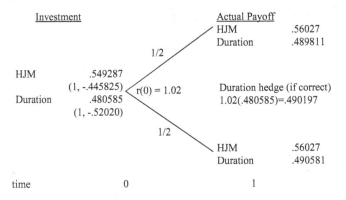

FIGURE 11.3   A Comparison of HJM Hedging versus Duration Hedging. The Bond Trading Strategy $(n_a(0), n_b(0))$ Is Given.

using the "b" bond. The delta gives the appropriate position in the "b" bond. The entire portfolio is then long 1 unit of bond "a" and short n_b units of bond "b," i.e.,

$$n_a = 1$$

and

$$n_b = -\left(\frac{P(1,2;u)-P(1,2;d)}{P(1,4;u)-P(1,4;d)}\right) = -.445835.$$

The initial investment in this hedged portfolio is:

$$n_u P(0,2) + n_b P(0,4) = 1(.961169) - .445835(.923845) = .549287.$$

If the portfolio is riskless, then to avoid arbitrage at time 1, its value should be the initial investment times the spot rate of interest over [0,1], i.e.,

$$.549287(1.02) = .56027.$$

The value of the HJM hedged portfolio at time 1 can be computed as follows:

Time 1, state u

$$n_a P(1,2;u) + n_b P(1,4;u) = 1(.982699) - .445835(.947497) = .56027.$$

Time 1, state d

$$n_a P(1,2;d) + n_b P(1,4;d) = 1(.978085) - .445835(.937148) = .56027.$$

The values are exactly as needed to generate a riskless portfolio. So the HJM hedge works!

The Duration Hedge

We now calculate the hedge based on modified duration. For this example it is easy to show that that duration of bonds "a" and "b" is:

$$D_a(0) = 2$$

and

$$D_b(0) = 4.$$

This follows because a zero-coupon bond's duration is always equal to its time to maturity.

Given that the forward rate curve is flat at 1.02, we have that the yield on both bonds "a" and "b" is identical and equal to $Y^a(0) = Y^b(0) = 1.02$. The bonds' modified durations are:

$$D_{M,a}(0) = 2/1.02$$

$$D_{M,b}(0) = 4/1.02.$$

The modified-duration hedge is determined from expression (3.13) in Chapter 3. It is given by

$$n_a = 1$$

$$n_b = \frac{-D_{M,a}(0)\mathcal{B}_a(0)}{D_{M,b}(0)\mathcal{B}_b(0)} = -\frac{2(.961169)/1.02}{4(.923845)/1.02} = -.52020$$

The initial investment in this portfolio is:

$$n_a P(0,2) + n_b P(0,4) = 1(.961169) - .52020(.923845) = .480585.$$

Again, if the portfolio is riskless, then to avoid arbitrage at time 1, its value should be the initial investment times the spot rate of interest over [0,1], i.e.,

$$(.480585)1.02 = .490197.$$

The value of the duration hedged portfolio at time 1 is:
Time 1, state u

$$n_a P(1,2;u) + n_b P(1,4;u) = 1(.982699) - .52020(.947497) = .489811.$$

Time 1, state d

$$n_a P(1,2;d) + n_b P(1,4;d) = 1(.978085) - .52020(.937148) = .490581.$$

This portfolio is not riskless! It earns more in the down state and less in the up state than the spot rate of interest.

This example illustrates that the duration hedge does not work.

This is not a pathological example, as the evolution in Figure 11.1 was calibrated (at the time of writing of the first edition) to market rate movements, where the time step corresponds to half a year. This completes the example. ☐

CHAPTER 12

Options on Bonds

THIS CHAPTER STUDIES THE arbitrage-free pricing of various types of options on bonds. It is the second application of the theory developed in Chapter 10. Two types of models are considered: distribution independent and distribution dependent. Distribution-independent methods are those that do not depend on a particular evolution for the term structure of interest rates, while distribution dependent structures do. Distribution-independent models generate very general results, often not specific enough to determine exact pricing and hedging descriptions. Usually, only pricing inequalities and bounds on synthetic constructions can be obtained. These results are often useful for building intuition, but for risk management applications, more specific results are required. Hence, the need for distribution dependent models. Distribution dependent models, as seen in Chapter 10, provide exact pricing and hedging prescriptions.

The first section of this chapter investigates the distribution-free relationships between the different types of options: calls, puts, European and American. The key result obtained is a put-call parity theorem, relating European calls and puts.

Distribution dependent models conclude the chapter. European options on zero-coupon bonds are studied next, followed by American options on coupon bonds, and finally callable coupon bonds. Without specifying an evolution for the term structure of interest rates, it is shown that a callable coupon bond is equivalent to an ordinary coupon bond plus a written American call option on the bond. Specifying a particular term structure model allows for a complete hedging analysis.

12.1 DISTRIBUTION-FREE OPTION THEORY

This section studies option pricing theory without invoking any assumptions relating to the probability distribution for the underlying asset(s). For the pricing of options on bonds, this means without invoking any assumptions on the evolution for the term structure of interest rates. For this reason, the insights obtained are general, not specific, and pertaining to the relations between the various types of options: calls and puts, both European and American. The most important relation derived is put-call parity.

The arguments employed use just the notion of arbitrage-free prices. We show that certain relations between the various options hold; otherwise an arbitrage opportunity exists. Consequently, to avoid arbitrage, we get our pricing relationships.

For this analysis, we need some notation. As in Chapter 2, for pedagogical reasons, we concentrate on options written on zero-coupon bonds. The analysis, however, is more general, and the results apply to all the other options studied in this book.

Let the underlying zero-coupon bond have maturity date T_2. Its time t price is denoted by $P(t,T_2)$.

Consider a European call option with strike price K and maturity date $T_1 \leq T_2$ written on this zero-coupon bond. Its time t price is denoted by $C(t,T_1,K:T_2)$. The otherwise-identical American call option's price is denoted by $A(t,T_1,K:T_2)$.

A European put option with strike K and maturity date $T_1 \leq T_2$ written on this same zero-coupon bond's time t price is denoted by $\mathcal{P}(t,T_1,K:T_2)$. The otherwise-identical American put's price is denoted by $\mathcal{A}(t,T_1,K:T_2)$.

12.1.1 Call Options

We first study some simple relations satisfied by call options.

To derive these relations, we show how to synthetically construct one type of option or portfolio using another type of option, sometimes throwing away value. Then, the price of the traded option or portfolio must be no greater than the synthetic option. It is no greater because value is often discarded in the synthetic construction. An example will help to clarify this approach.

First, we consider two identical European call options differing only in their strike prices, $K_1 \leq K_2$. The call option with the smaller strike price (K_1) is at least as valuable as the call option with the larger strike price (K_2). The reason is that the holder of the call option with the smaller strike (K_1)

can always discard value and create the larger strike call (K_2) by paying more upon exercise. Hence, it must be the case that

$$C(t,T_1,K_2:T_2) \leq C(t,T_1,K_1:T_2) \quad \text{if } K_1 \leq K_2. \qquad (12.1)$$

If this condition were violated, then $C(t,T_1,K_2:T_2) > C(t,T_1,K_1:T_2)$ holds. To create an arbitrage opportunity, at time t, buy the call with strike (K_1) and sell the call with strike (K_2). This has a strictly positive cash flow. At all future dates, by discarding value if necessary to create the (K_2) call with the (K_1) call, there is no future liability. Therefore, all future cash flows are zero (or positive if value is not discarded). This is an arbitrage opportunity. Thus, the only arbitrage-free price relation possible is expression (12.1).

As all of the subsequent arguments are identical, we will not provide the detailed description of the arbitrage opportunity to the extent that it is provided above. The details are straightforward and left to the reader as an exercise.

By an identical argument, we have that this relation also holds for American calls,

$$A(t,T_1,K_2:T_2) \leq A(t,T_1,K_1:T_2) \quad \text{if } K_1 \leq K_2. \qquad (12.2)$$

The next arbitrage-free pricing relation is that

$$P(t,T_2) - KP(t,T_1) \leq C(t,T_1,K:T_2). \qquad (12.3)$$

The underlying bond's price $P(t,T_2)$ less the present value of the strike $KP(t,T_1)$ must be less valuable than the call. This follows because the portfolio on the left side (holding the underlying bond and selling K zero-coupon bonds of maturity T_1) can be constructed from the call by (perhaps) discarding value. The procedure is to exercise the call at time T_1, even if the call is out-of-the-money (discarding value). Under this exercise strategy, the call creates this portfolio. Indeed, it is like owning the underlying and paying (for sure) K dollars at time T_1 to get it. Hence, expression (12.2) holds, or an arbitrage opportunity can be constructed.

Recall that an American call is identical to a European call except that it can be exercised any time from the date the contract is written through the maturity date of the option. As such, the holder of the American call can transform it into a European call by not exercising it early. Consequently, to avoid arbitrage, the European call must be worth less than the American call:

$$C(t,T_1,K:T_2) \leq A(t,T_1,K:T_2). \tag{12.4}$$

Our last relation concerns when the European call's price is equal to the American call's price.

If the commodity underlying the American call has no cash flows over the life of the option (as is the case for the T_2-maturity zero-coupon bond) and $P(t,T_1) < 1$ for $t < T_1$, then the American call and an otherwise-identical European call have equal value, i.e.,

$$C(t,T_1,K:T_2) = A(t,T_1,K:T_2). \tag{12.5}$$

This follows by showing that it is never optimal to exercise the American call early under the hypotheses of this theorem. Therefore, the European call is equivalent to the American call (it synthetically constructs it). The argument is quite simple. If exercised early at time t, the American call holder gets a cash flow equal to $P(t,T_2) - K$. But, from expressions (12.3) and (12.4) we see that

$$P(t,T_2) - KP(t,T_1) \leq C(t,T_1,K:T_2) \leq A(t,T_1,K:T_2).$$

But,

$$P(t,T_2) - K < P(t,T_2) - KP(t,T_1) \quad \text{because } P(t,T_1) < 1.$$

So,

$$P(t,T_2) - K < A(t,T_1,K:T_2).$$

This last inequality shows that the American call is always worth more alive; hence, it is never exercised early.

This is a powerful result. As such, it can often be misinterpreted. Suppose that the option is currently in-the-money, i.e., $P(t,T_2) - K > 0$. If exercised, one earns this difference. Suppose you also believe that interest rates will rise, so that at maturity, the T_2-maturity bond will trade for less than the strike K. If you wait until maturity, you believe that the American call option will have zero value.

Isn't this a case where the option should be exercised early? The answer is no. The result of expression (12.5) says that you should not exercise the option. Why? It is worth more if you sell it! The market price $A(t,T_1,K:T_2)$ exceeds the value if exercised $P(t,T_2) - K$. In summary, the result does not say that you should hold the option; the result only says that you should not exercise it.

12.1.2 Put Options

We now study some arbitrage-free relations satisfied by put options. As the arguments are similar, we just provide the relations.

First, we consider two identical European put options differing only in their strike prices, $K_1 \leq K_2$. The put option with the higher strike price (K_2) is at least as valuable as the put option with the smaller strike price (K_1). The reason is that the holder of the put option with higher strike (K_2) can always discard value and create the smaller strike call (K_1) by receiving less upon exercise. Hence, it must be the case that

$$\mathcal{P}(t,T_1,K_1:T_2) \leq \mathcal{P}(t,T_1,K_2:T_2) \quad \text{if } K_1 \leq K_2. \tag{12.6}$$

By an identical argument, we have that this relation also holds for American puts,

$$\mathcal{A}(t,T_1,K_1:T_2) \leq \mathcal{A}(t,T_1,K_2:T_2) \quad \text{if } K_1 \leq K_2. \tag{12.7}$$

The next arbitrage-free pricing relation is that

$$\mathcal{P}(t,T_1,K:T_2) \leq K. \tag{12.8}$$

The right side of expression (12.8) represents holding K dollars in the money market account. K dollars is the most that the put will ever be worth. Clearly, by discarding value (if necessary), one can synthetically construct the payoff to the European put. Hence, expression (12.8) must hold.

Because the holder of the American put can, by their own actions, transform it into a European put by not exercising it early, the European put must be worth less than the American put:

$$\mathcal{P}(t,T_1,K:T_2) \leq \mathcal{A}(t,T_1,K:T_2). \tag{12.9}$$

Unfortunately, there is no analogous result to expression (12.5) for put options. The reason is that if the underlying commodity's price gets low enough, it is optimal to exercise the American put early. This is in contrast to the situation for American calls where even if the underlying's price gets high, optimal exercise may not occur. Unfortunately, the proof of this statement is beyond the scope of this text and is left to independent reading (see Jarrow and Chatterjea [2]).

12.1.3 Put-Call Parity

This section generates an arbitrage-free pricing relation between otherwise-identical European calls and puts on zero-coupon bonds. As with the previous results, this relation applies more generally than just to options on zero-coupon bonds. A complete reference to this material is Jarrow and Chatterjea [2].

Put-call parity follows from one important observation. The observation is that a European call option with strike K and maturity T_1 on a T_2-maturity zero-coupon bond can be constructed synthetically with a buy and hold trading strategy using an otherwise-identical European put option, the underlying T_2-maturity zero-coupon bond, and a T_1-maturity zero-coupon bond. In particular, the synthetic call consists of:

(i) buying one unit of the otherwise-identical European put option,

(ii) selling K units of the T_1-maturity zero-coupon bond, and

(iii) buying one unit of the T_2-maturity zero-coupon bond.

As mentioned earlier, this is a buy and hold trading strategy. It is purchased at time t and held until the maturity date, T_1. The cash flows on this portfolio are detailed in Table 12.1.

TABLE 12.1   Cash Flow Comparison of a Call Option with a Synthetic Call Option

	Time T_1 Payoffs	
	$P(T_1,T_2) > K$	$P(T_1,T_2) \leq K$
Call	$P(T_1, T_2) - K$	0
Synthetic call		
Put	0	$K - P(T_1, T_2)$
$-K$ bonds with T_1-maturity	$-K$	$-K$
1 bond with T_2-maturity	$P(T_1, T_2)$	$P(T_1,T_2)$
Sum	$P(T_1,T_2) - K$	0

The top row gives the payoffs to the traded call. If $P(T_1,T_2) > K$, the call option is in-the-money, and it earns the difference $P(T_1,T_2) - K$. If $P(T_1,T_2) \leq K$, the call is out-of-the-money, and it earns zero.

The next set of rows provides the payoff to the synthetic call. The first row gives the payoffs to the traded put position. If $P(T_1,T_2) > K$, the put option is out-the-money, and it earns zero. If $P(T_1,T_2) \leq K$, the put is in-the-money, and it earns the difference $K - P(T_1,T_2)$. The next row is the payoff to selling K bonds, $-K$ dollars. The last row is the value of the position in the T_2-maturity bond. The sum of these three positions exactly duplicates the traded call's time T_1 payoffs. Thus, to avoid arbitrage, the price of the traded European call must equal the cost of constructing the synthetic call. This gives us the put-call parity theorem.

Given a European call and put both with strikes K and maturities T_1 on the same underlying commodity, and if the underlying commodity has no cash flows over the option's lives (as is the case with the T_2-maturity bond), then

$$C(t,T_1,K:T_2) = \mathcal{P}(t,T_1,K:T_2) + P(t,T_2) - KP(t,T_1). \qquad (12.10)$$

This is an important result. It implies that to price and hedge European puts, we need to only understand how to price and hedge European calls, and then apply put-call parity. Unfortunately, this result does not extend to American calls and puts. The best results available are upper and lower bounds for the American call's value in terms of the American put's value and a portfolio of zero-coupon bonds. These results are outside the scope of this text (see Jarrow and Chatterjea [2]).

12.2 EUROPEAN OPTIONS ON ZERO-COUPON BONDS

This section applies the theory developed in Chapter 10 to price a European call option on a zero-coupon bond. In contrast to the previous section, this theory utilizes assumptions regarding the probability distribution for the commodities underlying the options – the term structure of zero-coupon bond prices as detailed in Chapter 5.

Let the European call option's exercise date be at time τ^* for $0 \leq \tau^* \leq \tau$, and let its exercise price be $K > 0$. We assume that the call is written on the zero-coupon bond that matures at time $T \geq \tau^*$.

We simplify the notation for the value of the call option at time t to be $C(t)$. By definition, the boundary condition for the call option at the exercise date τ^* is:

$$C(\tau^*) = \max(P(\tau^*, T) - K, 0). \qquad (12.11)$$

If the T-maturity zero-coupon bond's price $P(\tau^*, T)$ exceeds the exercise price K, the option is exercised. It is said to be in-the-money, and its value is the difference $P(\tau^*, T) - K > 0$. Otherwise, it ends up out-of-the-money, unexercised, with zero value.

By the risk-neutral valuation procedure, its arbitrage-free value at time t is:

$$C(t) = \tilde{E}\left(\max\left(P\left(\tau^*, T\right) - K, 0\right) \middle/ B(\tau^*)\right) B(t). \qquad (12.12)$$

A synthetic call option can be constructed using the procedure described in Chapter 10.

We now illustrate this risk-neutral valuation procedure and the construction of the synthetic call option with an example.

Example: European Call Option Valuation and Synthetic Construction

We start with the evolution for the zero-coupon bond price curve as given in Figure 12.1. This evolution is arbitrage-free, because it is the same evolution given in Chapter 10, Figure 10.1. In that chapter it was shown that the pseudo probabilities satisfy the necessary and sufficient condition for an arbitrage-free evolution at each node in the tree.

Consider a European call option on the 4-period zero-coupon bond, with an exercise price $K = .961000$ and an exercise date $\tau^* = 2$. The value of the call option at its maturity date is $C(2; s_2) = \max[P(2, 4; s_2) - .961000, 0]$.

Substituting in the values from Figure 12.1 yields the following values:

at $(2; uu)$: max $[.967826 - 0.961000, 0] = 0.006826$

at $(2; ud)$: max $[.960529 - 0.961000, 0] = 0$

at $(2; du)$: max $[.962414 - 0.961000, 0] = 0.001414$

at $(2; dd)$: max $[.953877 - 0.961000, 0] = 0$

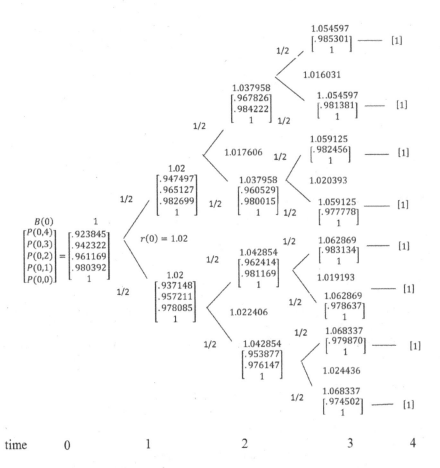

FIGURE 12.1   An Example of a One-Factor Bond Price Curve Evolution. The Money Market Account Values and Spot Rates Are Included on the Tree. Pseudo Probabilities Are Along Each Branch of the Tree.

These numbers are placed on the last node in Figure 12.2. To deduce the arbitrage-free value at time 1, we use the risk-neutral valuation procedure.

At time 1, state u:

$$C(1;u) = \tilde{E}_1(C(2;s_s))/r(1;u)$$

$$= \frac{(1/2)0.006826 + (1/2)0}{1.017606} = 0.003354$$

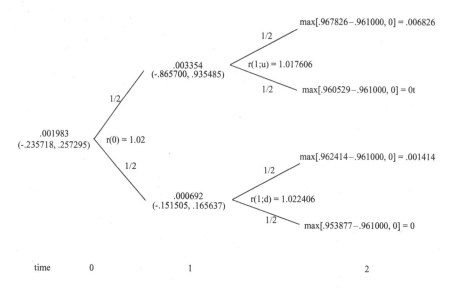

FIGURE 12.2 A European Call Option's Value $(C(t;s_t))$ on the 4-Period Bond $(P(t;4))$ with Exercise Price $K = .961000$ and Exercise Date 2; and the Synthetic Call Portfolio $(n_0(t;s_t), n_4(t;s_t))$. Pseudo Probabilities Are Along Each Branch of the Tree.

At time 1, state d:

$$C(1;d) = \tilde{E}_1(C(2;s_s))/r(1;d)$$

$$= \frac{(1/2)0.001414 + (1/2)0}{1.022406} = 0.000692$$

Finally, at time 0:

$$C(0) = \tilde{E}_0(C(1;s_1))/r(0)$$

$$= \frac{(1/2)0.003354 + (1/2)0.000693}{1.02} = 0.001983.$$

These numbers appear at the nodes on Figure 12.2.

If the call's time 0 value differs from its arbitrage-free value .001983, an arbitrage opportunity exists. The arbitrage opportunity involves creating the synthetic call. If the market price for the call exceeds .001983, short the traded call and hold the synthetic. Conversely, if the market price for the call is less than .001983, buy the traded call and short the synthetic.

To calculate the synthetic-call portfolio holdings in the money market account $(n_0(t;s_t))$ and the 4-period zero-coupon bond $(n_4(t;s_t))$, we use the delta approach of method 2 in Chapter 8. These calculations are as follows:

At time 1, state u:

$$n_4(1;u) = \frac{C(2;uu) - C(2;ud)}{P(2,4;uu) - P(2,4;ud)} = \frac{0.006826 - 0}{0.967826 - 0.960529}$$

$$= 0.935485$$

$$n_0(1;u) = \frac{C(1;u) - n_4(1;u)P(1,4;u)}{B(1)}$$

$$= \frac{0.003354 - (.935485)0.947497}{1.02} = -.865700$$

At time 1, state d:

$$n_4(1;d) = \frac{C(2;du) - C(2;dd)}{P(2,4;du) - P(2,4;dd)} = \frac{0.001414 - 0}{0.9621414 - 0.953877}$$

$$= 0.165637$$

$$n_0(1;d) = \frac{C(1;d) - n_4(1;d)P(1,4;d)}{B(1)}$$

$$= \frac{0.000692 - (.165637)0.937148}{1.02} = -.151505.$$

At time 0:

$$n_4(0) = \frac{C(1;u) - C(1;d)}{P(1,4;u) - P(1,4;d)} = \frac{0.003354 - 0.00692}{0.947497 - 0.937148}$$

$$= 0.257295$$

$$n_0(0) = \frac{C(0) - n_4(0)P(0,4)}{B(0)} = \frac{0.001983 - (.257295)0.923845}{1}$$

$$= -.235718$$

These portfolio holdings are given on Figure 12.2 under each node in the tree.

We see that a synthetic call consists of a short position in the money market account and a long position in the 4-period bond. These positions change across time and across states.

For this one-factor economy, it is also possible to create the synthetic call using the 3-period bond, because any zero-coupon bond and the money market account complete the market (prior to the zero-coupon bond's maturity).

The calculations follow similarly and are briefly recorded. Let the synthetic call positions using the money market account and the 3-period zero-coupon bond be denoted by $(n_0(t;s_t),n_3(t;s_t))$. They are calculated as follows:

At time 1, state u:

$$n_3(1;u) = \frac{C(2;uu) - C(2;ud)}{P(2,3;uu) - P(2,3;ud)}$$

$$= \frac{0.006826 - 0}{0.984222 - 0.980015} = 1.622534$$

$$n_0(1;u) = \frac{C(1;u) - n_3(1;u)P(1,3;u)}{B(1)}$$

$$= \frac{0.003354 - (1.622534)0.965127}{1.02} = -1.531958$$

At time 1, state d:

$$n_3(1;d) = \frac{C(2;du) - C(2;dd)}{P(2,3;du) - P(2,3;dd)} = \frac{0.001414 - 0}{0.981169 - 0.976147}$$

$$= 0.281561$$

$$n_0(1;d) = \frac{C(1;d) - n_4(1;d)P(1,3;d)}{B(1)}$$

$$= \frac{0.000692 - (.281561)0.957211}{1.02} = -.263550.$$

At time 0:

$$n_3(0) = \frac{C(1;u) - C(1;d)}{P(1,3;u) - P(1,3;d)} = \frac{0.003354 - 0.00692}{0.965127 - 0.957211}$$

$$= 0.336281$$

$$n_0(0) = \frac{C(0) - n_3(0)P(0,3)}{B(0)} = \frac{0.001983 - (.336281)0.942322}{1}$$

$$= -.314902$$

These holdings differ from those calculated for the 4-period zero-coupon bond. In general, a larger number of the 3-period bonds are needed to generate the synthetic call.

The calculation of the option's delta in terms of the 3-period zero-coupon bond could have been done differently. Note that

$$n_3(t;s_t) = n_4(t;s_t)\left(\frac{P(t+1,4;s_t u) - P(t+1,4;s_t d)}{P(t+1,3;s_t u) - P(t+1,3;s_t d)}\right).$$

In words, the option's delta in terms of the 3-period bond equals the option's delta in terms of the 4-period bond $n_4(0)$ multiplied by the delta of the 4-period bond in terms of the 3-period bond $[P(1,4;u) - P(1,4;d)]/[P(1,3;u) - P(1,3;d)]$.

For example,

$$n_3(0) = n_4(0)\left(\frac{P(1,4;u) - P(1,4;d)}{P(1,3;u) - P(1,3;d)}\right)$$

$$= 0.257295\left(\frac{0.947497 - 0.937148}{0.965127 - 0.957211}\right) = 0.336281.$$

The procedure for recalculating the option's delta in terms of a different underlying asset discussed in the previous example generalizes. It can be used, for example, to hedge the call option using another traded option or a futures contract. The choice of the underlying asset for use in the hedge is often determined by liquidity considerations, which are not formally addressed in the frictionless market model above. This completes the example. □

12.3 AMERICAN OPTIONS ON COUPON BONDS

This section illustrates the valuation procedure for American options by valuing an American call option on a coupon bond. Pricing an American call option on a zero-coupon bond is not instructive because both

American and European call options on a zero-coupon bond have identical values. This follows from expression (12.5) above because an American call option on an underlying asset with no cash flows will never be exercised early.

Here I will illustrate the American option valuation approach with an example.

Example: American Call Option Valuation

Consider the coupon bond analyzed in Chapter 11. For convenience, its cash flows are provided in Table 12.2. This coupon bond pays a coupon of 5 dollars at times 2 and 4 and a principal repayment of 100 dollars at time 4. The coupon bond's price is denoted by $\mathcal{B}(t)$.

We assume that the zero-coupon bond price curve evolves as in Figure 12.1. This is an arbitrage-free evolution.

Using the risk-neutral valuation procedure, the coupon bond price's evolution is computed (in Chapter 11) and given in Figure 12.3.

Consider an American call option on this coupon bond with a strike price of 101 and expiration date 2. The American feature implies that the option can be exercised at any date prior to expiration. Because of the discrete nature of this model, we need to make an assumption concerning the timing of early exercise and the receipt of the bond's coupon. We assume that if the option is exercised at time t, the option holder receives the next coupon payment at time $t+1$, not the time t payment. Any other convention concerning the timing of early exercise and the receipt of the coupon payment could be easily handled using an adjusted version of the following methodology. This assumption is a good approximation to actual practice.

We value the American option using risk-neutral valuation, by backward induction. The only complication is that at each node in the tree, besides valuing the future cash flows, we need to determine whether early exercise is optimal. This turns out to be a simple calculation. The argument is best illustrated by doing it!

TABLE 12.2   An Example of a Coupon Bond

	0	1	2	3	4	
						Time
Price	$\mathcal{B}(0)$					
Coupon			$5		$5	
Principal					$100	

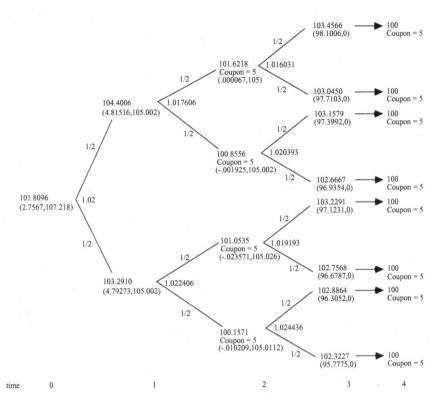

FIGURE 12.3 The Evolution of the Coupon Bond's Price for the Example in Table 12.2. The Coupon Payment at Each Date Is Indicated by the Nodes. The Synthetic Coupon-Bond Portfolio $(n_0(t;s_t), n_4(t;s_t))$ in the Money Market Account and 4-Period Zero-Coupon Bond Are Given under Each Node.

By the backward induction, at the maturity date 2, the American call's value is:

$$A(2;uu) = \max(\mathcal{B}(2;uu) - 101,0) = \max(101.6218 - 101,0) = 0.6218$$

$$A(2;ud) = \max(\mathcal{B}(2;ud) - 101,0) = \max(100.8557 - 101,0) = 0$$

$$A(2;du) = \max(\mathcal{B}(2;du) - 101,0) = \max(101.0535 - 101,0) = 0.0535$$

$$A(2;dd) = \max(\mathcal{B}(2;dd) - 101,0) = \max(100.1571 - 101,0) = 0$$

These numbers are tabulated in Figure 12.4.

At time 1, state u:

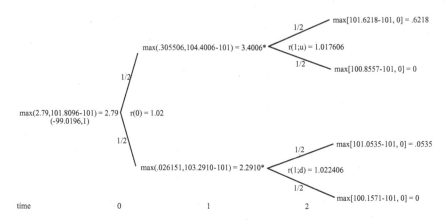

FIGURE 12.4   An Example of an American Call Option's Values with Constant Strike Price $K = 101$ and Maturity Date 2 on the Coupon Bond of Table 12.2. The Synthetic American Call Option's Portfolio in the Money Market Account $(n_0(t;s_t))$ and the Coupon Bond $(n_c(t;s_t))$ is Given under Each Node Where Relevant. The Symbol "*" Means That It Is Optimal to Exercise at the Time and State Identified.

$$A(1;u) = \max\left(\frac{(1/2)A(2;uu) + (1/2)A(2;ud)}{r(1;u)}, \mathcal{B}(1;u) - 101 \right)$$

$$= \max\left(\frac{(1/2)0.6218 + (1/2)0}{1.017606}, 104.4006 - 101 \right)$$

$$= \max(0.305506, 3.4006) = 3.4006$$

In valuing the American call at time 1 in state u we compare two values. The first,

$$\frac{(1/2)A(2;uu) + (1/2)A(2;ud)}{r(1;u)}$$

is the value of the call *if it remains unexercised.* This can be understood by recognizing that this quantity is a present value. It represents the present value of the future cash flows (and value) of the option, i.e., the value if not exercised. The second term,

$$\mathcal{B}(1;u) - 101$$

is the value of the call *if exercised*. It is the value of the coupon bond less the strike price. The optimal decision is that which provides the largest value (hence the "max").

This valuation implies that early exercise is optimal at time 1, state u giving a value of 3.4006. The synthetic call at time 1, state u, therefore, really does not exist, for it consists of zero units of the money market account and one coupon bond. Note that if the American call is not exercised, the holder loses (significant) value at time 2. It goes from 3.4006 to either .6218 if u occurs or 0 if d occurs.

At time 1, state d:

$$A(1;d) = \max\left(\frac{(1/2)A(2;du)+(1/2)A(2;dd)}{r(1;u)}, \mathcal{B}(1;d)-101 \right)$$

$$= \max\left(\frac{(1/2)0.0535+(1/2)0}{1.022406}, 103.2910-101 \right)$$

$$= \max(0.026151, 2.2910) = 2.2910$$

Again, early exercise is optimal at this node. The synthetic call at time 1, state d consists of zero units of the money market account and one unit of the coupon bond purchased at $K = 101$ dollars. If the American call is not exercised, it loses (significant) value at time 2. It goes from 2.2910 to either .0535 if u occurs or 0 if d occurs.

Finally, at time 0:

$$A(0) = \max\left(\frac{(1/2)A(1;u)+(1/2)A(1;d)}{r(0)}, \mathcal{B}(0)-101 \right)$$

$$= \max\left(\frac{(1/2)3.4006+(1/2)2.2910}{1.02}, 101.8096-101 \right)$$

$$= \max(2.79, 0.8096) = 2.79$$

Here it is optimal not to exercise the call, giving the call a value of 2.79.

To create the synthetic American call at time 0, we hold $n_0(0)$ units of the money market account and $n_c(0)$ units of the coupon bond determined as follows:

$$n_c(0) = \frac{A(1;u) - A(1;d)}{B(1;u) - B(1;d)} = \frac{3.4006 - 2.2910}{104.4006 - 103.2910} = 1$$

$$n_0(0) = \frac{A(0) - n_c(0)B(0)}{B(0)} = \frac{2.79 - 1(101.8096)}{1} = -99.0196$$

One needs to purchase the coupon bond ($n_c(0) = 1$) and short −99.0196 units of the money market account. The reason one buys one unit of the coupon bond is that the option is exercised for sure at time 1. Consequently, holding the option at time 0 is equivalent to holding the coupon bond, but paying for it at time 1!

Two observations about these calculations need to be made. First, the numerator in $n_c(0)$ contains the values for the option at time $t+1$ under the *optimal decision selections*, which in this case is exercising in both the up and the down states. Second, the denominator in $n_c(0)$ is the difference in the value of the coupon bond at time 1. There is no coupon payment at time 1. This completes the example. □

Given the previous example, we now provide the more formal and abstract analysis.

Coupon bonds were analyzed in Chapter 11. An arbitrary coupon bond can be represented as a sequence of known cash flows $C_1, C_2,..., C_T$ at times $0 \leq 1 \leq 2 \leq ... \leq T \leq \tau$, with an arbitrage-free value given by the risk-neutral valuation expression

$$B(t) = \sum_{i=t+1}^{T} C_i P(t,i). \tag{12.13}$$

By convention, the time t price of the coupon bond does not include the coupon payment made at time t, C_t. It represents the price ex-coupon.

For use in future applications, we study a more complex American call option than that given in the example. Here we analyze an American call option whose strike price can change across time. This is a slight generalization of the traditional definition of an American call. In this regard, we let the strike price be K_i dollars if the option is exercised on date i with $t \leq i \leq \tau^*$. The option's expiration date is $\tau^* \leq T$. For this option, the exercise price K_i changes across time according to this predetermined schedule.

The American feature implies that the call option can be exercised at any date prior to expiration. Because of the discrete nature of this model, we need to make an assumption concerning the timing of early exercise and the receipt of the bond's coupon. We assume that if the option is exercised at time t, the option holder receives the next coupon payment at time $t+1$, C_{t+1}, not the time t payment, C_t. Any other convention concerning the timing of early exercise and the receipt of the coupon payment could be easily handled using an adjusted version of the following methodology. This assumption is a reasonable approximation to practice.

The value of the American call at time t, under state s_t, is denoted by $A(t)$. To value this option, we use backward induction. We start at the expiration date of the option, time τ^*, and work backward in time, valuing the option at each intermediate date until we reach time t.

Let's begin. At the expiration date τ^* the American call's value, if not exercised before that date, is

$$A(\tau^*) = \max\left(\mathcal{B}(\tau^*) - K_{\tau^*}, 0\right). \tag{12.14}$$

At time $\tau^* - 1$, the value of the American option *if not exercised* is

$$\tilde{E}_{\tau^*-1}(A(\tau^*))/r(\tau^* - 1).$$

This represents the present value of the *future* payments on this option. This time $\tau^* - 1$ value does not include the coupon payment at time τ^*, since $\mathcal{B}(\tau^*)$ is the ex-coupon price of the bond. *If it is exercised*, its value is

$$\mathcal{B}(\tau^* - 1) - K_{\tau^* - 1}.$$

This exercised value includes within the bond's price the present value of the coupon payment received at time τ^*, C_{τ^*}. It does not include the coupon payment at time $\tau^* - 1$ because to get that payment, the option had to be exercised at time $\tau^* - 2$.

The time $\tau^* - 1$ value of the option is the largest of these, i.e.,

$$A(\tau^* - 1) = \max\left[\tilde{E}_{\tau^*-1}(A(\tau^*))\big/r(\tau^* - 1), \mathcal{B}(\tau^* - 1) - K_{\tau^* - 1}\right]. \tag{12.15}$$

Continuing, moving backward to time $\tau^* - 2$, the value of the option is

$$A(\tau^* - 2) = \max\left[\tilde{E}_{\tau^*-2}(A(\tau^* - 1))\big/r(\tau^* - 2), \mathcal{B}(\tau^* - 2) - K_{\tau^* - 2}\right]. \tag{12.16}$$

By induction,

$$A(t) = \max\left(\tilde{E}_t(A(t+1))/r(t), \mathcal{B}(t) - K_t\right).\qquad(12.17)$$

The synthetic American call option is obtained by duplicating the optimal exercise value as implicitly indicated in expression (12.17).

The above procedure for valuing the American call option is called *stochastic dynamic programming*. It is the standard procedure used for solving the American option valuation problem discussed in Chapter 9. Stochastic dynamic programming simultaneously determines both the optimal exercise policy and the maximum value of the option as indicated in the above derivation.[*] A good introductory reference to the mathematics underlying this technique can be found in Bertsekas and Shreve [1].

In the definition of the American call option given above, we allowed for an exercise price that changes across time, $\{K_i$ for $t \le i \le \tau^*\}$. This formulation includes two special cases worth mentioning.

The first is the typical American call option, in which the exercise price is a fixed constant K for all times, i.e., $K_i = K$ for all i. We studied this case in the preceding example.

The second case is a delayed-exercise American call option, in which exercise cannot occur prior to some future date $T_m > t$. Exercise can occur only between time T_m and the expiration date of the option, at time τ^*. The exercise price between these dates can change and is given by the schedule $\{K_i$ for $T_m \le i \le \tau^*\}$. This delayed-exercise American call option is a special case of the previous formulation. Indeed, to see this, let the exercise schedule for the entire life of the option $\{K_i$ for $t \le i \le \tau^*\}$ be given by

$$\text{Delayed exercise schedule} = \begin{cases} K_i \equiv M & \text{for } t \le i \le T_m \\ K_i & \text{for } T_m \le i \le t \end{cases}\qquad(12.18)$$

where M is selected to be a very large number, say $10^6 \gg C_{\tau^*}$.

Under this schedule we get the delayed-exercise American call. Technically, although the American call option under expression (12.18) allows exercise prior to time T_M, it will never be optimal to do so. This is seen by examining expression (12.17). The maximum on the right side

---

[*] In the notation of Chapter 9, the stochastic dynamic programming technique jointly determines the maximum $a^* \in A$ and the value under this a^* of $\sum\limits_{j=0}^{\tau^*} \tilde{E}_0\left(\dfrac{x(j,a^*)}{B(j)}\right)B(0)$.

of expression (12.17) is always $\tilde{E}_t(A(t+1))/r(t)$ prior to time T_m, because prior to time T_m, $\mathcal{B}(t) - K$ is a large negative number and $\tilde{E}_t(A(t+1))/r(t)$ is always nonnegative.

In summary, under the delayed exercise schedule (12.17), the American call option of expressions (12.14)–(12.17) is the delayed-exercise American call option. This delayed-exercise American call option is important in applications because it relates to the valuation of callable coupon bonds. This is discussed in the next section.

12.4 CALL PROVISIONS ON COUPON BONDS

This section studies the pricing and hedging of callable coupon bonds. Without specifying a particular evolution for the term structure of interest rates, it is first shown that a callable coupon bond is equivalent to an ordinary coupon bond plus a written *delayed-exercise* American call option on the ordinary coupon bond. Specifying a particular evolution, however, enables one to apply the techniques of Chapter 10.

We define a *callable coupon bond* to be any financial security with the following structure. It is similar to an ordinary coupon bond, but with an early retirement or payback provision that the issuer can invoke. This earlier payback provision is said to be a "call" provision. The callable coupon bond pays known cash flows of C_1, C_2, ..., C_T dollars at times $0 \leq 1 \leq 2 \leq ... \leq T \leq \tau$, where T is the maturity date of the bond. However, each cash flow is paid only if the bond is not retired (called) earlier than the payment date. The bond can be retired (called) under the following conditions. First, it cannot be called prior to time T_m. The bond is said to be *call protected* for T_m years. Second, between times T_m and the maturity, time T, it can be retired only at a cost of K_t dollars for $0 \leq t \leq T-1$. The cost schedule is allowed to change with time. After the bond is retired, no future cash flows are paid. This completes the description.

Thus, at the discretion of the entity issuing the bond, the callable bond can be retired at any time t for $T_m \leq t \leq T-1$ at a cost of K_t dollars. Retiring the bond at time t saves the entity the future cash flows of C_{t+1}, C_{t+2}, ..., C_T dollars, and this may be optimal if interest rates fall (relative to the time the bond was issued).

To value the callable coupon bond, we need to know the optimal call policy of the entity issuing the bond. To determine this, we assume that the objective of the entity issuing the bond is to minimize its liability; i.e., it desires to minimize the value of the callable coupon bond. Given this objective, we can use the stochastic dynamic programming approach introduced in the previous section to value and hedge this bond.

Prior to this demonstration, however, we need to define two other financial securities. First, consider an ordinary (noncallable) coupon bond with the identical cash flows C_1, C_2, ..., C_T and maturity date T as the callable coupon bond. From Chapter 11, we know that the price of this coupon bond at time t is given by

$$\mathcal{B}(t) = \sum_{i=t+1}^{T} C_i P(t,i) \tag{12.19a}$$

or

$$\mathcal{B}(t) = \tilde{E}_t(\mathcal{B}(t+1) + C_{t+1})/r(t) \quad \text{where } \mathcal{B}(T) = 0. \tag{12.19b}$$

Note that $\mathcal{B}(T) = 0$ because this represents the value of the bond, ex-coupon, at time T after all coupons and principal are repaid.

Second, consider a delayed-exercise American call option with expiration time $T-1$ on the ordinary coupon bond in expression (12.19). Let the exercise period start after time T_m, and let the exercise price schedule be $\{K_t$ for $T_m \le t \le T-1\}$. This delayed-exercise American call option was valued in the previous section, and its value is given by

$$A(t) = \max\left(\tilde{E}_t(A(t+1))/r(t), \mathcal{B}(t) - K_t\right). \tag{12.20}$$

where

$$A(T-1) = \max(\mathcal{B}(T-1) - K_{T-1}, 0).$$

In both expressions (12.19) and (12.20), the expectations are taken with respect to the pseudo probabilities from either a one-, two-, or $N \ge 3$-factor economy.

Let us denote the value of the callable coupon bond at time t as $D(t)$. This bond's price is ex-coupon.

It is intuitive that the callable coupon bond $D(t)$ is equivalent to an ordinary coupon bond $\mathcal{B}(t)$ less an American call option $A(t)$ written on the ordinary coupon bond, i.e.,

$$D(t) = \mathcal{B}(t) - A(t). \tag{12.21}$$

The American call option is sold (or written) because the issuer of the callable bond, not the holder, has the option to buy the bond back. The option to buy is a call. It is American since it can be exercised at any time after the protection period.

Given this decomposition of the callable bond, synthetically constructing the callable bond is straightforward. To construct a synthetic callable coupon bond, one needs to buy an ordinary (noncallable) coupon bond with the same characteristics and to short a delayed-exercise American call option on this ordinary coupon bond. If these do not trade, one can construct the synthetic callable coupon bond by synthetically constructing both the ordinary coupon bond (as described in Chapter 11) and the delayed-exercise American call option on this ordinary coupon bond (as described in the previous section). These constructions require a specific model for the evolution of the term structure of interest rates. Because examples of these constructions have been provided earlier in the text, no additional examples are provided here. This completes the discussion of callable coupon bonds except for the formal derivation of expression (12.21). This formal derivation follows.

Proof. The proof of expression (12.21) proceeds by backward induction. Starting at time T, the value of the callable coupon bond, if not called earlier, equals its value after the last cash payment, i.e.,

$$D(T) = 0. \tag{12.22a}$$

Using expression (12.19b), we can rewrite this as

$$D(T) = \mathcal{B}(T). \tag{12.22b}$$

At time $T-1$, according to the risk-neutral valuation procedure, the value of the callable coupon bond is

$$D(T-1) = \min\left(\tilde{E}_{T-1}(D(T) + C_T)/r(T-1), K_{T-1}\right). \tag{12.23a}$$

This follows because the entity issuing the bond will retire it if K_{T-1} is less than the value of the future cash flows if not retired ($\tilde{E}_{T-1}(D(T) + C_T)/r(T-1)$). The coupon payment at time $T-1$ is already owed, and it is paid regardless of the retirement decision at time $T-1$. Thus, it does not appear in this expression.

We perform some algebra on (12.23a). Substitution of (12.22a) gives

$$D(T-1) = \min\left(\tilde{E}_{T-1}(\mathcal{B}(T)+C_T)/r(T-1), K_{T-1}\right). \qquad (12.23b)$$

Next, using expression (12.19) yields

$$D(T-1) = \min[\mathcal{B}(T-1), \ K_{T-1}].$$

But this can be written as

$$D(T-1) = \mathcal{B}(T-1) - \max(\mathcal{B}(T-1) - K_{T-1}, 0).$$

Finally, using expression (12.20) gives

$$D(T-1) = \mathcal{B}(T-1) - A(T-1). \qquad (12.23b)$$

Expression (12.23b) shows that at time $T-1$, the callable coupon bond ($D(T-1)$) is equivalent to an ordinary coupon bond ($\mathcal{B}(T-1)$) less a delayed-exercise American call option on the ordinary coupon bond ($A(T-1)$). Continuing backward to time $T-2$,

$$D(T-2) = \min\left(\tilde{E}_{T-2}(D(T-1)+C_{T-1})/r(T-2), K_{T-2}\right). \qquad (12.24a)$$

We want to transform this expression to one similar to (12.23b). First, substitute (12.23b) into expression (12.24a) to obtain

$$D(T-2) = \min\left(\frac{\tilde{E}_{T-2}(\mathcal{B}(T-1)+C_{T-1})}{r(T-2)} - \frac{\tilde{E}_{T-2}(A(T-1)+C_{T-1})}{r(T-2)}, K_{T-2}\right).$$

Using expression (12.19) yields

$$D(T-2) = \min\left(\mathcal{B}(T-2) - \frac{\tilde{E}_{T-2}(A(T-1)+C_{T-1})}{r(T-2)}, K_{T-2}\right).$$

Algebra gives

$$D(T-2) = \mathcal{B}(T-2) - \max\left(\mathcal{B}(T-2) - K_{T-2}, \frac{\tilde{E}_{T-2}(A(T-1)+C_{T-1})}{r(T-2)}\right).$$

Finally, using expression (12.20) gives

$$D(T-2) = \mathcal{B}(T-2) - A(T-2). \qquad (12.24b)$$

Continuing backward in a similar fashion, at time t we get

$$D(t) = \min\left(\left[\tilde{E}_t\left(D(t+1)\right) + C_{t+1}\right]/r(t), K_t\right). \qquad (12.25a)$$

where $D(T) = 0$, or

$$D(t) = \mathcal{B}(t) - A(t). \qquad (12.25b)$$

Expression (12.25b) shows that the callable coupon bond ($D(t)$) is equivalent to an ordinary coupon bond ($\mathcal{B}(t)$) less a delayed-exercise American call option on the ordinary coupon bond ($A(t)$). This result is independent of any particular evolution of the term structure of interest rates. This completes the formal proof.

REFERENCES

1. Bertsekas, D., and S. Shreve, 1978. *Stochastic Optimal Control: The Discrete Time Case*, Academic, New York.
2. Jarrow, R., and A. Chatterjea, 2019. *An Introduction to Derivative Securities, Financial Markets, and Risk Management*, 2nd edition, World Scientific Press.

CHAPTER 13

Forwards and Futures

T HIS CHAPTER APPLIES THE general derivative pricing theory of Chapter 10 to forward contracts, futures contracts, and options on futures. Under stochastic interest rates, we will show that forward and futures contracts are distinct securities and that forward and futures prices are (usually) different quantities. The material in this chapter was motivated by the original insights of Cox, Ingersoll, and Ross [2] and of Jarrow and Oldfield [6].

It will be shown herein that the differences between forward and futures contracts are solely due to the differences in the timing of the cash flows from the two contracts. Although the cash flows occur at different times, the total cash flows received on both contracts are the same. Futures contracts receive cash flows every day due to marking-to-market, while forward contracts receive a cash flow only on the delivery date. Hence, understanding the differences in these two contracts is equivalent to understanding the reinvestment risk of random cash flows under stochastic interest rates. But, this is just a restatement of the canonical problem studied in Chapter 10 – determining the present value of random cash flows given stochastic interest rates. Hence, understanding the difference between forward and futures contracts is equivalent (conceptually) to understanding how to price interest rate derivatives.

This distinction between forward and futures contracts is not uniformly understood. There are still textbooks available that confuse these two contacts. One contract cannot be used (in a simple fashion) to arbitrage the other, contrary to occasional practice. Forward and futures contracts are different; therefore, synthetic forward and futures contracts will be constructed differently. Forward contracts can be constructed via buy and hold trading strategies, while futures contracts (in general) cannot. These distinctions and others will be clarified in this chapter.

The futures contracts analyzed here are simplified versions of the exchange-traded futures contracts (see Chapter 2). We simplified these contracts to facilitate understanding. References are provided at the end of the chapter for an analysis of the more complex exchange-traded Treasury futures contracts.

13.1 FORWARDS

We consider the simplified forward contract as defined in Chapter 4. Recall that this forward contract is issued on a T_2-maturity zero-coupon bond with delivery date T_1 where $T_1 \leq T_2$. The time t forward price on this contract is denoted by $F(t,T_1:T_2)$. The purpose of this section is to use the contingent claim valuation methodology of Chapter 10 to give an alternative characterization of the forward price.

Given there are no arbitrage opportunities and markets are complete, the subsequent analysis proceeds independently of the particular economy (with one, two, or $N \geq 3$ factors) studied. We use the risk-neutral valuation approach to value the forward contract. Unlike the previous chapters, this chapter proceeds by first studying the general case, rather than an example. This is done due to the simplicity of the analysis. An example follows the general analysis.

Consider the forward contract on the T_2-maturity zero-coupon bond with delivery date time T_1 and *initiated* at time t. Let $v(t)$ denote the value of this forward contract at time t.

To value this forward contract using the risk-neutral valuation procedure, we take the expected time T_1 payoff to the forward contract, using the pseudo probabilities, and discount it to the present.

In this regard, the time T_1 payoff is

$$P(T_1,T_2) - F(t,T_1:T_2).$$

This represents the price of the underlying T_2-maturity zero-coupon bond less the initial forward price on the contract. Applying the risk-neutral valuation formula gives

$$v(t) = \tilde{E}_t \left(\frac{P(T_1,T_2) - F(t,T_1:T_2)}{B(T_1)} \right) B(t). \tag{13.1}$$

Recall that by market convention, the forward price $F(t, T_1:T_2)$ is determined such that at initiation, the forward contract has zero value; i.e., no cash exchanges hands. Thus, the forward price is determined by setting expression (13.1) equal to zero; i.e.,

   the forward price $F(t, T_1:T_2)$ is determined such that

$$\tilde{E}_t \left(\frac{P(T_1, T_2) - F(t, T_1:T_2)}{B(T_1)} \right) B(t) = 0.$$

We now solve this equation. First, separating the various components we can rewrite this equation as:

$$\tilde{E}_t \left(\frac{P(T_1, T_2)}{B(T_1)} \right) B(t) - \tilde{E}_t \left(\frac{F(t, T_1:T_2)}{B(T_1)} \right) B(t) = 0. \tag{13.2}$$

But, we know from Chapter 9 that the underlying T_2-maturity zero-coupon bond's price is its expected value at time T_1 discounted to the present,

$$P(t, T_2) = \tilde{E}_t \left(\frac{P(T_1, T_2)}{B(T_1)} \right) B(t).$$

Similarly, the T_1-maturity zero-coupon bond's price is its expected payoff at time T_1 discounted to the present,

$$P(t, T_1) = \tilde{E}_t \left(\frac{1}{B(T_1)} \right) B(t).$$

We also know that $F(t, T_1:T_2)$ is a constant at time t. Combined, these facts imply that expression (13.2) can be rewritten as:

$$P(t, T_2) - F(t, T_1:T_2) P(t, T_1) = 0. \tag{13.3}$$

Rearranging terms gives the desired result:

$$F(t, T_1:T_2) = \frac{P(t, T_2)}{P(t, T_1)}. \tag{13.4}$$

The time t forward price with a T_1 expiration date contract on the T_2-maturity zero-coupon bond is the ratio of the two relevant zero-coupon bond prices.

This expression is easy to understand if one recognizes the ratio $1/P(t,T_1)$ as the time T_1 future value of a dollar received at time t. Then expression (13.4) can be interpreted as the future value at time T_1 of the T_2-maturity zero-coupon bond – hence, a forward price.

As an aside, from this expression we can obtain a relation between forward prices and forward rates:

$$F(t,T:T+1) = \frac{P(t,T+1)}{P(t,T)} = 1/f(t,T).$$

The forward price of a zero-coupon bond that matures at time $T+1$, one period after the forward contract's expiration date T, is the inverse of the time t forward rate for the same period $[T,T+1]$. This result makes sense since both quantities correspond to contracting today for a future date, and $F(t,T:T+1)$ is a price, while $f(t,T)$ is a rate.

The self-financing trading strategy that synthetically replicates a forward contract is easily determined. The synthetic forward contract is constructed with what is called a "cash and carry" strategy. The cash and carry strategy is to buy the spot (cash) commodity and carry (hold) it until the delivery date of the contract. The details of this strategy are *(i)* buy and hold until time T_1 the T_2-maturity zero-coupon bond and *(ii)* sell and hold until time T_1, the number $F(t,T_1:T_2)$ units of the T_1-maturity zero-coupon bond.

The time T_1 cash flows to this strategy are given in Table 13.1. The first row gives the payoff to the traded forward contract, $P(T_1,T_2) - F(t,T_1:T_2)$. The second set of rows gives the payoff to the synthetic forward contract. The first gives the payoff to holding the underlying T_2-maturity zero-coupon bond. The second gives the payoff to selling $F(t,T_1;T_2)$ of T_1-maturity zero-coupon bonds. The sum duplicates the forward contract's cash flows. This completes the proof.

TABLE 13.1 Cash Flows to a Cash and Carry Trading Strategy

	Time T_1 Cash Flows
Forward Contract	$P(T_1,T_2) - F(t,T_1:T_2)$
Synthetic Forward	
Buy T_2-maturity bond	$P(T_1,T_2)$
Sell $F(t,T_1:T_2)$	$-F(t, T_1:T_2) \cdot 1$
T_1-maturity bonds	
SUM	$P(T_1,T_2) - F(t, T_1:T_2)$

Note that after the initiation date, the forward contract can have non-zero value. Indeed, at a subsequent date $t < m \leq T_1$, the forward contract's value, denoted by $v(m)$, is

$$v(m) = \tilde{E}_m \left(\frac{P(T_1, T_2) - F(t, T_1 : T_2)}{B(T_1)} \right) B(m) = P(m, T_2) - F(t, T_1 : T_2) P(m, T_1)$$

$$(13.5)$$

which can differ from zero.

Example: Forward Price and Forward Contract Values

We illustrate the use of the valuation expressions (13.4) and (13.5) with an example.

Let the evolution of the zero-coupon bond price curve be as given in Figure 13.1. This is the same evolution as in Figure 10.1; hence, we know that the evolution is arbitrage-free.

Consider a forward contract with expiration date 2 on the 3-period zero-coupon bond.

Using the formula for the forward price given in expression (13.4), we compute the time 0 forward price to be:

$$F(0, 2:3) = P(0, 3)/P(0, 2) = .942322/.961169 = .980392.$$

The same formula generates forward prices at times 1 and 2 to be:

$$F(1, 2:3; u) = P(1, 3; u)/P(1, 2; u) = .965127/.982699 = .982119$$

$$F(1, 2:3; d) = P(1, 3; d)/P(1, 2; d) = .957211/.978085 = .978658$$

$$F(2, 2:3; uu) = P(2, 3; uu) = .984222$$

$$F(2, 2:3; ud) = P(2, 3; ud) = .980015$$

$$F(2, 2:3; du) = P(2, 3; du) = .981169$$

$$F(2, 2:3; dd) = P(2, 3; dd) = .976147.$$

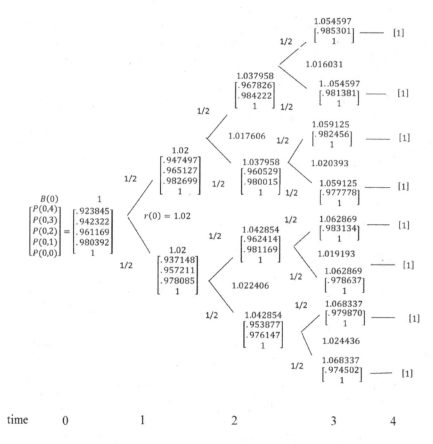

FIGURE 13.1 An Example of a One-Factor Bond Price Curve Evolution. The Money Market Account Values and Spot Rates Are Included on the Tree. Pseudo Probabilities Are Along Each Branch of the Tree.

These values are listed in Figure 13.2.

The value of the forward contract can now be determined. There are two approaches. One, we could use the second formula in expression (13.5). This just involves knowledge of the zero-coupon bond price evolution in Figure 13.1. Two, we could use the risk-neutral valuation approach, the first formula in expression (13.5). We choose the later.

The value of the forward contract at time 2 is the spot price of the 3-period zero-coupon bond at time 2 less the agreed-upon purchase price.

$$v(2; uu) = P(2, 3; uu) - F(0, 2:3) = .984222 - .980392 = .003830$$

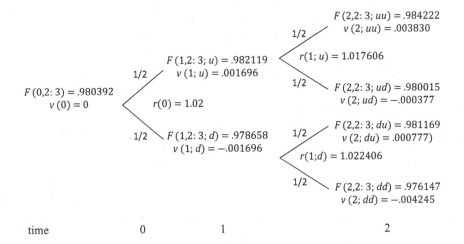

FIGURE 13.2 An Example of a Forward Contract Initiated at Time 0 on a 3-Period Zero-Coupon Bond. The Forward Contract Expires at Time 2, with Value ($v(t;s_i)$) and Forward Price ($F(t,2:3;s_i)$).

$$v(2;ud) = P(2,3;ud) - F(0,2:3) = .980015 - .980392 = -.000377$$

$$v(2;du) = P(2,3;du) - F(0,2:3) = .981169 - .980392 = .000777$$

$$v(2;dd) = P(2,3;dd) - F(0,2:3) = .976147 - .980392 = -.004245$$

The value of the forward contract at time 1 is obtained via the risk-neutral valuation procedure:

$$v(1;u) = \tilde{E}_1\left(v(2;s_2)\right)/r(1;u)$$

$$= \left[\left(\tfrac{1}{2}\right).003830 + \left(\tfrac{1}{2}\right)(-.000377)\right]/1.017606 = .001696$$

$$v(1;d) = \tilde{E}_1\left(v(2;s_2)\right)/r(1;d)$$

$$= \left[\left(\tfrac{1}{2}\right).000777 + \left(\tfrac{1}{2}\right)(-.004245)\right]/1.022406 = -.001696$$

Finally, at time 0:

$$v(0) = E_0\left(v(1;s_1)\right)/r(0) = \left[\left(\tfrac{1}{2}\right).001696 + \left(\tfrac{1}{2}\right)(-.001696)\right]/1.02 = 0.$$

The forward contract has zero value at initiation as it should.

The reader should verify that the values for the forward contract in Figure 13.2 satisfy the last equation in expression (13.5). Note also that in this example, the forward price changes over time and the forward contract's value changes over time with both positive and negative values.

Figure 13.3 illustrates these computations for a different forward contract initiated at time 0 with expiration date 3 on a 4-period zero-coupon bond. These calculations are left to the reader as an exercise. This completes the example. ☐

13.2 FUTURES

We consider the simplified futures contracts as defined in Chapter 4, which are issued on a T_2-maturity zero-coupon bond with expiration date T_1. The futures price was denoted by $\mathcal{F}(t,T_1;T_2)$. The purpose of this section is to use the contingent claim valuation methodology of Chapter 10 to give an alternative characterization for the futures price.

Given there are no arbitrage opportunities and complete markets, the subsequent analysis proceeds independently of the particular economy (with one, two, or $N \geq 3$ factors) studied. As in the previous section, we present the general case before discussing an example. This is due to the fact that the general analysis is simple enough to grasp by itself.

Consider the futures contract on the T_2-maturity zero-coupon bond with delivery date time T_1 and *initiated* at time t. By market convention, the futures price for this contract is determined such that the futures contract has zero value at initiation. Furthermore, marking-to-market guarantees that the futures contract also has zero value at each intermediate date over the contract's life. This is in contrast to the forward contract that can have nonzero value at intermediate dates.

To determine the futures price, we proceed in a backward-inductive fashion, using both the risk-neutral valuation procedure and the payoff to the futures contract. For easy comparison with the derivation of the forward price, let us denote the value of the futures contract at time t to be $V(t)$. We know by market convention, however, that $V(t) \equiv 0$ for all t.

At time $T_1 - 1$ the value of the futures contract is its discounted time T_1 payout. It is given by expression (13.6):

$$V(T_1 - 1) = \tilde{E}_{T_1-1}\left(\frac{P(T_1, T_2) - \mathcal{F}(T_1 - 1, T_1 : T_2)}{B(T_1)}\right)B(T_1 - 1). \qquad (13.6)$$

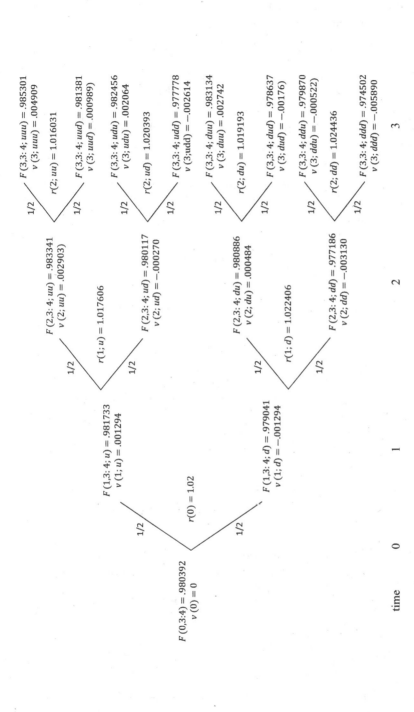

FIGURE 13.3 An Example of a Forward Contract Initiated Time 0 on a 4-Period Zero-Coupon Bond. The Forward Contract Expires at time 3, with Value ($v(t;s_t)$) and Forward Price ($F(t,3:4;s_t)$).

The market convention is that the futures price $F(T_1 - 1, T_1 : T_2)$ is determined such that the futures contract has zero value, i.e.,

the futures price $F(T_1 - 1, T_1 : T_2)$ *is determined such that*

$$\tilde{E}_{T_1 - 1}\left(\frac{P(T_1, T_2) - F(T_1 - 1, T_1 : T_2)}{B(T_1)}\right) B(T_1 - 1) = 0.$$

We now solve for the futures price. Recognizing that the money market account's value $B(T_1)$ is known at time $T_1 - 1$, it can be moved outside the expectations operator. Then, multiplying both the left and right side of this expression by $B(T_1)$ removes it from the expression. Finally, note that the futures price at expiration is the spot price, i.e., $F(T_1, T_1 ; T_2) = P(T_1, T_2)$. Substitution gives:

$$\tilde{E}_{T_1 - 1}\left(F(T_1, T_1 : T_2) - F(T_1 - 1, T_1 : T_2)\right) = 0. \tag{13.7}$$

Given that $F(T_1 - 1, T_1 : T_2)$ is known at time $T_1 - 1$, we get:

$$F(T_1 - 1, T_1 : T_2) = \tilde{E}_{T_1 - 1}\left(F(T_1, T_1 : T_2)\right). \tag{13.8}$$

The time $T_1 - 1$ futures price is the time $T_1 - 1$ expectation of its time T_1 value using the pseudo probabilities. We next show that this result is true for all remaining time periods.

Next, at time $T_1 - 2$, using risk-neutral valuation again yields:

$$V(T_1 - 2) = \tilde{E}_{T_1 - 2}\left(\frac{V(T_1 - 1) + [F(T_1 - 1, T_1 : T_2) - F(T_1 - 2, T_1 : T_2)]}{B(T_1 - 1)}\right) B(T_1 - 2). \tag{13.9a}$$

The time $T_1 - 2$ value of the futures contract equals its expected time $T_1 - 1$ *value plus cash flow* discounted. But, by market convention, the futures contract's value is always zero. So, we can rewrite expression (13.9a) as (13.9b).

$$0 = \tilde{E}_{T_1 - 2}\left(\frac{[F(T_1 - 1, T_1 : T_2) - F(T_1 - 2, T_1 : T_2)]}{B(T_1 - 1)}\right) B(T_1 - 2). \tag{13.9b}$$

The $F(T_1 - 2, T_1 : T_2)$ that solves this expression is the time $T_1 - 2$ futures price.

Because the money market account's value $B(T_1 - 1)$ is known at time $T_1 - 2$, following the same algebra that obtained expression (13.8) yields

$$\mathcal{F}(T_1 - 2, T_1 : T_2) = \tilde{E}_{T_1 - 2}\left(\mathcal{F}(T_1 - 1, T_1 : T_2)\right). \tag{13.10}$$

Continuing the same argument inductively backward in time generates our final result:

$$\mathcal{F}(t, T_1 : T_2) = \tilde{E}_t\left(\mathcal{F}(t + 1, T_1 : T_2)\right). \tag{13.11}$$

This says that under the pseudo probabilities, *futures prices are martingales*. This is an important result. It facilitates the computation of futures prices, and it gives the pseudo probabilities their fourth name: *futures price martingale* probabilities.*

Using the law of iterated expectations (backward substitution), this martingale property implies that

$$\mathcal{F}(t, T_1 : T_2) = \tilde{E}_t\left(P(T_1, T_2)\right). \tag{13.12}$$

The futures price is the time t expectation of the underlying T_2-maturity zero-coupon bond's price trading at time T_1. This is an important result as it simplifies the computation of the futures price to just taking an expected value of the spot price at the delivery date using the pseudo probabilities. We illustrate this computation with an example.

Example: Futures Price Computations

Let the evolution of the zero-coupon bond price curve again be as given in Figure 13.1. This is the same evolution as in Figure 10.1; hence, we know that the evolution is arbitrage-free.

To illustrate the determination of the futures price across time, consider a futures contract on a 3-period zero-coupon bond with delivery date time 2.

By definition, at time 2, the futures price is the spot price, i.e.,

$$\mathcal{F}(2, 2:3; uu) = P(2, 3; uu) = .984222$$

---

* Other names for these futures price martingale probabilities are risk-neutral probabilities, martingale probabilities, and pseudo probabilities.

$$F(2,2{:}3;ud) = P(2,3;ud) = .980015$$

$$F(2,2{:}3;du) = P(2,3;du) = .981169$$

$$F(2,2{:}3;dd) = P(2,3;dd) = .976147.$$

Using expression (13.11), we obtain

$$F(1,2{:}3;u) = \left[(1/2).984222 + (1/2).980015\right] = .982119$$

and

$$F(1,2{:}3;d) = \left[(1/2).981169 + (1/2).976147\right] = .978658.$$

The time 0 futures price is

$$F(0,2{:}3) = \left[(1/2).982119 + (1/2).978658\right] = .980388.$$

The futures price is seen to change randomly over time. These values are depicted in Figure 13.4. The futures contract has zero value at each time and state. The cash flow paid out each period, due to marking-to-market, is also indicated below each node. The cash flow at time t, state s_t is the change in the futures price, i.e.,

$$F(t,2{:}3;st) - F(t-1,2{:}3;st-1).$$

For example, at time 2, state uu,

cash flow $= F(2,2:3;uu) - F(1,2:3;u) = .984222 - .982119 = .002104.$

The cash flows at the up and down nodes should differ only in sign, because the expected value of the cash flow is zero. This can be verified by examining Figure 13.4.

To create a synthetic futures contract, we need to create the cash flows in Figure 13.4. The only difference between this procedure and what was previously done is that here we are duplicating cash flows instead of values. Otherwise, it is the same.

The procedure for computing the synthetic futures contract is to use the delta approach. At time 1, state u we choose $n_3(1;u)$ units of the 3-period zero-coupon bond via

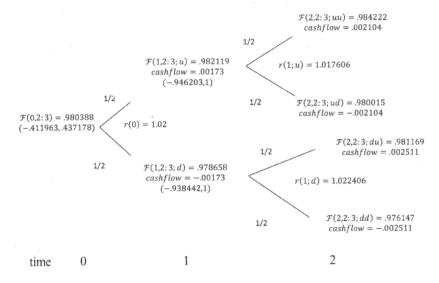

FIGURE 13.4   An Example of a Futures Contract with Expiration Date 2 on the 3-Period Zero-Coupon Bond. Futures prices ($\mathcal{F}(t,2{:}3)$) Are Given at Each Node. Pseudo Probabilities Are on Each Branch of the Tree. The Synthetic Futures Contract ($n_0(t;s_t)$, $n_3(t;s_t)$) in the Money Market Account and the 3-Period Bond Are Also Provided.

$$n_3(1;u) = \frac{\text{cash flow}(2;uu) - \text{cash flow}(2;ud)}{P(2,3;uu) - P(2,3;ud)}$$

$$= \frac{.002104 - (-.002104)}{.984222 - .980015} = 1$$

and the position in the money market account is:

$$n_0(1;u) = \left[0 - n_3(1;u)P(1,3;u)\right]/B(1)$$

$$= -1(.965127)/1.02 = -.946203.$$

The first zero in the expression for $n_0(1;u)$ represents the value of the futures contract at time 1, state u.

Continuing, at time 1, state d,

$$n_3(1;d) = \frac{\text{cash flow}(2;du) - \text{cash flow}(2;dd)}{P(2,3;du) - P(2,3;dd)}$$

$$= \frac{.002511 - (-.002511)}{.981169 - .976147} = 1$$

and

$$n_0(1;d) = [0 - n_3(1;d)P(1,3;d)]/B(1)$$

$$= -1(.957211)/1.02$$

$$= -.938442.$$

Finally, at time 0,

$$n_3(0) = \frac{\text{cash flow}(1;u) - \text{cash flow}(1;d)}{P(1,3;u) - P(1,3;d)}$$

$$= \frac{.00173 - (-.00173)}{.965127 - .957211}$$

$$= .437178$$

and

$$n_0(0) = 0 - n_3(0)P(0,3)$$

$$= 0 - (.437178)(.942322)$$

$$= -.411963.$$

This completes the method for determining the synthetic futures contract. Note that the cost of forming this portfolio is:

$$n_0(1;u)B(1) + n_3(1;u)P(1,3;u) = -.946203(1.02) + 1(.965127) = 0.$$

This is the time 0 arbitrage-free value of the futures contract, as it should be!

The futures prices for a 3-period futures contract on the 4-period zero-coupon bond are provided in Figure 13.5. The calculations are left to the reader as an exercise. This completes the example. ☐

13.3 THE RELATIONSHIP BETWEEN FORWARD AND FUTURES PRICES

This section clarifies and formalizes the intuition given in Chapter 4 relating forward and futures prices. We first demonstrate through an example that when interest rates are random, futures prices are (not equal to and) less than forward prices. This is due to the negative correlation between interest rates and futures prices on zero-coupon bonds.

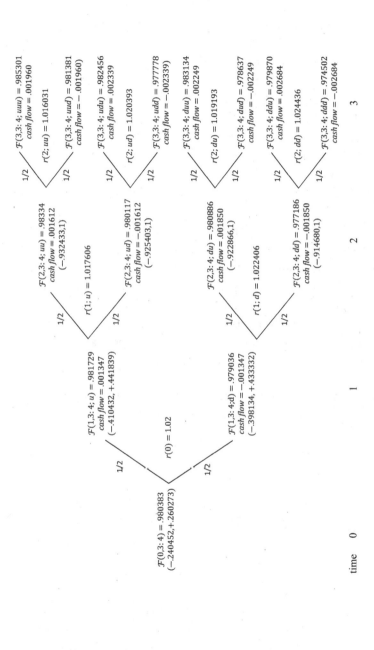

FIGURE 13.5 An Example of a Futures Contract with Expiration Date 3 on a 4-Period Zero-Coupon Bond. Futures Prices ($\mathcal{F}(t,3:4)$) Are Given at Each Node. Pseudo Probabilities Are Along the Branches of the Tree. The Synthetic Futures Contract ($n_0(t;s_t)$, $n_4(t;s_t)$) in the Money Market Account and 4-Period Bond Is Also Provided.

Example: Relationship Between Forward and Futures Prices

This example uses the forward contracts and futures contracts contained in Figures 13.2 through 13.5.

The forward and futures prices for contracts with a delivery date of time 2 on a 3-period zero-coupon bond are given in Figures 13.1 and 13.2. For comparison, these prices are listed in Table 13.2.

We see that at the expiration date 2, forward prices, futures prices, and spot prices are equal. This is as it should be since this is true by the definition of the contracts. At time 1, forward prices equal futures prices. This is because 1 day before delivery, both contracts are identical. To find any difference, one must proceed at least 2 days before delivery. Finally, 2 days before delivery at time 0, the forward price (.980392) differs from the futures price (.980388). The forward price exceeds the futures price as was to be shown.

Table 13.3 provides a comparison of forward and futures prices for contracts with expiration date 3 on the 4-period zero-coupon bond given in Figures 13.3 and 13.4. As before, both the time 2 and time 3 forward and futures prices are equal. These correspond to the day before expiration and the expiration date of the contracts. However, prior to that date, forward and futures prices differ. At both earlier dates, futures prices are less than the forward prices. The differences at time 0 are greater than the differences at time 1. This completes the example. □

We can now use the theory of the preceding sections to relate forward and futures prices. From expression (13.4), using the fact that

$$P(t,T_2) = \tilde{E}_t\left(\frac{P(T_1,T_2)}{B(T_1)}\right)B(t),$$

TABLE 13.2   A Comparison of Forward and Futures Prices for a 2-Period Contract on the 3-Period Zero-Coupon Bond. Spot Prices Are Also Included

(Time, State)	Forward Price	Futures Price	Spot Price
0	.980392	.980388	.942322
(1,u)	.982119	.982119	.965127
(1,d)	.978658	.978658	.957211
(2,uu)	.984222	.984222	.984222
(2,ud)	.980015	.980015	.980015
(2,du)	.981169	.981169	.981169
(2,dd)	.976147	.976147	.976147

TABLE 13.3   A Comparison of Forward and Futures Prices for a 3-Period
Contract on the 4-Period Zero-Coupon Bond. Spot Prices Are Also Included

(Time, State)	Forward Price	Futures Price	Spot Price
0	.980392	.980383	.923845
(1,u)	.981733	.981729	.947497
(1,d)	.979041	.979036	.937148
(2,uu)	.983341	.983341	.967826
(2,ud)	.980117	.980117	.960529
(2,du)	.980886	.980886	.962414
(2,dd)	.977186	.977186	.953877
(3,uuu)	.985301	.985301	.985301
(3,uud)	.981381	.981381	.981381
(3,udu)	.982456	.982456	.982456
(3,udd)	.977778	.977778	.977778
(3,duu)	.983134	.983134	.983134
(3,dud)	.978637	.978637	.978637
(3,ddu)	.97987	.97987	.97987
(3,ddd)	.974502	.974502	.974502

we obtain

$$F\left(t,T_1:T_2\right)=\tilde{E}_t\left(\frac{P\left(T_1,T_2\right)}{B(T_1)}\right)\frac{B(t)}{P(t,T_1)}.$$

Consider the following property of expectations: if x and y are random variables,

$$\tilde{E}(xy)=\tilde{E}(x)\tilde{E}(y)+\text{co}\tilde{v}(x,y)\text{ where }\text{co}\tilde{v}(x,y)\equiv\tilde{E}\left[(x-\tilde{E}(x))(y-\tilde{E}(y))\right].$$

Using this property yields

$$F\left(t,T_1:T_2\right)=\tilde{E}_t\left(P\left(T_1,T_2\right)\right)\tilde{E}_t\left(\frac{1}{B(T_1)}\right)\frac{B(t)}{P(t,T_1)}$$

$$+\text{co}\tilde{v}\left(P\left(T_1,T_2\right),\frac{1}{B(T_1)}\right)\frac{B(t)}{P(t,T_1)}.$$

(13.14)

Finally, using expression (13.12) gives

$$F\left(t,T_1:T_2\right)=\mathcal{F}\left(t,T_1:T_2\right)+\text{co}\tilde{v}\left(P\left(T_1,T_2\right),\frac{1}{r(t)\cdots r(T_1-1)}\right)\frac{1}{P(t,T_1)}.\quad(13.15)$$

The forward price equals the futures price plus an adjustment term. The adjustment term reflects the covariance between the T_2-maturity zero-coupon bond's price and the spot rates over the time period $[t, T_1]$. When this covariance is zero, forward and futures prices are identical. When the covariance is positive, forward prices exceed futures prices. The covariance is positive, in general, when interest rates are negatively correlated with zero-coupon bond prices. Expression (13.15) finally formalizes the simple intuition introduced in the discussion of futures contracts in Chapter 4.

We note that this covariance term is zero, in general, only on the day before delivery at time $T_1 - 1$. At time $T_1 - 1$, expression (13.15) becomes

$$F\left(T_1 - 1, T_1 : T_2\right) = \mathcal{F}\left(T_1 - 1, T_1 : T_2\right) + \text{cõv}\left(P(T_1, T_2), \frac{1}{r(T_1 - 1)}\right) \frac{1}{P\left(T_1 - 1, T_1\right)}.$$

$$(13.16)$$

This last term is zero because $1/r(T_1 - 1)$ is not random when viewed at time $T_1 - 1$. Thus, on the day before the forward and futures contracts expire, the forward price equals the futures price. This makes sense, because on this day alone over the life of the contracts, the two contracts are guaranteed to have identical cash flows.

13.4 OPTIONS ON FUTURES

This section analyzes the arbitrage-free valuation of options on futures. Let us consider a futures contract with delivery date T_1 on a T_2-maturity zero-coupon bond with $T_1 \le T_2$.

Next, consider a European call option on this futures contract. Let the European call option have a maturity date of $\tau^* < T_1$ and an exercise price of $K > 0$. By definition, the payoff to the option contract at maturity is

$$\max(F(t^*, T_1 : T_2) - K, 0).$$

If it finishes in the money, the call owner receives a cash payment equal to the futures price $\mathcal{F}(\tau^*, T_1 : T_2)$ minus the exercise price K. Otherwise, it has zero value.

Using the risk-neutral valuation procedure and denoting the call's value at time t as $C(t)$, we obtain

$$C(t) = \tilde{E}_t\left(\max\left(\mathcal{F}\left(\tau^*, T_1 : T_2\right) - K, 0\right) \big/ B(\tau^*)\right) B(t). \qquad (13.17)$$

This call can be created synthetically using any T-maturity bond with $T > \tau^*$ and the money market account. It can also be created synthetically using the underlying futures contract and the money market account. This procedure is illustrated through the following example.

Example: Options on Futures

Let the evolution of the zero-coupon bond price curve again be as given in Figure 13.1. This is the same evolution as in Figure 10.1; hence, we know that the evolution is arbitrage-free.

Consider the futures contract of Figure 13.5. This is a 3-period futures contract on a 4-period zero-coupon bond. Let us value a European call option on this futures price with maturity date 2 and strike price $K = .981000$.

The payoffs to the option at date 2 are

$$C(2; uu) = \max(.983341 - .981000, 0) = .002341$$

$$C(2; ud) = \max(.980117 - .981000, 0) = 0$$

$$C(2; du) = \max(.980886 - .981000, 0) = 0$$

$$C(2; dd) = \max(.977186 - .981000, 0) = 0.$$

The values at time 1 are

$$C(1; u) = [(1/2).002341 + (1/2)0]/1.017606 = .001150$$

$$C(1; d) = [(1/2)0 + (1/2)0]/1.022406 = 0.$$

Finally, its value at time 0 is

$$C(0) = [(1/2).001150 + (1/2)0]/1.02 = .000564.$$

The synthetic futures options can be generated by investing in the 3-period zero-coupon bond $(n_3(t; s_t))$ and the money market account $(n_0(t; s_t))$ as follows:

$$n_3(1;u) = \frac{C(2;uu) - C(2;ud)}{P(2,3;uu) - P(2,3;ud)} = \frac{.0023410 - 0}{.984222 - .980015} = .556460$$

$$n_0(1;u) = \left[C(1;u) - n_3(1;u)P(1,3;u)\right]/B(1)$$

$$= \left[.001150 - .556460(.965127)\right]/1.02$$

$$= -.525400$$

At time 1, state u, hold .556460 units of the 3-period zero-coupon bond and short .525400 units of the money market account.

At time 1, state d, hold zero units of both since the call has zero value.

At time 0,

$$n_3(0) = \frac{C(1;u) - C(1;d)}{P(1,3;u) - P(1,3;d)} = \frac{.001150 - 0}{.965127 - .957211} = .145324$$

$$n_0(0) = C(0) - n_3(0)P(0,3)$$

$$= .000564 - .145324(.942322) = -.136378.$$

That is, hold .145324 units of the 3-period zero-coupon bond and short .136378 units of the money market account.

Instead, we could have hedged this option using the underlying futures contract. The synthetic position in the futures contract $(n_F(t;s_t))$ and the money market account $(n_0(t;s_t))$ can be obtained as follows:

At time 1, state u,

$$n_F(1;u) = \frac{C(2;uu) - C(2;ud)}{\text{cash flow}(2;uu) - \text{cash flow}(2;ud)}$$

$$= \frac{.002341 - 0}{.001612 - (-.001612)}$$

$$= .726117$$

and

$$n_0(1;u) = \left[(C(1;u) - n_F(1;u) \cdot 0\right]/B(1)$$

$$= [.001150]/1.02$$

$$= .001127.$$

We hold .726117 units of the futures contract and .001127 units of the money market account; both are positive positions.

At time 1, state d, hold zero units of both assets as the option has zero value.

At time 0,

$$n_F(0) = \frac{C(1;u) - C(1;d)}{\text{cash flow}(1;u) - \text{cash flow}(1;d)}$$

$$= \frac{.001150 - 0}{.001347 - (-.001347)}$$

$$= .426875$$

and

$$n_0(0) = C(0) - n_F(0) \cdot 0$$

$$= .000564.$$

We hold .426875 units of the futures contract and .000564 units of the money market account.

This completes the synthetic construction using the futures contract. A general result appears in this construction. When futures contracts are being used as the hedging instrument, the dollar position in the money market account in the synthetic option is always equal to the value of the option being hedged. This is because the value of a futures contract is always zero.

The above calculations are summarized in Figure 13.6. This completes the example. ☐

13.5 EXCHANGE-TRADED TREASURY FUTURES CONTRACTS

The futures contracts analyzed in this chapter are simplified versions of the Treasury futures contracts actually traded on organized exchanges. Exchange-traded Treasury futures contracts have various imbedded options related to the delivery procedure. These imbedded options, known as *(i)* the delivery option, *(ii)* the wildcard option, and *(iii)* the quality option, can significantly influence the futures price. To value and hedge these futures contracts with their imbedded options, one can utilize the theory of Chapter 10 in conjunction with an explicit description of the exchange-traded Treasury futures contracts. This analysis is straightforward and is left to outside reading. For a description of the imbedded

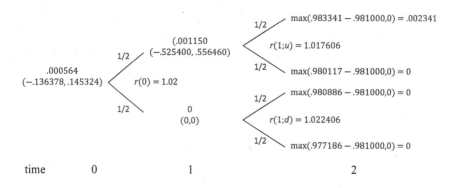

FIGURE 13.6 An Example of a European Call Option with Maturity Date 2 and Exercise Price $K = .981000$ on the Futures Price from Figure 13.5. The Synthetic Call Position in the Money Market Account $(n_0(t;s_t))$ and the 3-Period Zero-Coupon Bond $(n_3(t;s_t))$ Are Given under Each Node. Pseudo Probabilities Are Along the Branches of the Tree.

options, see Jarrow and Chatterjea [5]. For an analysis with respect to the influence of the imbedded options on the futures price, see Cohen [1], Gay and Manaster [3, 4], and Kane and Marcus [7].

REFERENCES

1. Cohen, H., 1995. "Isolating the Wild Card Option." *Mathematical Finance* 5 (2), 155–166.
2. Cox, J., J. Ingersoll, and S. Ross, 1981. "The Relation between Forward Prices and Futures Prices." *Journal of Financial Economics* 9 (4), 321–346.
3. Gay, G., and S. Manaster, 1984. "The Quality Option Implicit in Futures Contracts." *Journal of Financial Economics* 13, 353–370.
4. Gay, G., and S. Manaster, 1986. "Implicit Delivery Options and Optimal Delivery Strategies for Financial Futures Contracts." *Journal of Financial Economics* 16, 41–72.
5. Jarrow, R., and A. Chatterjea, 2019. *An Introduction to Derivative Securities, Financial Markets, and Risk Management*, 2nd edition, World Scientific Press.
6. Jarrow, R., and G. Oldfield, 1981. "Forward Contracts and Futures Contracts." *Journal of Financial Economics* 9 (4), 373–382.
7. Kane, A., and A. Marcus, 1986. "Valuation and Optimal Exercise of the Wild Card Option in Treasury Bond Futures Markets." *Journal of Finance* 41 (1), 195–207.

CHAPTER 14

Swaps, Caps, Floors, and Swaptions

S WAPS, CAPS, FLOORS, AND swaptions are very useful interest rate securities. Imagine yourself the treasurer of a large corporation who has borrowed funds from a bank using a floating-rate loan. A floating-rate loan is a long-term debt instrument whose interest payments vary (float) with respect to the current rates for short-term borrowing. Suppose the loan was taken when interest rates were low, but now rates are high. Rates are projected to move even higher. The current interest payments on the loan are high and if they go higher, the company could face a cash flow crisis, perhaps even bankruptcy. The company's board of directors is concerned.

Is there a way you can change this floating-rate loan into a fixed-rate loan, without retiring the debt and incurring large transaction costs (and a loss on your balance sheet)? That's where swaps, caps, floors, and swaptions apply. The solution is to enter into a fixed for floating-rate swap or simultaneously purchase caps and floors with predetermined strikes. If you had thought about this earlier, you could have entered into a swaption at the time the loan was made to protect the company from such a crisis.

Understanding this solution is the motivation for studying this chapter. To understand the solution, one needs to understand the pricing and hedging of swaps, caps, floors, and swaptions. This chapter applies the contingent claims valuation theory developed in Chapter 10 to do just this. Prior to this demonstration, however, it is necessary to understand fixed-rate and floating-rate loans. Fortunately for us, we have already studied fixed- and floating-rate loans in this book. At that time, however, they were analyzed from a different perspective and given different names. Fixed-rate loans correspond to (noncallable) coupon bonds, and floating-rate loans correspond to shorting the money market account.

The analysis in this chapter is presented from two related perspectives. The first perspective takes advantage of the simple structure of the swap contract and the previously stated analogy between fixed-rate loans/coupon bonds and floating-rate loans/money market accounts to derive the results quickly and efficiently. The second perspective is based on a detailed cash flow analysis, using the risk-neutral valuation procedure of Chapter 10 to obtain present values. The same results are obtained from both perspectives. This redundancy facilitates understanding. In addition, the detailed cash flow analysis is the method employed for analyzing the exotic interest rate swaps discussed in Chapter 15.

14.1 FIXED-RATE AND FLOATING-RATE LOANS

This chapter studies fixed- and floating-rate loans. Fixed- and floating-rate loans are common methods used for borrowing funds. Floating-rate loans are "long-term" debt contracts whose interest payments vary (float) with respect to short-term interest rates. When short-term rates are low, the interest rate payments are low. When short-term rates are high, the interest rate payments on the loan are high. In contrast, fixed-rate loans are "long-term" debt contracts whose interest rate payments are fixed (constant) at the time the loan is made. If short-term rates change, the interest rate payment on the loan does not. Consumers and corporations use both types of loans. Consistent with the previous theory, we only study default-free loans in this chapter. Although credit risk is an important feature of both consumer and corporate borrowing, a discussion of this topic is postponed to Chapter 18.

In our simple discrete time model, the short-term rate of interest corresponds to the spot rate $r(t)$, and each period in the model requires an interest payment. We define a *floating-rate loan* for L dollars (the principal) with maturity date T to be a debt contract that obligates the borrower to pay the spot rate of interest times the principal L every period, up to and including the maturity date, time T. At time T, the principal of L dollars is also repaid.

In our frictionless and default-free setting, this floating-rate loan is equivalent to shorting L units of the money market account and distributing the gains (paying out the spot rate of interest times L dollars) every period.

Paying out the interest as a cash flow maintains the value of the short position in the money market account at L dollars. At time T the short position is closed out.

A typical floating-rate loan of 1 dollar is depicted in Table 14.1. Examining Table 14.1 we see that the initial borrowing (the principal) is 1 dollar. At the end of each period, the interest payment owed on the dollar, $r(t) - 1$, is paid. The outstanding borrowing (principal) remains at 1 dollar. At the termination date (time T), the interest plus principal is paid, and it is equal to $r(T-1)$ dollars. To get a floating-rate loan of larger principal, say L dollars, one simultaneously enters into L of these unit-value floating-rate loans.

As the floating-rate loan is market determined, it costs 0 dollars to enter into a floating-rate loan contract. This is identical to saying that it costs 0 dollars to short the money market account in a frictionless and competitive economy. Note that after its initiation, the floating-rate loan is continually reset on the interest payment dates to have a 1-dollar value.

Computing the present value of the cash flows paid on a floating-rate loan with a dollar principal and maturity date T makes this same point. Let the value of the floating-rate loan at time t given by $V_r(t)$. Using the risk-neutral valuation procedure from Chapter 10, the present value of the cash flows to the floating-rate loan is:

$$V_r(t) = \tilde{E}_t \left(\sum_{j=t}^{T-1} \frac{[r(j)-1]}{B(j+1)} \right) B(t) + \tilde{E}_t \left(\frac{1}{B(T)} \right) B(t) = 1 \qquad (14.1)$$

where the expectations are taken with respect to the pseudo probabilities in either the one-, two-, or $N \geq 3$-factor case. The derivation of expression (14.1) is given below.

TABLE 14.1   Cash Flow from a Floating-Rate Loan of a Dollar (the Principal), with Maturity Date T

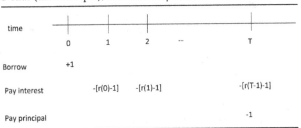

Expression (14.1) shows that the value of the cash flows from the floating-rate loan at time t equals 1 dollar, which is the amount borrowed.

Derivation of Expression (14.1)

Looking at the cash flows, we get

$$\frac{\text{cash}}{\text{flows}} = \sum_{j=t}^{T-1} \frac{(r(j)-1)}{B(j+1)} + \frac{1}{B(T)} = \sum_{j=t}^{T-2} \frac{(r(j)-1)}{B(j+1)} + \frac{r(T-1)}{B(T)}.$$

But $B(T) = B(T-1)r(T-1)$. So

$$\frac{\text{cash}}{\text{flows}} = \sum_{j=t}^{T-2} \frac{(r(j)-1)}{B(j+1)} + \frac{1}{B(T-1)} = \sum_{j=t}^{T-3} \frac{(r(j)-1)}{B(j+1)} + \frac{r(T-2)}{B(T-1)}.$$

Continuing backward in this fashion, one gets

$$\text{cash flows} = \frac{r(t)1}{B(t+1)} + \frac{1}{B(t+1)} = \frac{r(t)}{B(t+1)} = \frac{1}{B(t)}.$$

Thus,

$$V_r(t) = \tilde{E}_t(\text{cash flows})B(t) = \tilde{E}_t(1) = 1.$$

This completes the proof. □

We define a *fixed-rate loan* with interest rate c (one plus a percent) for L dollars (the principal) and with maturity date T to be a debt contract that obligates the borrower to pay $(c-1)$ times the principal L every period, up to and including the maturity date, time T. At time T, the principal of L dollars is also repaid.

A *fixed-rate loan* of $\mathcal{B}(0)$ dollars at fixed rate (C/L) and maturity T, in our frictionless and default-free setting, is equivalent to shorting the coupon bond described in Chapter 11. Shorting the coupon bond consists of receiving the value of the bond $\mathcal{B}(0)$ at time 0 (the principal) and paying out a sequence of equal and fixed coupon payments, C dollars at equally spaced times $0 \leq 1 \leq 2 \leq \ldots \leq T-1 \leq \tau$. The last payment represents an interest payment of C dollars plus the principal L. This cash flow is depicted

TABLE 14.2   Cash Flow to a Fixed-Rate Loan with Coupon C, Principal L, and Maturity Date T

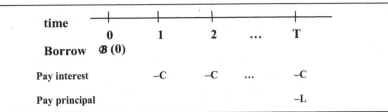

time	0	1	2	...	T
Borrow	$\mathcal{B}(0)$				
Pay interest		$-C$	$-C$	...	$-C$
Pay principal					$-L$

in Table 14.2. The (coupon) rate on the loan is defined to be $(1+C/L)$ per period.

Computing the present value of the cash flows paid on the fixed-rate loan can make the same point. Let the value of the fixed-rate loan at time t given state s_t be denoted by $V_c(t)$. To make the coupon rate compatible with the rate convention used in this book, we define $c \equiv 1 + C/L$.

Using the risk-neutral valuation procedure from Chapter 10,

$$V_c(t) = \tilde{E}_t \left(\sum_{j=t}^{T-1} \frac{C}{B(j+1)} \right) B(t) + \tilde{E}_t \left(\frac{L}{B(T)} \right) B(t)$$

$$= \sum_{j=t}^{T-1} CP(t, j+1) + LP(t, T) \tag{14.2}$$

$$= \mathcal{B}(t)$$

The second equality in expression (14.2) follows from the fact that $P(t,T) = \tilde{E}_t(1/B(T))B(t)$.

The third equality in expression (14.2) follows from synthetically constructing the coupon bond using a portfolio of zero-coupon bonds (see Chapter 11).

Expression (14.2) shows that the value of the cash flows to a fixed-rate loan at time t equals $\mathcal{B}(t)$, which is the amount borrowed.

The difference between floating- and fixed-rate loans can now be clarified. Floating-rate loans have a fixed market value at each date, but randomly changing interest rate payments $r(t) - 1$. In contrast, the fixed-rate loan has changing market values $\mathcal{B}(t)$, but fixed interest rate payments of C. Thus, these two types of loans exhibit opposite and symmetric risks. Floating-rate loans have no capital gains risk, only interest rate payment risk.

Conversely, fixed-rate loans have capital gains risk, but no interest rate payment risk.

14.2 INTEREST RATE SWAPS

An *interest rate swap* is a financial contract that obligates the holder to receive fixed-rate loan payments and pay floating-rate loan payments (or vice versa). We study interest rate swaps in this section. We first study the valuation of interest rate swaps, then the definition of the swap rate, and finally the construction of synthetic swaps.

14.2.1 Swap Valuation

To motivate the use of a swap, consider an investor who has a fixed-rate loan with interest rate c (one plus a percent), a principal of L dollars and a maturity date T. The cash payment at every intermediate date t is $C = (c - 1)L$. The principal repaid at time T is L dollars. The investor wants to exchange this fixed-rate loan for a floating-rate loan with principal L dollars, maturity date T, and floating interest payments of $L(r(t - 1) - 1)$ dollars per period. He does this by entering into a swap receiving fixed and paying floating. Figure 14.1 illustrates the cash flow streams from this swap transaction. The net payment from the fixed-rate loan plus the swap is equivalent to a floating-rate loan. Given this motivation for the use of swaps, we now consider a swap's valuation.

The cash flows from a swap receiving fixed and paying floating are depicted in Table 14.3. The swap holder receives fixed-rate payments $C = (c - 1)L$ and pays floating at $(r(t - 1) - 1)L$. The principals cancel and are never exchanged. The swap contract, therefore, has the value $\mathcal{B}(0) - L$. This equals the present value of the fixed-rate payments (plus principal) less the floating-rate payments (plus principal). Note that the present values of the principals cancel in this difference.

Let $S(t)$ represent the value of the swap at time t. Then the value of the swap at any period t is given by $S(t) = \mathcal{B}(t) - L$.

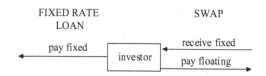

FIGURE 14.1   An Illustration of a Swap Changing a Fixed-Rate Loan into a Floating-Rate Loan.

TABLE 14.3   The Cash Flows and Values from a Swap Receiving Fixed and Paying Floating

time	0	1	2	...	T - 1	T
Floating Payments		$-[r(0)-1]L$	$-[r(1)-1]L$	...	$-[r(T-2)-1]L$	$-[r(T-1)-1]L$
Fixed Payments		$+C$	$+C$	...	$+C$	$+C$
Net Payments	0	$C-[r(0)-1]L$	$C-[r(1)-1]L$	...	$C-[r(T-2)-1]L$	$C-[r(T-1)-1]L$
Swap Value	$B(0)-L$	$B(1)-L$	$B(2)-L$		$B(T-1)-L$	$B(T)-L$

Computing the present value of the cash flows from the swap generates this same value. Using the risk-neutral valuation procedure of Chapter 10,

$$S(t) = \tilde{E}_t \left(\sum_{j=t}^{T-1} \frac{[C-(r(j)-1)L]}{B(j+1)} \right) B(t). \qquad (14.3)$$

Defining $c \equiv 1 + C/L$ to be one plus the coupon rate on the fixed-rate loan, we can rewrite this as

$$S(t) = \tilde{E}_t \left(\sum_{j=t}^{T-1} \frac{[c-r(j)]L}{B(j+1)} \right) B(t) \qquad (14.4)$$

$$= B(t) - L.$$

The derivation of expression (14.4) is given below.

Expression (14.4) shows that the swap's time 0 value is equal to the discounted expected cash flows from receiving fixed payments at rate c and paying floating payments at $r(t)$. The swap value as given in expression (14.4) is independent of any particular model for the evolution of the term structure of interest rates. This is because expression (14.4) can be determined just knowing the price of the coupon bond, and the coupon bond's price can be determined using the time t zero-coupon bond prices.

The derivation of expression (14.4) now follows for interested readers. Expression (14.4) will prove useful in the next chapter when we study exotic interest rate swaps.

Derivation of Expression (14.4)

To get the first equality, substitute $C = (c-1)L$ into expression (14.3) and simplify.

To get the second equality, write

$$\tilde{E}_t \left(\sum_{j=t}^{T-1} \frac{[C-(r(j)-1)L]}{B(j+1)} \right) B(t) = \tilde{E}_t \left(\sum_{j=t}^{T-1} \frac{C}{B(j+1)} \right) B(t)$$

$$-\tilde{E}_t \left(\sum_{j=0}^{T-1} \frac{(r(j)-1)}{B(j+1)} \right) B(t)L$$

$$= \tilde{E}_t \left(\sum_{j=t}^{T-1} \frac{C}{B(j+1)} + \frac{L}{B(T)} \right) B(t)$$

$$- \left[\tilde{E}_t \left(\sum_{j=t}^{T-1} \frac{(r(j)-1)}{B(j+1)} + \frac{1}{B(T)} \right) \right] B(t)L.$$

This last inequality follows by adding and subtracting $\tilde{E}_t(1/B(T))B(t)L$. Finally, using expressions (14.1) and (14.2) gives the result. This completes the proof. □

14.2.2 The Swap Rate

The *swap rate* is defined to be that coupon rate C/L such that the swap has zero value at time 0, i.e., such that $S(0) = 0$ or $\mathcal{B}(0) = L$. In other words, this rate makes the time 0 coupon bond's value at par (at L), and the swap fairly priced at zero dollars.

It is important to emphasize that this determination of the swap rate is under the assumption of no default risk for either counterparty to the swap contract. Default risk and credit risk spreads are important elements in the actual application of these techniques to the swap market. These extensions require a generalization of the model presented, and they are available in Jarrow and Turnbull [2].

Using the procedure described in Chapter 10, we can synthetically create the swap contract. An example will illustrate these computations.

Example: Swap Valuation

We use the evolution of the zero-coupon bond price curve given in Figure 14.2. This evolution is arbitrage-free as it is the same zero-coupon bond price curve evolution studied in Chapter 10.

Consider a swap receiving fixed and paying floating with maturity date $T = 3$ and principal $L = 100$.

First, we need to determine the swap rate. To do this, we need to find the coupon payment C per period such that the value of the swap is zero, i.e., $S(0) = 0$.

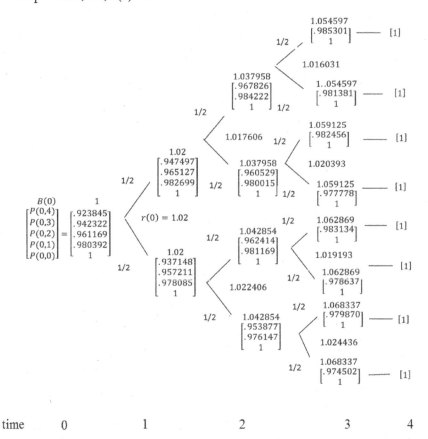

FIGURE 14.2 An Example of a One-Factor Bond Price Curve Evolution. The Money Market Account Values and Spot Rates Are Included on the Tree. Pseudo Probabilities Are Along Each Branch of the Tree.

We first compute the swap's value for an arbitrary coupon payment of C:

$$S(0) = \mathcal{B}(0) - 100$$

$$= CP(0,1) + CP(0,2) + (C+100)P(0,3) - 100$$

$$= C[.980392 + .961169 + .942322] + 100(.942322) - 100$$

$$= C(2.8838) - 5.7678.$$

Setting $S(0) = 0$ and solving for C yields

$$C = 5.7678 / 2.8838 = 2.$$

The swap rate is $C/L = 2/100 = .02$.

The cash flows and values of this swap, with swap rate .02, are shown in Figure 14.3. We receive fixed and pay floating. The calculations are as follows.

At time 3, for each possible state:

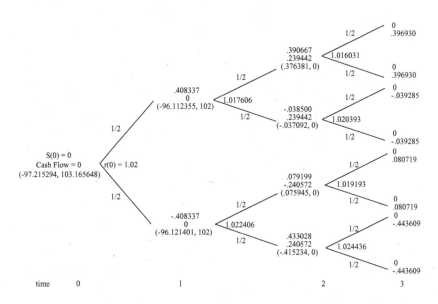

FIGURE 14.3 An Example of a Swap Receiving Fixed and Paying Floating with Maturity Time 3, Principal $100, and Swap Rate .02. Given First Is the Swap's Value, then the Swap's Cash Flow. The Synthetic Swap Portfolio in the Money Market Account and 3-Period Zero-Coupon Bond $(n_0(t;s_t), n_3(t;s_t))$ Is Given under Each Node.

$$S(3; uuu) = L - L = 100 - 100 = 0$$

$$\text{cash flow}(3; uuu) = C - (r(2; uu) - 1)L = 2 - 1.60307 = .39693$$

$$S(3; uud) = 100 - 100 = 0$$

$$\text{cash flow}(3; uud) = 2 - (r(2; uu) - 1)100 = .39693$$

$$S(3; udu) = 100 - 100 = 0$$

$$\text{cash flow}(3; udu) = 2 - (r(2; ud) - 1)100 = -.039285$$

$$S(3; udd) = 100 - 100 = 0$$

$$\text{cash flow}(3; udd) = 2 - (r(2; ud) - 1)100 = -.039285$$

$$S(3; duu) = 100 - 100 = 0$$

$$\text{cash flow}(3; duu) = 2 - (r(2; du) - 1)100 = .080719$$

$$S(3; dud) = 100 - 100 = 0$$

$$\text{cash flow}(3; dud) = 2 - (r(2; du) - 1)100 = .080719$$

$$S(3; ddu) = 100 - 100 = 0$$

$$\text{cash flow}(3; ddu) = 2 - (r(2; dd) - 1)100 = -.443609$$

$$S(3; ddd) = 100 - 100 = 0$$

$$\text{cash flow}(3; ddd) = 2 - (r(2; dd) - 1)100 = -.443609$$

Continuing backward through the tree:

$$S(2; uu) = \mathcal{B}(2; uu) - L = 102P(2,3; uu) - 100$$

$$= 102(.984222) - 100 = .390667$$

$$\text{cash flow}(2; uu) = C - [r(1; u) - 1]L = 2 - 1.76056 = .239442$$

$$S(2; ud) = 102P(2,3; ud) - 100 = -.038500$$

$$\text{cash flow}(2; ud) = 2 - (r(1; u) - 1)100 = .239442$$

$$S(2;du) = 102P(2,3;du) - 100 = .079199$$

$$\text{cash flow}(2;du) = 2 - (r(1;d) - 1)100 = -.240572$$

$$S(2;dd) = 102P(2,3;dd) - 100 = -.433028$$

$$\text{cash flow}(2;dd) = 2 - (r(1;d) - 1)100 = -.240572.$$

Finally, at time 1:

$$S(1;u) = \mathcal{B}(1;u) - L = 2P(1,2;u) + 102P(1,3;u) - 100$$

$$= 2(.982699) + 102(.965127) - 100 = .408337$$

$$\text{cash flow}(1;u) = C - [r(0) - 1]L = 2 - 2 = 0$$

$$S(1;d) = 2P(1,2;d) + 102P(1,3;d) - 100 = -.408337$$

$$\text{cash flow}(1;d) = 2 - (r(0) - 1)100 = 0.$$

From Figure 14.3 we see that the cash flow from the swap can be positive or negative. Similarly, the value of the swap can be positive or negative as well, depending upon the movements of the spot rate of interest. This completes the example. ☐

14.2.3 Synthetic Swaps

There are three basic ways of creating a swap synthetically. The first is to use a *buy and hold* strategy. This method is to short the money market account (pay floating) and to synthetically create the coupon bond as a portfolio of zero-coupon bonds. Then, the value of this combined position at each date and state will match the values of the swap and its cash flow. This synthetic swap is independent of any particular model for the evolution of the term structure of interest rates. Unfortunately, synthetically constructing the swap via a portfolio of zero-coupon bonds has two practical problems. One, not all zero-coupon bonds may trade. Two, the initial transaction costs will be high.

A second method for synthetically creating this swap is to use a portfolio of forward contracts written on the spot rate at future dates (analogous to FRAs, see Chapter 1). This approach is also independent of a particular specification of the evolution of the term structure of interest rates.

A third method for synthetically creating this swap is to use a dynamic portfolio consisting of a single zero-coupon bond (for a one-factor model) and the money market account. This approach requires a specification of the evolution of the term structure of interest rates. Prior to illustrating this third approach through an example, we first discuss the synthetic construction of swaps using forward contracts written on the spot rate.

14.2.3.1 Swap Construction Using FRAs

This section explains the synthetic construction of swaps using forward contracts written on the spot rate of interest, called forward rate agreements or FRAs.

We define an FRA on the spot rate of interest with delivery date T, contract rate c (one plus a percent), and principal L to be that contract that has a certain payoff of $[r(T-1)-c]L$ dollars at time T.

Notice that the spot rate in this FRA's payoff at time T is spot rate from time $T-1$. By construction, the payoff to the FRA at its delivery date is the time $T-1$ spot rate less the contract rate c multiplied by the principal L. The contract rate c is set at the date the contract is initiated, say at time 0. It is set by mutual consent of the counter parties to the contract. At initiation, the contract rate need not give the FRA zero initial value (however, a typical FRA sets the rate at initiation such that the contract has zero value, see Jarrow and Chatterjea [1]). In the case where the value of the contract at initiation is non-zero, the counter parties would sign the contract and the fair value of the FRA is exchanged in cash.

Let us denote the time t value of an FRA with delivery date T and contract rate c with principal 1 dollar as $V_f(t,T; c)$. Using the techniques of Chapter 13, the time t value of this FRA is:

$$V_f(t,T;c) = \tilde{E}_t\left(\frac{r(T-1)-c}{B(T)}\right)B(t).$$

But, $r(T-1)/B(T) = 1/B(T-1)$, so

$$V_f(t,T;c) = \tilde{E}_t\left(\frac{1}{B(T-1)}\right)B(t) - c\tilde{E}_t\left(\frac{1}{B(T)}\right)B(t).$$

Recalling that $P(t,T) = \tilde{E}_t(1/B(T))B(t)$, substitution gives:

$$V_f(t,T;c) = P(t,T-1) - cP(t,T).$$

At initiation, the FRA's value would be: $V_f(0,T;c) = P(0,T-1) - cP(0,T)$. Hence, the long position in the contract would receive this value in cash at the signing of the contract, and the short would pay this amount to the long.

To construct a synthetic swap, note that from Table 14.3 the third row, the net payment to the swap at time T is identical to the payoff from being *short* a single FRA with delivery date T, contract rate c, and principal L. Hence, a synthetic swap can be constructed at time 0 by shorting a portfolio of FRAs: all with contract rate c and principal L, but with differing delivery dates. The delivery dates included in the collection of short FRAs should be times 1, 2, ... , T.

The value of this collection of short FRAs is:

$$-\sum_{t=1}^{T} V_f(0,t;c)L = -\sum_{t=1}^{T} L[P(0,t-1) - cP(0,t)]$$

$$= -L + L\sum_{t=0}^{T} [c-1]P(0,t) + LP(0,T)$$

$$= -L + \mathcal{B}(0) = S(0).$$

This is the value of the swap with maturity T and principal L receiving fixed and paying floating at time 0, as expected! This completes our synthetic construction of swaps using FRAs.

14.2.3.2 Synthetic Swap Construction using a Dynamic Trading Strategy

The section explains the construction of a synthetic swap using a dynamic trading strategy involving a single zero-coupon bond and the money market account. An example illustrates the procedure.

Example: Synthetic Swap Construction

Consider the swap contract in Figure 14.3.

The third method discussed above for constructing the swap is to use a dynamic self-financing trading strategy in the 3-period zero-coupon bond and the money market account employing the delta approach of Chapter 8. This construction is now detailed.

At time 2, state *uu* the value of the swap and its cash flow are known for sure.

The swap can be synthetically created by holding none of the 3-period zero-coupon bond,

$$n_3(2;uu) = 0,$$

and

$$n_0(2;uu) = \left[S(2;uu) - n_3(2;uu)P(2,3;uu)\right]/B(2;u)$$

$$= [.390667]/1.037958 = .376381$$

units of the money market account.

The calculations for the remaining states are similar:

$$n_3(2;ud) = 0$$

$$n_0(2;ud) = -.038500/1.037958 = -.037092$$

$$n_3(2;du) = 0$$

$$n_0(2;du) = .079199/1.042854 = .075945$$

$$n_3(2;dd) = 0$$

$$n_0(2;dd) = -.433028/1.042854 = -.415234.$$

At time 1, state u the number of 3-period zero-coupon bonds held is

$$n_3(1;u) = \frac{(S(2;uu) + \text{cash flow}(2;uu)) - (S(2;ud) + \text{cash flow}(2;ud))}{P(2,3;uu) - P(2,3;ud)}$$

$$= \frac{.630109 - .200942}{.984222 - .980015} = 102.$$

The number of units of the money market account held is

$$n_0(1;u) = [S(1;u) - n_3(1;u)P(1,3;u)]/B(1)$$

$$= [.408337 - 102(.965127)]/1.02 = -96.112355.$$

At time 1, state d the calculations are

$$n_3(1;d) = \frac{(S(2;du) + \text{cash flow}(2;du)) - (S(2;dd) + \text{cash flow}(2;dd))}{P(2,3;du) - P(2,3;dd)}$$

$$= \frac{-.161373 - (-.6736)}{.981169 - .976149} = 102$$

and

$$n_0(1;d) = [S(1;d) - n_3(1;d)P(1,3;d)]/B(1)$$

$$= [-.408337 - 102(.957211)]/1.02 = -96.121401.$$

At time 0,

$$n_3(0) = \frac{(S(1;u) + \text{cash flow}(1;u)) - (S(1;d) + \text{cash flow}(1;d))}{P(1,3;u) - P(1,3;d)}$$

$$= \frac{.408337 - (-.408337)}{.965127 - .957211} = 103.165648$$

$$n_0(0) = [S(0) - n_3(0)P(0,3)]$$

$$= 0 - 103.165648(.942322) = -97.215294.$$

The time 0 synthetic swap consists of a long position of 103.165 3-period zero-coupon bonds and short −97.215 units of the money market account. This makes sense since this swap is long fixed borrowing and short floating.

Rather than using the 3-period zero-coupon bond, the swap could be synthetically constructed using any other interest rate sensitive security, e.g., a futures or option contract on the 3-period zero-coupon bond. These alternate constructions are left to the reader as an exercise. This completes the example. □

14.3 INTEREST RATE CAPS

This section values interest rate caps. A simple *interest rate cap* is a provision often attached to a floating-rate loan that limits the interest paid per period to a maximum amount, $k - 1$, where k is 1 plus a percentage. For example, if $r(t) > k$ occurs, the holder of the floating-rate loan with an interest rate cap pays only k dollars (not $r(t)$). Obviously, an interest rate cap is to the advantage of the borrower and to the disadvantage of the lender. So, if the floating-rate loan in Table 14.1 has zero value, a floating-rate loan with an interest rate cap attached may have positive value. That is, the holder of a floating-rate loan with an interest rate cap may be willing to pay a positive amount to enter into the contract, because the interest rate cap potentially saves on interest payments in future periods. The purpose of this section is to value simple interest rate caps.

Interest rate caps trade separately from floating-rate loans. Consider an interest rate cap with cap rate k and maturity date τ^*. We can decompose this cap into the sum of τ^* caplets.

A *caplet* is defined to be an interest rate cap specific to only a single time period. Specifically, it is equivalent to a European call option on the spot interest rate with strike k and maturity the specific date of the single time period. For example, a caplet with maturity T and a strike k has a time T cash flow equal to:

$$\max(r(T-1)-k,0).$$

This cash flow is known at time $T-1$ because the spot rate is known at time $T-1$. This cash flow can be interpreted as the interest rate difference $(r(T-1)-k)$ earned on a principal of 1 dollar over the time period $[T-1, T]$, but only if this difference is positive.

The arbitrage-free value of the T-maturity caplet at time t, denoted by $c(t,T)$, is obtained using the risk-neutral valuation procedure:

$$c(t,T) = \tilde{E}_t \left(\max\left(r(T-1)-k,0\right)/B(T)\right)B(t). \tag{14.5}$$

An interest rate cap is then the sum of the values of the caplets from which it is composed. Let $I(t, \tau^*)$ denote the value of an interest rate cap at time t with strike k and expiration date τ^*, then

$$I(t,\tau^*) = \sum_{j=1}^{\tau^*} c(t,t+j). \tag{14.6}$$

The interest rate cap can be created synthetically by using the procedure described in Chapter 10.

Example: Cap Valuation

To illustrate this computation, we utilize the zero-coupon bond price curve evolution contained in Figure 14.1. We know that this evolution is arbitrage-free so that the techniques of Chapter 10 can be applied.

Consider an interest rate cap with maturity date $\tau^*=3$ and a strike of $k=1.02$. This interest rate cap can be decomposed into three caplets: one at time 1, one at time 2, and one at time 3. We value and discuss the synthetic construction of each caplet in turn.

From Figure 14.1 we see that the caplet at time 1, $c(0,1)$, has zero value. Formally,

$$c(0,1) = \tilde{E}_0\left(\max\left(r(0) - 1.02, 0\right)\right)/r(0)$$

$$= \left[(1/2)(0) + (1/2)(0)\right]/1.02 = 0.$$

Hence, this caplet can be excluded from further consideration.

Next, consider the caplet with maturity at time 2. By expression (14.5), at time 2 its value under each state is as follows:

$$c(2,2;uu) = \max(r(1;u) - 1.02, 0) = \max(1.017606 - 1.02, 0) = 0$$

$$c(2,2;ud) = \max(r(1;u) - 1.02, 0) = 0$$

$$c(2,2;du) = \max(r(1;d) - 1.02, 0) = \max(1.022406 - 1.02, 0) = .002406$$

$$c(2,2;dd) = \max(r(1;d) - 1.02, 0) = .002406$$

These numbers can be found in Figure 14.4. Continuing backward through the tree,

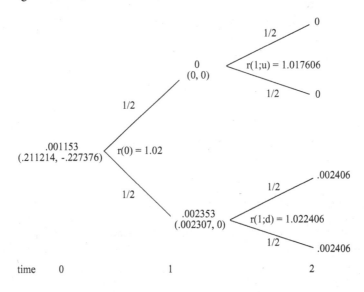

FIGURE 14.4 An Example of a 2-Period Caplet with a 1.02 Strike. The Synthetic Caplet Portfolio in the Money Market Account and 3-Period Zero-Coupon Bond $(n_0(t;s_t), n_3(t;s_t))$ Is Given under Each Node.

$$c(1,2;u) = \left[(1/2)0 + (1/2)0\right]/r(1;u) = 0$$

$$c(1,2;d) = \left[(1/2).002406 + (1/2).002406\right]/1.022406 = .002353.$$

The denominator in the calculation of $c(1,2;d)$ is $r(1;d) = 1.022406$.
Finally, at time 0, the caplet's value is

$$c(0,2) = \left[(1/2)0 + (1/2).002353\right]/1.02$$

$$= .001153.$$

These numbers appear in Figure 14.4.

We can synthetically create this 2-period caplet with the money market account and a 3-period zero-coupon bond.

At time 1, state u no position is required. At time 1, state d the number of 3-period zero-coupon bonds is

$$n_3(1;d) = \frac{c(2,2;du) - c(2,2;dd)}{P(2,3;du) - P(2,3;dd)} = \frac{.002406 - .002406}{.981169 - .976147} = 0.$$

The number of units of the money market account held is:

$$n_0(1;d) = \left[c(1,2;d) - n_3(1;d)P(1,3;d)\right]/B(1)$$

$$= \left[.002353 - 0(.957211)\right]/1.02$$

$$= .002307.$$

At time 0, the calculations are

$$n_3(0) - \frac{c(1,2;u) - c(1,2;d)}{P(1,3;u) - P(1,3;d)} = \frac{0 - .002353}{.965127 - .957211} = -.227376,$$

and

$$n_0(0) = \left[c(0,2) - n_3(0)P(0,3)\right]$$

$$= .001153 + .227376(.942322)$$

$$= .211214.$$

The time 0 synthetic 2-period caplet consists of .211214 units of the money market account and *short* .227376 units of the 3-period zero-coupon bond. This makes sense since the 3-period zero-coupon bond declines as interest rates rise. This inverse price movement necessitates a short position in the 3-period zero-coupon bond to duplicate the caplet.

The calculations for the value of the 3-period caplet and the synthetic 3-period caplet positions in the money market account and the 4-period zero-coupon bond are presented in Figure 14.5. These calculations are similar to those described above, and they are left to the reader to verify as an exercise.

The interest rate cap's value is the sum of the three separate caplets' values, i.e.,

$$I(0,3) = c(0,1) + c(0,2) + c(0,3)$$

$$= 0 + .001153 + .001131 = .002284 \text{ dollars.}$$

The synthetic interest rate cap can be obtained as the sum of the three separate synthetic caplets. These positions are easily read off Figures 14.4 and 14.5, and they are left to the reader to aggregate. The position involves the money market account, the 3-period bond, and the 4-period bond. This completes the example. □

14.4 INTEREST RATE FLOORS

This section values interest rate floors. An *interest rate floor* is a provision often associated with a floating-rate loan that guarantees that a minimum interest payment of $k - 1$ is made, where k is 1 plus a percentage. Unlike an interest rate cap, this provision benefits the lender. It therefore reduces the value of a floating-rate loan below zero. That is, the holder of a floating-rate loan plus a floor must be paid some positive amount to enter into the loan. The purpose of this section is to value interest rate floors.

Interest rate floors trade separately. Consider an interest rate floor with floor rate k and maturity date τ^*. This interest rate floor can be decomposed into the sum of τ^* floorlets.

A *floorlet* is an interest rate floor specific to only a single time period. Equivalently, the floorlet is a European put on the spot interest rate with strike price k and maturity the date of the single time period. For example, a floorlet with maturity T and strike k has a time T cash flow of

$$\max(k - r(T-1), 0).$$

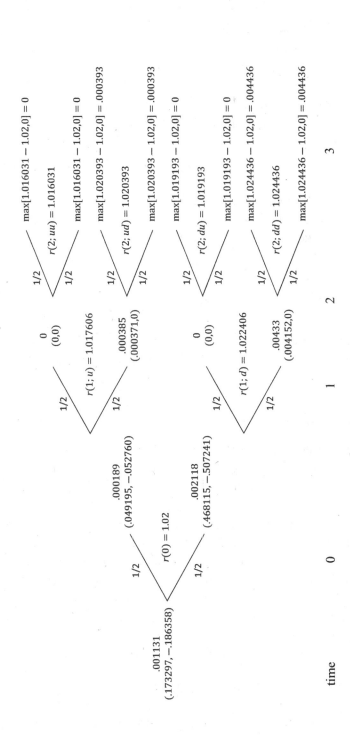

FIGURE 14.5 An Example of a 3-Period Caplet with a 1.02 Strike. The Synthetic Caplet Portfolio in the Money Market Account and 4-Period Zero-Coupon Bond $(n_0(t;s_t),\, n_4(t;s_t))$ Is Given under Each Node.

This cash flow is known at time $T-1$ because the spot rate is known at time $T-1$. This cash flow can be interpreted as the interest rate difference $(k-r(T-1))$ earned on a principal of 1 dollar over the time period $[T-1, T]$, but only if this difference is positive.

The arbitrage-free value of the T-maturity floorlet at time t, denoted by $d(t,T)$, is obtained using the risk-neutral valuation procedure:

$$d(t,T) = \tilde{E}_t \left(\max\left(k - r(T-1), 0\right)/B(T)\right) B(t). \tag{14.7}$$

An interest rate floor equals the sum of the values of the τ^* floorlets of which it is composed. Let $J(t, \tau^*)$ denote the value of an interest rate floor at time t with strike k and expiration date τ^*, then

$$J(t, \tau^*) = \sum_{j=1}^{\tau^*} d(t, t+j). \tag{14.8}$$

Using the procedure described in Chapter 10 we can synthetically construct the interest rate floor. We illustrate these computations with an example.

Example: Floor Valuation

To illustrate this computation, we utilize the zero-coupon bond price curve evolution given in Figure 14.2. As before, we know that this evolution is arbitrage-free due to the existence and properties of the pseudo probabilities.

Consider an interest rate floor with maturity date $\tau^* = 3$ and strike $k = 1.0175$. This interest rate floor can be decomposed into three floorlets: one at time 1, one at time 2, and one at time 3. We value and discuss the synthetic construction of each floorlet in turn.

From Figure 14.2 we see that the floorlet at time 1, $d(0,1)$, has zero value. Formally,

$$d(0,1) = \tilde{E}_0 \left(\max\left(1.0175 - r(0), 0\right)\right)/r(0)$$

$$= \left[1/2(0) + (1/2)0\right]/1.02 = 0.$$

Hence, this floorlet can be excluded from further consideration.

Next, consider the floorlet with maturity at time 2. By expression (14.7) its value at time 2 is zero under all states, i.e.,

$$d(2,2;uu) = \max(1.0175 - r(1;u),0) = \max(1.0175 - 1.017606,0) = 0$$

$$d(2,2;ud) = \max(1.0175 - r(1;u),0) = 0$$

$$d(2,2;du) = \max(1.0175 - r(1;d),0) = \max(1.0175 - 1.022406,0) = 0$$

$$d(2,2;dd) = \max(1.0175 - r(1;d),0) = 0.$$

Hence, at time 1 and time 0 its value is also zero. Consequently, this floorlet can also be excluded from future consideration.

The calculations for the remaining 3-period floorlet are contained in Figure 14.6. The time 3 payoffs to the floorlet, using expression (14.7), are

$$d(3,3;uuu) = \max(1.0175 - r(1;uu),0) = \max(1.0175 - 1.016031,0) = .001469$$

$$d(3,3;uud) = \max(1.0175 - r(1;uu),0) = .001469$$

$$d(3,3;udu) = \max(1.0175 - r(1;ud),0) = \max(1.0175 - 1.020393,0) = 0$$

$$d(3,3;udd) = \max(1.0175 - r(1;ud),0) = 0$$

$$d(3,3;duu) = \max(1.0175 - r(1;du),0) = \max(1.0175 - 1.019193,0) = 0$$

$$d(3,3;dud) = \max(1.0175 - r(1;du),0) = 0$$

$$d(3,3;ddu) = \max(1.0175 - r(1;dd),0) = \max(1.0175 - 1.024436,0) = 0$$

$$d(3,3;ddd) = \max(1.0175 - r(1;dd),0) = 0$$

The floorlet has a positive value only at time 2, state uu. Its value is

$$d(2,3;uu) = [(1/2).001469 + (1/2).001469]/1.016031 = .001446.$$

Continuing backward in the tree, the floorlet has a positive value only at time 1, state u:

$$d(1,3;u) = [(1/2).001446 + (1/2)0]/1.017606 = .000711.$$

Finally, its time 0 value is

$$d(0,3) = [(1/2).000711 + (1/2)0]/1.02 = .000348.$$

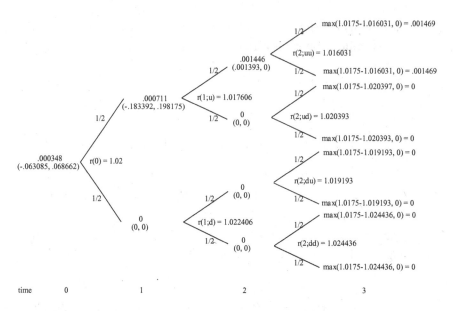

FIGURE 14.6 An Example of a 3-Period Floorlet with a 1.0175 Strike. The Synthetic Floorlet Portfolio in the Money Market Account and 4-Period Zero-Coupon Bond $(n_0(t;s_t), n_4(t;s_t))$ Is Given under Each Node.

To synthetically construct the floorlet, we use the 4-period zero-coupon bond and the money market account. The calculations are as follows:

At time 2, state uu,

$$n_4(2;uu) = \frac{d(3,3;uuu) - d(3,3;uud)}{P(3,4;uuu) - P(3,4;uud)} = \frac{.001469 - .001469}{.985301 - .981381} = 0$$

$$n_0(2;uu) = \left[d(2,3;uu) - n_4(2;uu)P(2,4;uu)\right]/B(2;uu)$$

$$= \left[.001446 - 0(.984222)\right]/1.037958 = .001393.$$

At time 1, state u,

$$n_4(1;u) = \frac{d(2,3;uu) - d(2,3;ud)}{P(2,4;uu) - P(2,4;ud)} = \frac{.001446 - 0}{.967826 - .960529} = .198175$$

$$n_0(1;u) = \left[d(1,3;u) - n_4(1;u)P(1,4;u)\right]/B(1;u)$$

$$= \left[.000711 - (.198175).947497\right]/1.02$$

$$= -.183392.$$

Finally, at time 0,

$$n_4(0) = \frac{d(1,3;u) - d(1,3;d)}{P(1,4;u) - P(1,4;d)} = \frac{.000711 - 0}{.947497 - .937148} = .068662$$

$$n_0(0) = d(0,3) - n_4(0)P(0,4)$$

$$= .000348 - .068662(.923845) = -.063085.$$

To duplicate the 3-period floorlet at time 0, one must go long .068662 units of the 4-period zero-coupon bond and go short .063085 units of the money market account.

The interest rate floor's value is the sum of the three separate floorlets' values, i.e.,

$$J(0,3) = d(0,1) + d(0,2) + d(0,3)$$

$$= 0 + 0 + .000348 = .000348 \text{ dollars.}$$

The synthetic interest rate floor can be obtained by summing the positions in the three synthetic floorlets. This completes the example. ☐

We can now revisit the situation discussed in the introduction to this chapter. Given a floating-rate loan with maturity T, one can synthetically construct a fixed-rate loan via buying a cap with strike k and maturity T, and simultaneously selling a floor with strike k and maturity T. The combined position is equivalent to a fixed-rate loan with coupon rate k.

14.5 SWAPTIONS

This section values swaptions, which are options issued on interest rate swaps. An interest rate swap changes floating- to fixed-rate loans or vice versa. Swaptions, then, are "insurance contracts" issued on the decision to switch into a fixed-rate or floating-rate loan at a future date.

To understand this statement, consider the decision concerning which type of loan to take. This decision depends upon one's view of future interest rate movements and projected cash flows. For example, if short-term rates are low relative to fixed rates, and they are expected to remain low, a floating-rate loan looks good. So, suppose one enters into a floating-rate loan. Now, also suppose that this decision goes bad; i.e., rates move up unexpectedly. This situation can cause a severe cash flow crisis as the interest payments rise.

Now, consider how a swaption purchased at the time the loan was issued affects this situation. It would be the perfect "insurance contract." Indeed, it would allow the floating-rate loan holder to switch back to a fixed-rate loan and, most importantly, *at a fixed rate set at the time the loan was made!* The cash flow crisis is avoided. Of course, as with all insurance contracts, the swaption has a cost. Determining the fair price of the swaption is the purpose of this section.

Consider the swap receiving fixed and paying floating discussed earlier in this chapter. This swap has a swap rate C/L, a maturity date T, and a principal equal to L dollars. Its time t, state s_t value is denoted by $S(t)$ and is given in expression (14.4).

This simplest type of swaption is a European call option on this swap. A *European call option* on the swap $S(t)$ with an expiration date $T^* \le T$ and a strike price of K dollars is defined by its payoff at time T^*, which is equal to max $[S(T^*) - K, 0]$.

The arbitrage-free value of the swaption, denoted by $O(t)$, is obtained using the risk-neutral valuation procedure, i.e.,

$$O(t) = \tilde{E}_t \left(\max[S(T^*) - K, 0] / B(T^*) \right) B(t). \tag{14.9}$$

Using the procedure described in Chapter 10 we can synthetically create the European call option.

Before that, however, a simple manipulation of expression (14.9) generates an important insight. Recall that a swap can be viewed as a long position in a coupon-bearing bond and a short position in the money market account. Substitution of expression (14.4), that is based on this relation, into expression (14.9) generates the following expression:

$$O(t) = \tilde{E}_t \left(\max[\mathcal{B}(T^*) - (L + K), 0] / B(T^*) \right) B(t) \tag{14.10}$$

Expression (14.10) shows that:

> *a European call option with strike K and expiration T* on a swap receiving fixed and paying floating with maturity T, principal L, and swap rate C/L is equivalent to a European call option with a strike L + K and an expiration date of T* on a (noncallable) coupon bond $\mathcal{B}(t;s_t)$ with maturity T, coupon C, and principal L.*

The pricing and synthetic construction of these bond options was discussed in Chapter 12. Thus, we have already studied the pricing and synthetic construction of swaptions. There is nothing new to add! Nonetheless, for completeness, we give an example based on expression (14.9).

Example: European Call Option On A Swap

We illustrate the calculation of the swaption value in expression (14.9) using the swap example given in Figure 14.3. Recall that the swap in this example is receiving fixed and paying floating. It has a swap rate $C/L=.02$, a maturity date $T=3$, and a principal $L=100$. The evolution of the zero-coupon bond price curve is as given in Figure 14.2.

Consider a European call option on this swap. Let the maturity date of the option be $T^*=1$, and let the strike price be $K=0$.

Using the risk-neutral valuation procedure, the value of the swaption is as follows:

Time 1, state u:

$$O(1;u) = \max\left[S(1;u)-K,0\right] = \max\left[.408337,0\right] = .408337$$

Time 1, state d:

$$O(1;d) = \max\left[S(1;d)-K,0\right] = \max\left[-.408337,0\right] = 0$$

Time 0:

$$O(0) = [(1/2)O(1;u)+(1/2)O(1;d)]/r(0)$$

$$= [(1/2)(.408337)+(1/2)0]/1.02 = .200165.$$

We can synthetically create this swaption with the money market account and a 3-period zero-coupon bond. At time 0 the calculations are as follows:

$$n_3(0) = \frac{O(1;u)-O(1;d)}{P(1,3;u)-P(1,3;d)} = \frac{.408337-0}{.965127-.957211} = 51.5838$$

$$n_0(0) = [O(0)-n_3(0)P(0,3)]/B(0)$$

$$= .200165-(51.5838).942322 = -48.4083.$$

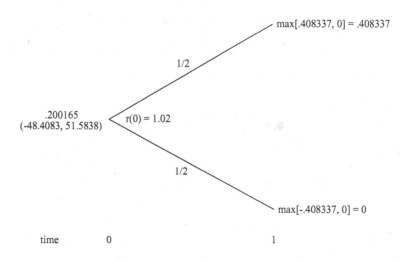

FIGURE 14.7 An Example of a European Call Option with Strike 0 and Expiration Date 1 on the Swap in Figure 14.3. The Synthetic Swaption in the Money Market Account and 3-Period Zero-Coupon Bond $(n_0(t;s_t), n_3(t;s_t))$ Is Given under Each Node.

The synthetic option consists of 51.5838 units of the 3-period zero-coupon bond, and it is short 48.4083 units of the money market account. The swaption's value is .200165. These numbers are presented in Figure 14.7. This completes the example. □

More complex or "exotic" swaptions can be valued using the techniques of Chapter 10. For example, *American* swaptions (calls and puts) can be valued and hedged using the stochastic dynamic programming approach illustrated previously in Chapter 12.

REFERENCES

1. Jarrow, R., and A. Chatterjea, 2019. *An Introduction to Derivative Securities, Financial Markets, and Risk Management,* 2nd edition, World Scientific Press.
2. Jarrow, R., and S. Turnbull, 1995. "Pricing Derivatives on Financial Securities Subject to Credit Risk." *Journal of Finance* 50 (1), 53–85.

CHAPTER 15

Interest Rate Exotics

T HE PURPOSE OF THIS chapter is to show the "power" of the HJM model by pricing interest rate *exotic options*. These interest rate options are called exotic because they are more complex than ordinary call and put options. We study digital options, range notes, and index-amortizing swaps. Even though they are more complex, interest rate exotics are popular securities. Digital options are used because they provide similar "insurance" to caps and floors, but they are cheaper. Range notes provide a partial hedge for floating rate loans with caps and floors attached. Finally, index-amortizing swaps provide a partial hedge for the prepayment option embedded in mortgage-backed securities.

15.1 SIMPLE INTEREST RATES

To study exotic options, we must first define the notion of a simple interest rate. A *simple interest rate of maturity* $T - t$, denoted by $R(t,T)$, is defined in terms of a zero-coupon bond price $P(t,T)$ as

$$R(t,T) \equiv \left(\frac{1}{P(t,T)} - 1 \right) \bigg/ T - t. \qquad (15.1)$$

Alternatively,

$$P(t,T) = \frac{1}{1 + R(t,T)(T - t)}. \qquad (15.2)$$

Unlike the other rates in this book, $R(t,T)$ represents a *percentage*; i.e., it is a number greater than -1. Recall that all other rates in this book are defined to be one plus a percentage (a number greater than zero).

Simple interest rates measure the holding-period return on a zero-coupon bond. In this sense they are similar to the yield on a zero-coupon bond, except that yields include compounding, while simple interest rates do not.*

Example: Simple Interest Rates

Consider the zero-coupon bond prices as given in Figure 15.1. Using these zero-coupon bond prices at time 0, the corresponding simple interest rates are

$$R(0,1) = \frac{1}{P(0,1)} - 1 = \frac{1}{.980392} - 1 = .020000$$

$$R(0,2) = \left(\frac{1}{P(0,2)} - 1\right)\Big/2 = \left(\frac{1}{.961169} - 1\right)\Big/2 = .020200$$

$$R(0,3) = \left(\frac{1}{P(0,3)} - 1\right)\Big/3 = \left(\frac{1}{.942322} - 1\right)\Big/3 = .020403$$

$$R(0,4) = \left(\frac{1}{P(0,4)} - 1\right)\Big/4 = \left(\frac{1}{.923845} - 1\right)\Big/4 = .020608$$

Note that the simple interest rate of maturity one is $R(0,1) = r(0) - 1 = .02$. This completes the example. □

By their definitions, we see that the spot rate at time t equals one plus the simple interest rate with unit maturity; i.e.,

$$r(t) = 1/P(t,t+1) = 1 + R(t,t+1). \tag{15.3}$$

Digital options and range notes are typically written on simple interest rates.

---

* Recall that the yield $y(t,T)$ is defined in Chapter 4 as $P(t,T) = 1/[y(t, T)]^{(T-t)}$. This expression explicitly incorporates compounding of interest (interest earned on interest).

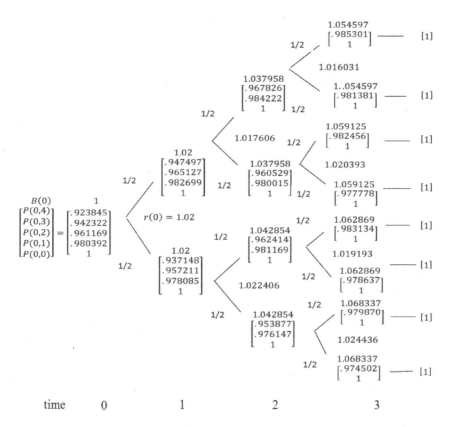

time   0   1   2   3

FIGURE 15.1 An Example of a One-Factor Bond Price Curve Evolution. The Money Market Account Values and Spot Rates Are Included on the Tree. Pseudo Probabilities Are Along Each Branch of the Tree.

15.2 DIGITAL OPTIONS

Digital options are used in the market because they provide similar "insurance" to call and put options, but they are much cheaper to purchase. Digitals are cheaper because their payoffs are constant, and they do not increase linearly with the underlying's value. Call and put option's payoffs do. We study the simplest digital option, a European call digital option.

A *European call digital option* with expiration date T and strike price k on the simple interest rate with time to maturity T^* is defined by its payoff at expiration. The digital's time T payoff is defined by

$$D(T) = \begin{cases} 1 & \text{if } R(T, T + T^*) > k \\ 0 & \text{if } R(T, T + T^*) \le k \end{cases} \tag{15.4}$$

The digital call option pays a dollar if the simple interest rate with time to maturity T^* exceeds the strike rate k. Otherwise, it pays nothing. It is called a digital because the payoff is a unit (digit).

Denote the digital's time t value as $D(t)$. Using the risk-neutral valuation procedure of Chapter 10, the value of the digital option at time t is

$$D(t) = \tilde{E}_t\left(\frac{D(T)}{B(T)}\right)B(t). \tag{15.5}$$

It is interesting to rewrite expression (15.4) using expression (15.1). Note that

$$R(T,T+T^*) > k \Leftrightarrow 1/P(T,T+T^*) > 1+kT^* \Leftrightarrow P(T,T+T^*) < 1/(1+kT^*).$$

This allows us to rewrite expression (15.4) as

$$D(T) = \begin{cases} 1 \text{ if } P(T,T+T^*) < 1/(1+kT^*) \\ 0 \text{ if } P(T,T+T^*) \geq 1/(1+kT^*) \end{cases} \tag{15.6}$$

Thus, a European digital *call* option on *the simple interest rate* is equivalent to a European digital *put* option on *the zero-coupon bond* with maturity date $(T+T^*)$. Expression (15.6) is often more useful for analysis because it depends only on the evolution of the term structure of zero-coupon bond prices (and not the term structure of simple interest rates).

A synthetic digital option can be valued and synthetically constructed by using the techniques of Chapter 10.

Example: Digital Call Valuation

This example is based on the zero-coupon bond price curve evolution as given in Figure 15.1. This evolution is arbitrage-free because it is the same evolution as used in Figure 10.1 in Chapter 10.

Consider a European call digital option with expiration date $T=2$ and strike price $k=.02$ on the simple interest rate with time to maturity $T^*=2$.

To value this option, we use expression (15.6). The calculations are as follows.

The modified strike is

$$1/(1+kT^*) = 1/(1+.02(2)) = .961538.$$

The payoffs are
Time 2, state uu:

$$D(2;uu) = 0 \quad \text{because} \quad P(2,4;uu) = .967826 > .961538$$

Time 2, state ud:

$$D(2;ud) = 1 \quad \text{because} \quad P(2,4;ud) = .960529 < .961538$$

Time 2, state du:

$$D(2;du) = 0 \quad \text{because} \quad P(2,4;du) = .962414 > .961538$$

Time 2, state dd:

$$D(2;dd) = 1 \quad \text{because} \quad P(2,4;dd) = .953877 < .961538$$

Time 1, state u:

$$D(1;u) = \frac{(1/2)0 + (1/2)1}{1.017606} = .49135$$

The synthetic digital in the 4-period zero-coupon bond and the money market account is

$$n_4(1;u) = \frac{D(2;uu) - D(2;ud)}{P(2,4;uu) - P(2,4;ud)} = \frac{0-1}{.967826 - .960529} = -137.0426$$

$$n_0(1;u) = [D(1;u) - n_4(1;u)P(1,4;u)]/B(1)$$

$$= [.49135 + (137.0426).947497]/1.02 = 127.7831.$$

Time 1, state d:

$$D(1;d) = \frac{(1/2)0 + (1/2)1}{1.022406} = .48904.$$

The synthetic digital in the 4-period zero-coupon bond and the money market account is

$$n_4(1;d) = \frac{D(2;du) - D(2;dd)}{P(2,4;du) - P(2,4;dd)} = \frac{0-1}{.962414 - .953877} = -117.1372$$

$$n_0(1;d) = [D(1;d) - n_4(1;d)P(1,4;d)]/B(1)$$

$$= [.48904 + (117.1372).937148]/1.02 = 108.1019.$$

Time 0:

$$D(0) = \frac{(1/2).49135 + (1/2).48904}{1.02} = .48058.$$

The synthetic digital in the 4-period zero-coupon bond and the money market account is

$$n_4(0) = \frac{D(1;u) - D(1;d)}{P(1,4;u) - P(1,4;d)} = \frac{.49135 - .48904}{.947497 - .937148} = .22321$$

$$n_0(0) = [D(0) - n_4(0)P(0,4)]/B(0)$$

$$= [.48058 + (.22321).923845]/1.02 = .27437.$$

These numbers are contained in Figure 15.2. The synthetic digital construction shows the difficulty in replicating the option (considering transaction costs). As the expiration date approaches, the delta in the 4-period zero becomes quite large. It goes from +.22321 units at time 0 to either −137.0426 at time 1 in the up state or −117.1372 at time 1 in the down state. This completes the example. □

The above techniques can be extended to European digital *put* options. *American* digital options can be valued using the stochastic dynamic programming technique discussed in Chapter 12.

15.3 RANGE NOTES

Range notes provide a partial hedge against floating rate loans with caps and floors attached. This section studies the pricing and hedging of range notes.

A *range note* is a financial security with a principal of L dollars and a maturity date T that pays the spot rate* of interest $r(t) - 1$ times L on any

---

* A generalization is to allow the note to pay based on a second simple interest rate of maturity τ^* instead of the spot rate of interest.

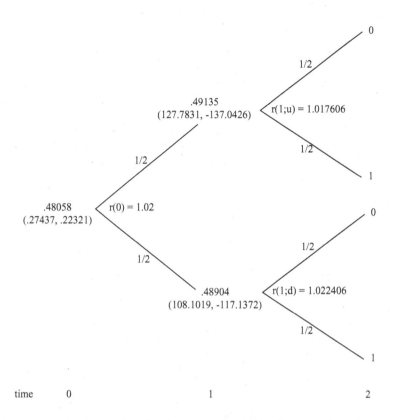

FIGURE 15.2   An Example of a European Digital Call Option's Values with Strike $k = .02$ and Expiration Date $T = 2$ on the Simple Interest Rate with Time to Maturity $T^* = 2$. The Synthetic Option Portfolio in the Money Market Account and the 4-Period Zero-Coupon Bond $(n_0(t;s_t), n_4(t;s_t))$ Is Given under Each Note.

date t over the life of the contract where the simple interest rate with maturity T^*, $R(t, t + T^*)$ lies within the range k_l to k_u. The range limits k_l and k_u are called the lower and upper bound, respectively. The cash flow is paid at time $t + 1$.

This description is clarified by introducing the notation for an *index function*:

$$1_{\{k_l < R(j, j+T^*) < k_u\}} = \begin{cases} 1 & \text{if } k_l < R(t, t+T^*) < k_u \\ 0 & \text{otherwise} \end{cases} \tag{15.7}$$

Using this new index function, we can write the cash flows to a range note from time $t+1$ until maturity as

$$\text{cashflows from } t+1 \text{ to } T = \sum_{j=t}^{T-1} [r(j)-1]L\,1_{\{k_l<R(j,j+T^*)<k_u\}}. \tag{15.8}$$

Expression (15.8) has the first cash flow at time $t+1$, $[r(t)-1]Ll_{\{k_l<R(t,t+T^*)<k_u\}}$. The last payment is made at time T, and it is determined by the simple interest rate at time $T-1$.

Let $N(t)$ denote the value of the range note at time t. Then using the risk-neutral valuation procedure gives

$$N(t) = \tilde{E}_t \left(\sum_{j=t}^{T-1} [r(j)-1]L\,1_{\{k_l<R(j,j+T^*)<k_u\}} \,/\, B(j+1) \right) B(t). \tag{15.9}$$

Given this description, we can now understand why a range note can be used as a partial hedge against a floating rate note with a cap and floor attached. If the floating rate note has principal L and pays based on the spot rate, the payments are $[r(t)-1]L$. If the cap has rate k_u and the floor has rate k_l then these payments only occur in the range (k_l, k_u). These payments are represented by expression (15.9). Outside of this range, either k_lL or k_uL is received. It is only these fixed payments outside of the range that are not included in expression (15.9).

A synthetic range note can be constructed by using the techniques of Chapter 10.

Example: Range Note Valuation

This example is based on the zero-coupon bond price curve evolution as given in Figure 15.1. This evolution is arbitrage-free as a quick verification of the pseudo probabilities reveals.

Consider a range note with the following provisions. Its maturity is $T=3$ with principal $L=100$. Let the lower bound be $k_l=.018$ and the upper bound be $k_u=.022$ on the simple interest rate with maturity $T^*=2$.

To value this range note, we first compute the evolution of the simple interest rate $R(t, t+2) = [1/P(t, t+2)-1]/2$. This evolution is given in Figure 15.3. We only need the evolution up to time 2 because

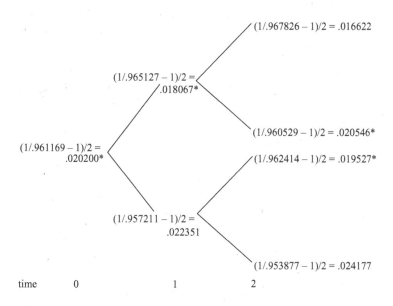

FIGURE 15.3 An Example of the Evolution of a Simple Interest Rate of Maturity 2. An Asterisk "*" Denotes That the Simple Interest Rate Lies between $k_1 = .018$ and $k_u = .022$.

the time 3 payment is based on the simple interest rate at time 2. A sample calculation, at time 2, state uu, is

$$R(2,4;uu) = \left[\frac{1}{P(2,4;uu)} - 1\right]\Big/2$$

$$= [1/.967826 - 1]/2 = .016622.$$

From this evolution, we can compute the cash flows to the range note at each time and state. Sample calculations for time 3 are

$$\text{cash flow}(3;uuu) = 100[r(2;uu) - 1]\,1_{\{.018 < R(2,4;uu) < .02\}} = 0$$

because $R(2,4;uu) = .016622$ is below the lower bound, and

$$\text{cash flow}(3;udu) = 100[r(2;ud) - 1]\,1_{\{.018 < R(2,4;ud) < .022\}} = 100(1.020393 - 1)$$

$$= 2.0393$$

because $R(2,4;ud) = .020546$ is between the lower and upper bounds.

The remaining calculations are left as exercises for the reader. The results are presented in Figure 15.4.

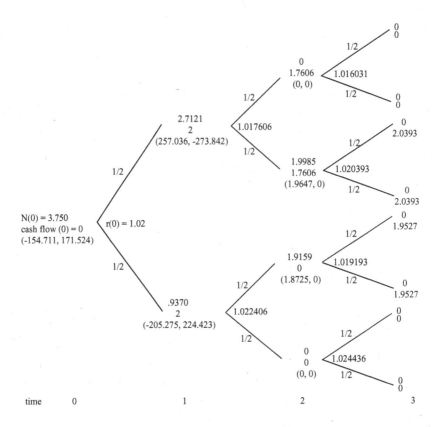

FIGURE 15.4 An Example of a Range Note with Maturity $T=3$, Principal $L=100$, Lower Bound $k_1=.018$, Upper Bound $k_u=.022$ on the Simple Interest Rate with Maturity $T^*=2$. At Each Node: The First Number Is the Value $(N(t;s_t))$, the Second Number is the Cash Flow (cash flow$(t;s_t)$). The Synthetic Range Note in the Money Market Account and the 4-Period Zero-Coupon Bond $(n_0(t;s_t), n_4(t;s_t))$ Is Given under Each Node.

Using the risk-neutral valuation procedure, we can compute the value of the range note at time t from these cash flows as follows:

$$N(t;s_t) = \big((1/2)[N(t+1;s_tu)+\text{cash flow}(t+1;s_tu)]$$
$$+(1/2)[N(t+1;s_td)+\text{cash flow}(t+1;s_td)]\big)\big/r(t;s_t) \tag{15.10}$$

where $N(3;s_3)\equiv 0$ for all s_3.

A sample calculation for time 1, state u is

$$N(1;u)=[1/2(0+1.7606)+1/2(1.9985+1.7606)]/1.017606=2.7121.$$

This occurs because $N(2;uu) = 0$, cash flow$(2;uu) = 1.7606$, $N(2;ud) = 1.9985$, cash flow$(2;ud) = 1.7606$, and $r(1;u) = 1.017606$.

The remaining calculations are left as exercises for the reader. The results are contained in Figure 15.4.

Finally, we can construct this range note synthetically using the money market account and the 4-period zero-coupon bond via the equations

$$n_4(t;s_t) = \frac{[N(t+1;s_t u) + \text{cash flow}(t+1;s_t u)] - [N(t+1;s_t d) + \text{cash flow}(t+1;s_t d)]}{[P(t+1,4;s_t u) - P(t+1,4;s_t d)]}$$

(15.11a)

and

$$n_0(t;s_t) = [N(t;s_t) - n_4(t;s_t)P(t,4;s_t)] / B(t;s_{t-1}).$$ (15.11b)

A sample calculation for time 1, state u is

$$n_4(1;u) = \frac{(0+1.7606) - (1.9985+1.7606)}{.967827 - .960529} = -273.842,$$

and

$$n_0(1;u) = [2.7121 + 273.842(.947497)] / 1.02 = 257.036.$$

This follows because

$N(2;uu) = 0,$
cash flow$(2;uu) = 1.7606,$
$N(2;ud) = 1.9985,$
cash flow$(2;ud) = 1.7606,$
$P(2,4;uu) = .967827,$
$P(2,4;ud) = .960529,$
$N(1;u) = 2.7121,$
$P(1,4;u) = .947497,$ and
$B(1) = 1.02.$

The remaining calculations are left as exercises for the reader. The results are given in Figure 15.4.

It is interesting to observe that the range note goes from being long the 4-period zero-coupon bond at time 0 (171.524 units) to short the 4-period zero-coupon bond at time 1, state u (−273.842 units). This whiplash effect makes hedging range notes very difficult (considering transaction costs). This completes the example. ☐

15.4 INDEX-AMORTIZING SWAPS

Index-amortizing swaps are interest rate swaps in which the principal declines (amortizes) when interest rates decline. Consequently, these instruments are useful as (partial) hedges against prepayment risk in mortgage-backed securities.* Unlike the "plain vanilla" swaps discussed in Chapter 14, these exotic swaps are difficult to value because their cash flows (in general) depend on the entire history of spot interest rates. These instruments are commonly thought of as one of the most complex interest rate derivatives.

Formally, an *index-amortizing swap* is a swap (say, receive fixed and pay floating) in which the principal is reduced by an amortizing schedule based on the spot rate of interest. The amortizing schedule does not apply until after a prespecified *lockout period* has passed.

In symbols, let T be the maturity of the index-amortizing swap with initial principal L_0 receiving fixed at rate c (one plus a percent) and paying floating at $r(t)$. Let the lockout period be T^* years. For $t \leq T^*$ the principal is fixed at L_0. For $t > T^*$ the following change in the principal occurs:

$$L(t) = L(t-1)[1 - a(t)] \text{ for } t > T^* \qquad (15.12)$$

where $L(T^*) = L_0$ and $a(t)$ is the amortizing amount that occurs at time t.

This expression states that the principal remaining at time t, $L(t)$, equals the principal at time $t-1$, $L(t-1)$, reduced by the time t amortizing schedule amount $a(t)$.

A typical amortizing schedule looks like the following:

---

* Mortgages are sometimes prepaid when the level of interest rates declines. This prepayment of a mortgage's principal is similar to the reduction in the principal on an index-amortizing swap. Mortgage-backed securities and prepayment risk is not discussed further in this textbook. Two useful references are McConnell and Singh [1] and Titman and Torous [2].

$$a(t) \equiv \begin{cases} 0 & \text{if } r(t) > k_0 \\ b_0 & \text{if } k_0 \geq r(t) > k_1 \\ b_1 & \text{if } k_1 \geq r(t) > k_2 \\ b_2 & \text{if } k_2 \geq r(t) > k_3 \\ b_3 & \text{if } k_3 \geq r(t) > k_4 \\ b_4 & \text{if } k_4 \geq r(t) > k_5 \\ 1 & \text{if } k_5 \geq r(t). \end{cases} \tag{15.13}$$

where $k_0 > k_1 > k_2 > k_3 > k_4 > k_5$ and $0 < b_0 < b_1 < b_2 < b_3 < b_4 < 1$ are positive constants determined at the initiation of the swap.

The amortizing schedule in expression (15.13) depends on the spot interest rate at time t. If the spot interest rate is larger than k_0, no reduction in principal occurs: $a(t) = 0$. If the spot interest rate lies between k_0 and k_1, a reduction of $a(t) = b_0$ percent of the principal occurs; if the spot interest rate lies between k_1 and k_2, a reduction of $a(t) = b_1$ percent of the principal occurs; and so forth down the schedule. Although we have only written seven different levels for the amortization, it is easy to augment this schedule to accommodate an arbitrary number of levels.

Example: Amortizing Schedule

An example of an amortizing schedule is

$$a(t) \equiv \begin{cases} 0 & \text{if } r(t) > 1.01 \\ .25 & \text{if } 1.01 \geq r(t) > 1.0075 \\ .5 & \text{if } 1.0075 \geq r(t) > 1.005 \\ .75 & \text{if } 1.005 \geq r(t) > 1.0025 \\ 1 & \text{if } 1.0025 > r(t). \end{cases}$$

This completes the example. □

The time t cash flow to the (receive fixed, pay floating) index-amortizing swap can be written as

$$\text{cash flow}(t) = [c - r(t-1)]L(t-1). \tag{15.14}$$

The cash flow at time t is determined at time $t-1$. The cash flow is the sum of receiving fixed $(c-1)L(t-1)$ and paying floating $[r(t-1) - 1]L(t-1)$.

The principal on which the 1-dollar interest payments are based is determined according to expressions (15.12) and (15.13).

Let $IA(t)$ denote the time t value of the index-amortizing swap. Using the risk-neutral valuation procedure of Chapter 10, we have that

$$IA(t) = \tilde{E}_t \left(\sum_{j=t+1}^{T} [c - r(j-1)]L(j-1)/B(j) \right) B(t). \qquad (15.15)$$

Expression (15.15) represents the present value of the cash flows to the index-amortizing swap from time $t+1$ through its maturity at time T. This value differs from the plain vanilla swap of expression (14.4) only by the replacement of a constant principal L with the amortizing principal $L(j-1)$. The complexity in valuation occurs because of this small difference. Note that the principal $L(j-1)$ and spot interest rate $r(j-1)$ are correlated. Furthermore, the principal depends (in general) on the history of the interest rate process prior to time $j-1$. This implies, for example, that the principal at time $T-1$ will also be correlated with the spot interest rate occurring earlier at time t. Consequently, valuation becomes complex. We illustrate this procedure through an example.

Example: Index-Amortizing Swap Valuation

This example is based on the zero-coupon bond price curve evolution in Figure 15.1. This figure has been proven to be arbitrage-free.

To illustrate the computations, we consider a very simple index-amortizing swap. Let the swap receive fixed at rate $c = 1.02$ and pay floating. Let the swap's maturity be $T = 3$ years, and let the initial principal be $L_0 = 100$. Let the lockout period be $T^* = 1$ year. Let the amortizing schedule be given by

$$a(t) = \begin{cases} 0 & \text{if } r(t) \geq 1.018 \\ .5 & \text{if } r(t) < 1.018. \end{cases}$$

That is, the swap amortizes 50 percent of its principal each date the spot interest rate lies below 1.018.

This example is identical to the plain vanilla swap example studied in Chapter 14, with the exception of the amortizing principal. Recall that the plain vanilla swap had zero value at time 0. We see below that the index-amortizing swap has a negative value at time 0. This is because the principal decreases on the index-amortizing swap exactly when the payment stream is potentially the largest (when fixed minus floating is largest). This is to the disadvantage of the holder of the index-amortizing swap (relative to a plain vanilla swap).

We first compute the cash flows to the index-amortizing swap. These are given in Figure 15.5.

A sample set of calculations are as follows:
At time 1, state u the cash flow is

$$[c-r(0)]L_0 = [1.02-1.02]100 = 0.$$

At time 2, state uu the cash flow is

$$[c-r(1;u)]L_1 = [1.02-1.017606]50 = .1197$$

because $r(1;u) = 1.017606 < 1.018$ gives $a(1;u) = 0.5$, and therefore

$$L_1 = L_0[1-a(1;u)] = 100[1-.5] = 50.$$

At time 3, state uuu the cash flow is

$$[c-r(2;uu)]L_2 = [1.02-1.016031]25 = .099225$$

because

$$L_2 = L_1[1-a(2;uu)] = 50[1-.5] = 25$$

as $a(2;u) = 0.5$ since $r(2;uu) = 1.016031 < 1.018$.

The remaining calculations are left as exercises for the reader. Note that the lower branches of this tree are identical to those in Chapter 14 for the plain vanilla swap.

To value the index-amortizing swap, we use the risk-neutral valuation procedure. The value at time t is computed as follows:

$$IA(t;s_t) = [(1/2)(IA(t+1;s_tu)+ \text{cash flow}(t+1;s_tu))$$
$$+ (1/2)(IA(t+1;s_td)+ \text{cash flow}(t+1;s_td))]/r(t;s_t) \quad (15.16)$$

where

$$IA(3;s_3) \equiv 0 \text{ for all } s_3.$$

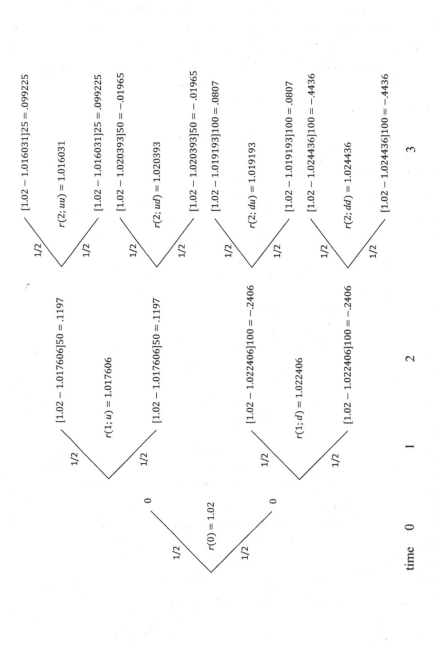

FIGURE 15.5   An Example of the Cash Flows from an Index-Amortizing Swap with Maturity $T = 3$, Initial Principal $L_0 = 100$, Lockout Period $T^* = 1$, Which Amortizes 50 Percent of the Principal if $r(t; s_t) < 1.018$.

For example, at time 2, state uu:

$$A(2;uu) = [(1/2)(0+.099225)+(1/2)(0+.099225)]/1.016031$$

$$= .097659,$$

and at time 1, state u:

$$IA(1;u) = [(1/2)(.0977+.1197)+(1/2)(-.0193+.1197)]/1.017606$$

$$= .1562$$

The remaining calculations are similar and are left as exercises for the reader. They are contained in Figure 15.6. Notice that the index-amortizing swap's value at time 0, $IA(0) = -.1236$. This is less than the plain vanilla swap's value at time 0 that is zero.

Finally, we compute the synthetic index-amortizing swap portfolio using the money market account and the 3-period bond via the equations

$$n_3(t;s_t) = \frac{[(IA(t+1;s_tu)+\text{cash flow}(t+1;s_tu)) - (IA(t+1;s_td)+\text{cash flow}(t+1;s_td))]}{P(t+1,3;s_tu)-P(t+1,3;s_td)}$$

and

$$n_0(t;s_t) = [IA(t;s_t)-n_3(t;s_t)P(t,3;s_t)]/B(t) \qquad (15.17)$$

for $t=0$ and 1.

For time 2, since the money market account and the 3-period bond are identical, we arbitrarily place all the investment in the money market account. Some sample calculations are as follows:
Time 2, state uu:

$$n_3(2;uu) = 0$$

$$n_0(2;uu) = IA(2;uu)/B(2;u) = .0977/1.037958 = .094088$$

Time 1, state u:

$$n_3(1;u) = \frac{[IA(2;uu)+\text{cash flow}(2;uu)]-[IA(2;ud)+\text{cash flow}(2;ud)]}{P(2,3;uu)-P(2,3;ud)}$$

$$= \frac{(.0977+.1197)-(-.0193+.1197)}{.984222-.980015} = 27.8108$$

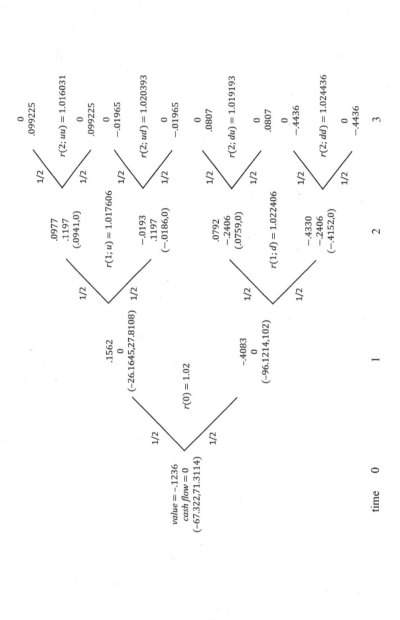

FIGURE 15.6 An Example of an Index-Amortizing Swap with Maturity $T = 3$, Initial Principal $L_0 = 100$, Lockout Period $T^* = 1$, Which Amortizes 50 Percent of the Principal if $r(t; s_t) < 1.018$. The First Number Is the Value, and the Second Is the Cash Flow. The Synthetic Index-Amortizing Swap Portfolio in the Money Market Account and 3-Period Zero-Coupon Bond $(n_0(t; s_t), n_3(t; s_t))$ Is Given under Each Node.

$$n_0(1;u) = [IA(1;u) - n_3(1;u)P(1,3;u)] / B(1)$$

$$= [.1562 - (27.8108).965127] / 1.02 = -26.1615.$$

The remaining calculations are left as exercises for the reader. All calculations are presented in Figure 15.6. This completes the example. □

The above techniques for valuing exotic interest rate options are easily modified to incorporate more complexities, such as early exercise considerations. The procedure requires a detailed specification of the interest rate option's cash flows given an evolution of the term structure of interest rates. Once this is understood, the procedures of Chapter 10 are easily applied.

REFERENCES

1. McConnell, J., and M. Singh, 1994. "Rational Prepayments and the Valuation of Collateralized Mortgage Obligations." *Journal of Finance* 49, 891–921.
2. Titman, S., and W. Torous, 1989. "Valuing Commercial Mortgages: An Empirical Investigation of the Contingent Claims Approach to Valuing Risky Debt." *Journal of Finance* 44, 345–373.

IV

Implementation/Estimation

CHAPTER 16

Continuous-Time Limits

THIS CHAPTER DISCUSSES THE computer implementation of the interest rate option models developed in the previous chapters. As a discrete-time model, its approximation to reality is quite good when the number of periods (τ) is large. In this case, the discrete-time model is approximating the continuous trading limit. In fact, for purposes of empirical estimation, it is convenient to reparameterize the discrete-time model in terms of its continuous-time limit. The actual implementation of computer code is then done under this reparameterization.

The primary purpose of this chapter is to study this reparameterization and the resulting continuous-time limit. A secondary purpose is to demonstrate how to construct arbitrage-free zero-coupon bond price evolutions such as those used in the examples of the previous chapters.

Some words of caution are needed. This chapter and the next contain material that is more abstract than that contained in the preceding chapters. This is because these chapters transform the material from easily understood examples to actual applications. "Applications" mean computer code and empirical estimation. These are necessarily more complex and abstract topics.

16.1 MOTIVATION

This section discusses the intuition behind the construction of the discrete time approximation to the continuous-time limit economy.

To parameterize the forward rate process,[*][†] in terms of its continuous limit, we need to change the time scale in the discrete-time model. As it is currently constructed, there are τ time periods $t = 0, 1, 2, ..., \tau$. These time periods are arbitrarily specified. In order to take limits, let us fix a future date $\overline{\tau}$ (say, January 1, 2050), and divide the time horizon 0 to $\overline{\tau}$ into sub-periods of equal length Δ. Thus, in terms of calendar time, the discrete periods $0, 1, 2, ..., \tau$ correspond to the dates $0, \Delta, 2\Delta, 3\Delta, ..., \tau\Delta \equiv \overline{\tau}$.

We are interested in studying the various discrete-time economies when the number of trading dates becomes large (i.e., $\tau \to \infty$) or, equivalently, when the time between trades becomes small (i.e., $\Delta \to 0$).

The discrete-time forward rate is denoted by $f_\Delta(t,T)$. Recall that this represents "one plus the percentage" forward rate at time t for the future time period $[T, T+\Delta]$. The continuously compounded (continuous-time) forward rate $\tilde{f}(t,T)$ corresponds to that rate such that

$$f_\Delta(t,T) \approx e^{\tilde{f}(t,T)\Delta}.$$

It is that rate, compounded continuously for Δ units of time, that equals the discrete rate. Note that this expression implies that $\tilde{f}(t,T)$ is a *percentage*, i.e., a number between -1 and $+1$.

More formally, (assuming the limit exists)

$$\tilde{f}(t,T) \equiv \lim_{\Delta \to 0} \log\left(f_\Delta(t,T)\right)/\Delta.$$

In constructing the continuous-time economy, we are concerned with changes in continuously compounded forward rates, i.e.,

$$\tilde{f}(t+\Delta,T) - \tilde{f}(t,T) \approx \frac{\log f_\Delta(t+\Delta,T)}{\Delta} - \frac{\log f_\Delta(t,T)}{\Delta}.$$

From the continuous-time perspective, the evolution of *observed* zero-coupon bond prices and forward rates is generated by a *continuous time empirical economy* with parameters[‡] *(i)* $\mu^*(t,T)$, the expected change in the

---

[*] For the bond trading models of Chapter 9, one could estimate the parameters of the bond process directly. This estimation is problematic, however, because the parameters of the bond process are (in general) nonstationary.

[†] This chapter's analysis follows Heath, Jarrow, and Morton [5].

[‡] In mathematical notation, to be defined later, $d\tilde{f}(t,T) = \mu^*(t,T)dt + \sigma(t,T)dW(t)$ where both $\sigma(t,T)$ and $dW(t)$ can be vector-valued processes for multiple-factor economies.

continuously compounded forward rates per unit time, and *(ii)* $\sigma(t,T)$, the standard deviation of changes in the continuously compounded forward rates per unit time. The standard deviation is sometimes called the forward rate's *volatility*. The word "empirical" is intended to distinguish the "actual" or "empirical" economy from the transformation used for valuation, called the "pseudo" economy.

We would like to construct an approximating discrete-time empirical economy such that as the step size shrinks ($\Delta \to 0$), the discrete-time economy approaches the continuous-time economy. This is illustrated at the top of Figure 16.1.

Recall that a discrete-time empirical economy is characterized by *(i)* the probability of movements of forward rates $q_t^\Delta(s_t)$ and *(ii)* the (one plus) percentage changes in forward rates across the various states ($\alpha_\Delta(t,T;s_t)$, $\beta_\Delta(t,T;s_t)$ for the one-factor case). It is a fact from probability theory (see He [4]) that under mild technical conditions, this approximation can be obtained by choosing

$$\frac{E_t\left\{\dfrac{\log f_\Delta(t+\Delta,T)}{\Delta} - \dfrac{\log f_\Delta(t,T)}{\Delta}\right\}}{\Delta} \to \mu^*(t,T) \quad \text{as } \Delta \to 0,$$

and

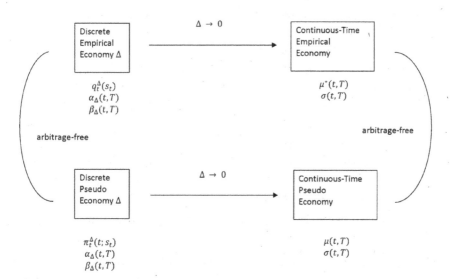

FIGURE 16.1 Discrete Approximation System to the Empirical- and Pseudo-Continuous-Time Economies.

$$\frac{\mathrm{Var}_t \left\{ \dfrac{\log f_\Delta (t+\Delta, T)}{\Delta} - \dfrac{\log f_\Delta (t, T)}{\Delta} \right\}}{\Delta} \to \sigma(t, T)^2 \quad \text{as } \Delta \to 0,$$

where the expectations and variance are obtained using the actual probabilities $q_t^\Delta(s_t)$. Under these conditions, for small Δ, the two economies will be similar, and the discrete-time empirical economy will be a good approximation to the continuous-time empirical economy (and conversely).

We now study the bottom part of Figure 16.1. Given the discrete-time empirical economy constructed above, as in Chapter 10, the assumption of no arbitrage gives the existence of *unique* pseudo probabilities $\pi_\Delta(t;s_t)$, which are used for the valuation of contingent claims. The pseudo economy is represented by the lower part of Figure 16.1.

The discrete-time pseudo economy is characterized by *(i)* the probability of movements of forward rates $\pi_\Delta(t;s_t)$ and *(ii)* the (one plus) percentage changes in forward rates across the various states ($\alpha_\Delta(t,T;s_t)$, $\beta_\Delta(t,T;s_t)$ for the one-factor case). The percentage changes in forward rates are *identical* across the two discrete-time economies; only the likelihoods of the movements differ. However, we need $\pi_\Delta(t;s_t) > 0$ if and only if $q_t^\Delta(s_t) > 0$. Those states with positive probability in the pseudo economy must have positive probability in the empirical economy, and conversely. This is called the *equivalent probability* condition. The arbitrage-free link between the two discrete-time economies (empirical and pseudo) is depicted on the left side of Figure 16.1.

Analogous to the discrete-time case, the assumption of no arbitrage in the continuous-time model gives the existence of unique pseudo probabilities, which are used for the valuation of contingent claims.* These no-arbitrage restrictions imply that the limit pseudo economy has parameters *(i)* $\mu(t,T)$, the expected change in the continuously compounded forward rates per unit time, and *(ii)* $\sigma(t,T)$, the standard deviation of changes in the continuously compounded forward rates per unit time. The standard deviations of changes in forward rates are identical across the two limit economies; only the likelihoods (and, therefore, the expected changes of forward rates) can differ.† This is the equivalent probability condition.

---

* Heath, Jarrow, and Morton [6] provide the technical details.
† It can be shown that $\mu'(t,T) = \mu(t,T) - \sigma(t,T)\phi(t)$ where $\phi(t)$ is a risk premium. This is discussed later in this chapter. The probability measures are equivalent (agree on zero probability events) because $\sigma(t,T)$ is identical across the two economies.

This arbitrage-free link between the two limit economies is depicted on the right side of Figure 16.1.

The construction is complete if the discrete-time pseudo economy also converges to the limit pseudo economy, because then for small Δ, contingent claim values as computed in the discrete-time model will be good approximations to contingent claim values as computed in the continuous-time model (and conversely). This is illustrated in the bottom part of Figure 16.1. This construction can be obtained if we choose

$$\frac{\tilde{E}_t\left\{\dfrac{\log f_\Delta(t+\Delta,T)}{\Delta} - \dfrac{\log f_\Delta(t,T)}{\Delta}\right\}}{\Delta} \to \mu(t,T) \quad \text{as } \Delta \to 0,$$

and

$$\frac{\tilde{\text{Var}}_t\left\{\dfrac{\log f_\Delta(t+\Delta,T)}{\Delta} - \dfrac{\log f_\Delta(t,T)}{\Delta}\right\}}{\Delta} \to \sigma(t,T)^2 \quad \text{as } \Delta \to 0,$$

where the expectation and variance are obtained using $\pi_\Delta(t;s_t)$.

There are numerous ways of constructing the discrete-time economies such that Figure 16.1 is satisfied. This follows because we require only that the limiting systems match. From among these constructions, we select the one that makes the computation of contingent claim values as simple as possible. This simplicity of computation occurs, for example, if the pseudo probabilities satisfy $\pi_\Delta(t,;s_t) = 1/2$ for all Δ, t, and s_t. This is the identification that has been used throughout this text. We show below that such a construction is always possible.

In summary, the purpose of this chapter is to construct a system of discrete-time and limit economies satisfying Figure 16.1 and such that the pseudo probabilities satisfy $\pi_\Delta(t,;s_t) = 1/2$ for all Δ, t, and s_t. This is the task to which we now turn.

16.2 ONE-FACTOR ECONOMY

Consider the one-factor economy of Chapter 5. The forward rate process can be characterized as:

$$f_\Delta\left(t+\Delta,T;s_{t+\Delta}\right)=\begin{cases}\alpha_\Delta\left(t,T;s_t\right)f_\Delta\left(t,T;s_t\right) & \text{if } s_{t+\Delta}=s_t u \\ \\ \quad\text{with probability } q_t^\Delta(s_t)>0 \\ \\ \beta_\Delta\left(t,T;s_t\right)f_\Delta\left(t,T;s_t\right) & \text{if } s_{t+\Delta}=s_t d \\ \\ \quad\text{with probability } 1-q_t^\Delta(s_t)\end{cases}\qquad(16.1)$$

where $\tau\Delta-\Delta\geq T\geq t+\Delta$, and both t and T are integer multiples of Δ.

The forward rates and the actual probabilities are indexed by Δ. If u occurs, forward rates change proportionately by $\alpha_\Delta(t,T;s_t)$, and if d occurs, forward rates change proportionately by $\beta_\Delta(t,T;s_t)$. The actual probabilities of the up and down movements are given by $q_t^\Delta(s_t)$ and $1-q_t^\Delta(s_t)$, respectively. Expression (16.1) is expression (5.5) of Chapter 5 rewritten to include the new time notation.

Let us now reparameterize expression (16.1) in terms of three new stochastic processes, $\mu(t,T;s_t)$, $\sigma(t,T;s_t)$, and $\phi(t;s_t)$, implicitly defined as follows:

$$\alpha_\Delta\left(t,T;s_t\right)\equiv e^{\left[\mu(t,T;s_t)\Delta-\sigma(t,T;s_t)\sqrt{\Delta}\right]\Delta}$$

$$\beta_\Delta\left(t,T;s_t\right)\equiv e^{\left[\mu(t,T;s_t)\Delta+\sigma(t,T;s_t)\sqrt{\Delta}\right]\Delta}\qquad(16.2)$$

$$q_t^\Delta\left(s_t\right)\equiv(1/2)+(1/2)\phi\left(t;s_t\right)\sqrt{\Delta}$$

It is important to note the powers of the Δ's in expression (16.2). This parameterization was selected with Figure 16.1 in mind. To explain, substitution of expression (16.1) into (16.2) gives

$$f_\Delta\left(t+\Delta,T;s_{t+\Delta}\right)=\begin{cases}f_\Delta\left(t,T;s_t\right)e^{\left[\mu(t,T;s_t)\Delta-\sigma(t,T;s_t)\sqrt{\Delta}\right]\Delta} & \text{if } s_{t+\Delta}=s_t u \\ \\ \quad\text{with probability } (1/2)+(1/2)\phi\left(t;s_t\right)\sqrt{\Delta} \\ \\ f_\Delta\left(t,T;s_t\right)e^{\left[\mu(t,T;s_t)\Delta+\sigma(t,T;s_t)\sqrt{\Delta}\right]\Delta} & \text{if } s_{t+\Delta}=s_t d \\ \\ \quad\text{with probability } (1/2)-(1/2)\phi\left(t;s_t\right)\sqrt{\Delta}.\end{cases}\qquad(16.3)$$

Next, take natural logarithms of both sides of expression (16.3) to obtain

$$\frac{\log f_\Delta\left(t+\Delta,T;s_{t+\Delta}\right)}{\Delta} - \frac{\log f_\Delta\left(t,T;s_t\right)}{\Delta}$$

$$= \begin{cases} \mu\left(t,T;s_t\right)\Delta - \sigma\left(t,T;s_t\right)\sqrt{\Delta} & \text{if } s_{t+\Delta} = s_t u \\ \qquad \text{with probability } (1/2)+(1/2)\phi\left(t;s_t\right)\sqrt{\Delta} \\ \mu\left(t,T;s_t\right)\Delta + \sigma\left(t,T;s_t\right)\sqrt{\Delta} & \text{if } s_{t+\Delta} = s_t d \\ \qquad \text{with probability } (1/2)-(1/2)\phi\left(t;s_t\right)\sqrt{\Delta}. \end{cases} \qquad (16.4)$$

The mean and variance of the changes in forward rates can be computed to be

$$E_t\left\{\frac{\log f_\Delta(t+\Delta,T)}{\Delta} - \frac{\log f_\Delta(t,T)}{\Delta}\right\} = [\mu(t,T)-\phi(t)\sigma(t,T)]\Delta, \quad (16.5a)$$

and

$$\text{Var}_t\left\{\frac{\log f_\Delta(t+\Delta,T)}{\Delta} - \frac{\log f_\Delta(t,T)}{\Delta}\right\} = \sigma(t,T)^2\Delta - \phi(t)^2\sigma(t,T)^2\Delta^2. \quad (16.5b)$$

Derivation of Expression (16.5)

For simplicity of notation, define $\Delta \log f_\Delta(t,T) \equiv \log f_\Delta(t+\Delta,T) - \log f_\Delta(t,T)$. Using expression (16.4),

$$E_t\left(\frac{\Delta\log f_\Delta(t,T)}{\Delta}\right) = (1/2+\phi(t)\sqrt{\Delta}/2)(\mu(t,T)\Delta - \sigma(t,T)\sqrt{\Delta})$$

$$+(1/2-\phi(t)\sqrt{\Delta}/2)(\mu(t,T)\Delta + \sigma(t,T)\sqrt{\Delta})$$

$$= \mu(t,T)\Delta - \phi(t)\sigma(t,T)\Delta.$$

$$\text{Var}_t\left(\frac{\Delta\log f_\Delta(t,T)}{\Delta}\right) = E_t\left(\left(\frac{\Delta\log f_\Delta(t,T)}{\Delta}\right)^2\right) - \left(E_t\left(\frac{\Delta\log f_\Delta(t,T)}{\Delta}\right)\right)^2$$

$$= (\mu(t,T)\Delta - \sigma(t,T)\sqrt{\Delta})^2(1/2+\phi(t)\sqrt{\Delta}/2)$$

$$+(\mu(t,T)\Delta + \sigma(t,T)\sqrt{\Delta})^2(1/2-\phi(t)\sqrt{\Delta}/2)$$

$$-(\mu(t,T)\Delta - \phi(t)\sigma(t,T)\Delta)^2$$

$$= [\mu(t,T)^2\Delta^2 - 2\mu(t,T)\Delta\sigma(t,T)\sqrt{\Delta} + \sigma(t,T)^2\Delta](1/2+\phi(t)\sqrt{\Delta}/2)$$

$$+[\mu(t,T)^2\Delta^2 + 2\mu(t,T)\Delta\sigma(t,T)\sqrt{\Delta} + \sigma(t,T)^2\Delta](1/2-\phi(t)\sqrt{\Delta}/2)$$

$$-[\mu(t,T)^2\Delta^2 - 2\mu(t,T)\Delta\phi(t)\sigma(t,T)\Delta + \phi(t)^2\sigma(t,T)^2\Delta^2]$$

Canceling like terms gives

$$\text{Var}_t\left(\frac{\Delta \log f_\Delta(t,T)}{\Delta}\right) = -2\mu(t,T)\Delta\sigma(t,T)\sqrt{\Delta}\phi(t)\sqrt{\Delta}/2 - 2\mu(t,T)\Delta\sigma(t,T)\sqrt{\Delta}\phi(t)\sqrt{\Delta}/2$$

$$+2\mu(t,T)\Delta\phi(t)\sigma(t,T)\Delta + \sigma(t,T)^2\Delta - \phi(t)^2\sigma(t,T)^2\Delta^2$$

$$= \sigma(t,T)^2\Delta - \phi(t)^2\sigma(t,T)^2\Delta^2.$$

This completes the derivation. □

Dividing expression (12.5) by Δ and taking limits of these quantities as Δ gives

$$\lim_{\Delta \to 0}\frac{E_t\left\{\dfrac{\log f_\Delta(t+\Delta,T)}{\Delta} - \dfrac{\log f_\Delta(t,T)}{\Delta}\right\}}{\Delta} = \mu(t,T) - \phi(t)\sigma(t,T), \qquad (16.6a)$$

and

$$\lim_{\Delta \to 0}\frac{\text{Var}_t\left\{\dfrac{\log f_\Delta(t+\Delta,T)}{\Delta} - \dfrac{\log f_\Delta(t,T)}{\Delta}\right\}}{\Delta} = \sigma(t,T)^2. \qquad (16.6b)$$

Because the expectations are taken under the empirical probabilities, expression (16.6a) represents the drift $\mu^*(t,T) \equiv \mu(t,T) - \phi(t)\sigma(t,T)$ of the empirical process and $\sigma(t,T)$ represents the volatility of the empirical process. Both $\mu^*(t,T)$ and $\sigma(t,T)$ can in principle be estimated from past observations of forward rates.

The stochastic process $\phi(t)$ can be interpreted as a risk premium, i.e., a measure of the excess expected return (above the spot rate) per unit of standard deviation for the zero-coupon bonds.

The parameterization of the empirical discrete-time process as given in expression (16.2), therefore, is an approximation to the empirical continuous economy with drift $\mu^*(t,T) \equiv \mu(t,T) - \phi(t)\sigma(t,T)$ and volatility $\sigma(t,T)$. This identification corresponds to the top right of Figure 16.1.

We next investigate the discrete-time pseudo economy implied by expression (16.2) and the assumption of no arbitrage.

16.2.1 Arbitrage-Free Restrictions

This section studies the restrictions implied by no arbitrage on the above system. From Chapter 10, this system is arbitrage-free if and only if there exist unique pseudo probabilities $\pi_\Delta(t;s_t)$ strictly between 0 and 1 such that

$$\pi_\Delta\left(t;s_t\right) = \frac{r_\Delta\left(t;s_t\right) - d_\Delta\left(t,T;s_t\right)}{u_\Delta\left(t,T;s_t\right) - d_\Delta\left(t,T;s_t\right)} \tag{16.7}$$

for all s_t, $0 \le t < T - \Delta$, and $T \le \tau\Delta$, where both t and T are integer multiples of Δ.

In Chapter 5, expression (5.10), we characterized the zero-coupon bond price evolution's parameters in terms of the forward rate evolution's parameters. We use that expression now. Substitution of expression (16.2) into (5.10) gives*:

$$u_\Delta\left(t,T;s_t\right) = r_\Delta\left(t;s_t\right)e^{\left[-\sum_{j=t+\Delta}^{T-\Delta}\mu(t,j;s_t)\Delta + \sum_{j=t+\Delta}^{T-\Delta}\sigma(t,j;s_t)\sqrt{\Delta}\right]\Delta} \tag{16.8a}$$

and

$$d_\Delta\left(t,T;s_t\right) = r_\Delta\left(t;s_t\right)e^{\left[-\sum_{j=t+\Delta}^{T-\Delta}\mu(t,j;s_t)\Delta - \sum_{j=t+\Delta}^{T-\Delta}\sigma(t,j;s_t)\sqrt{\Delta}\right]\Delta}. \tag{16.8b}$$

Substitution of these expressions into (16.7) and simplification yield

$$\pi_\Delta\left(t;s_t\right) = \frac{1 - e^{\left[-\sum_{j=t+\Delta}^{T-\Delta}\mu(t,j;s_t)\Delta - \sum_{j=t+\Delta}^{T-\Delta}\sigma(t,j;s_t)\sqrt{\Delta}\right]\Delta}}{e^{\left[-\sum_{j=t+\Delta}^{T-\Delta}\mu(t,j;s_t)\Delta + \sum_{j=t+\Delta}^{T-\Delta}\sigma(t,j;s_t)\sqrt{\Delta}\right]\Delta} - e^{\left[-\sum_{j=t+\Delta}^{T-\Delta}\mu(t,j;s_t)\Delta - \sum_{j=t+\Delta}^{T-\Delta}\sigma(t,j;s_t)\sqrt{\Delta}\right]\Delta}} \tag{16.9}$$

for all s_t and $0 \le t < T - \Delta$ and $T \le \tau\Delta$, where both t and T are integer multiples of Δ.

Expression (16.9) gives the cross-restrictions on the drifts and volatilities of the forward rate process that are both necessary and sufficient for the existence of the pseudo probabilities.

Under these pseudo probabilities the change in the logarithm of forward rates is represented as

---

* The summations in expression (16.8) are for steps of size Δ, so $j = t + \Delta$, $t + 2\Delta$, ... , $T - \Delta$.

$$\frac{\log f_\Delta\left(t+\Delta,T;s_{t+\Delta}\right)}{\Delta} - \frac{\log f_\Delta\left(t,T;s_t\right)}{\Delta}$$

$$=\begin{cases} \mu(t,T;s_t)\Delta - \sigma(t,T;s_t)\sqrt{\Delta} & \text{if } s_{t+\Delta} = s_t u \\ \qquad\text{with probability } \pi(t;s_t) \\ \mu(t,T;s_t)\Delta + \sigma(t,T;s_t)\sqrt{\Delta} & \text{if } s_{t+\Delta} = s_t d \\ \qquad\text{with probability } 1 - \pi(t;s_t). \end{cases}$$

(16.10)

The mean and the variance of the changes in forward rates can be computed under the pseudo probabilities:

$$\tilde{E}_t\left\{\frac{\log f_\Delta(t+\Delta,T)}{\Delta} - \frac{\log f_\Delta(t,T)}{\Delta}\right\} = \mu(t,T)\Delta + (1 - 2\pi_\Delta(t))\sigma(t,T)\sqrt{\Delta},$$

(16.11a)

and

$$\tilde{\text{Var}}\left\{\frac{\log f_\Delta(t+\Delta,T)}{\Delta} - \frac{\log f_\Delta(t,T)}{\Delta}\right\} = 4\sigma(t,T)^2\,\Delta\pi_\Delta(t)(1 - \pi_\Delta(t)). \quad (16.11b)$$

Derivation of Expression (16.11)

For simplicity of notation, define $\Delta\log f_\Delta(t,T) \equiv \log f_\Delta(t+\Delta,\ T) - \log f_\Delta(t,T)$. Using expression (16.10),

$$\tilde{E}_t\left(\frac{\Delta\log f_\Delta(t,T)}{\Delta}\right) = \pi_\Delta(t)\left(\mu(t,T)\Delta - \sigma(t,T)\sqrt{\Delta}\right) + (1 - \pi_\Delta(t))\left(\mu(t,T)\Delta + \sigma(t,T)\sqrt{\Delta}\right)$$

$$= \mu(t,T)\Delta + (1 - 2\pi_\Delta(t))\sigma(t,T)\sqrt{\Delta}.$$

$$\tilde{\text{Var}}_t\left(\frac{\Delta\log f_\Delta(t,T)}{\Delta}\right) = \tilde{E}_t\left(\left(\frac{\Delta\log f_\Delta(t,T)}{\Delta}\right)^2\right) - \left(\tilde{E}_t\left(\frac{\Delta\log f_\Delta(t,T)}{\Delta}\right)\right)^2$$

$$= (\mu(t,T)\Delta - \sigma(t,T)\sqrt{\Delta})^2\pi_\Delta(t) + (\mu(t,T)\Delta - \sigma(t,T)\sqrt{\Delta})^2(1 - \pi_\Delta(t))$$

$$- \left(\mu(t,T)\Delta + (1 - 2\pi_\Delta(t))\sigma(t,T)\sqrt{\Delta}\right)^2$$

Expanding the squares gives

$$\tilde{V}ar_t\left(\frac{\Delta \log f_\Delta(t,T)}{\Delta}\right) = (\mu(t,T)^2\Delta^2 - 2\mu(t,T)\sigma(t,T)\sqrt{\Delta} + \sigma(t,T)^2\Delta)\pi_\Delta(t)$$

$$+(\mu(t,T)^2\Delta^2 + 2\mu(t,T)\sigma(t,T)\sqrt{\Delta} + \sigma(t,T)^2\Delta)(1-\pi_\Delta(t))$$

$$-(\mu(t,T)^2\Delta^2 + 2\mu(t,T)(1-2\pi_\Delta(t))\sigma(t,T)\sqrt{\Delta} + \sigma(t,T)^2\Delta(1-2\pi_\Delta(t))^2)$$

Canceling like terms and simplifying yields

$$\tilde{V}ar_t\left(\frac{\Delta \log f_\Delta(t,T)}{\Delta}\right) = 4\sigma(t,T)^2\Delta\pi_\Delta(t)(1-\pi_\Delta(t))$$

This completes the derivation.   □

In summary, to construct the lower part of Figure 16.1, we require three conditions. The first is that expression (16.9) regarding the pseudo probabilities holds. Second, we also require that as $\Delta\to 0$,

$$\lim_{\Delta\to 0}\frac{\tilde{E}_t\left\{\frac{\log f_\Delta(t+\Delta,T)}{\Delta} - \frac{\log f_\Delta(t,T)}{\Delta}\right\}}{\Delta} = \mu(t,T) \qquad (16.12a)$$

and third,

$$\lim_{\Delta\to 0}\frac{\tilde{V}ar_t\left\{\frac{\log f_\Delta(t+\Delta,T)}{\Delta} - \frac{\log f_\Delta(t,T)}{\Delta}\right\}}{\Delta} = \sigma(t,T)^2. \qquad (16.12b)$$

These three conditions imply

$$\pi_\Delta(t;s_t) - (1/2) + o\left(\sqrt{\Delta}\right) \qquad (16.13)$$

where

$$o\left(\sqrt{\Delta}\right) \text{ is defined by } \lim_{\Delta\to 0} o\left(\sqrt{\Delta}\right)/\sqrt{\Delta} = 0.$$

Derivation of Expression (16.13)

From expressions (16.11) and (16.12) one obtains

$$\lim_{\Delta\to 0}\frac{(1-2\pi_\Delta(t))\sigma(t,T)}{\sqrt{\Delta}} = 0$$

and

$$\lim_{\Delta \to 0} 4\pi_\Delta(t)(1 - \pi_\Delta(t)) = 1.$$

The first of these gives (16.13), since it implies that

$$\frac{(1/2 - \pi_\Delta(t))}{\sqrt{\Delta}} = 0.$$

The second equality is also satisfied by (16.13). This completes the derivation. □

For computational efficiency, it is convenient to set the pseudo probabilities $\pi_\Delta(t; s_t) = 1/2$ for all t and s_t. This is a special case of expression (16.13).

With this restriction, we get convergence to the continuous time economy and the pseudo probabilities from expression (16.9) satisfy

$$(1/2)\left[e^{\left[-\sum\limits_{j=t+\Delta}^{T-\Delta} \mu(t,j;s_t)\Delta + \sum\limits_{j=t+\Delta}^{T-\Delta} \sigma(t,j;s_t)\sqrt{\Delta}\right]\Delta} - e^{\left[-\sum\limits_{j=t+\Delta}^{T-\Delta} \mu(t,j;s_t)\Delta - \sum\limits_{j=t+\Delta}^{T-\Delta} \sigma(t,j;s_t)\sqrt{\Delta}\right]\Delta} \right]$$

$$= \left[1 - e^{\left[-\sum\limits_{j=t+\Delta}^{T-\Delta} \mu(t,j;s_t)\Delta - \sum\limits_{j=t+\Delta}^{T-\Delta} \sigma(t,j;s_t)\sqrt{\Delta}\right]\Delta} \right]$$

(16.14)

This is true if and only if

$$e^{\left[\sum\limits_{j=t+\Delta}^{T-\Delta} \mu(t,j;s_t)\Delta \right]\Delta} = (1/2)\left[e^{\left[+\sum\limits_{j=t+\Delta}^{T-\Delta} \sigma(t,j;s_t)\sqrt{\Delta}\right]\Delta} + e^{\left[-\sum\limits_{j=t+\Delta}^{T-\Delta} \sigma(t,j;s_t)\sqrt{\Delta}\right]\Delta} \right]$$

(16.15)

$$\equiv \cosh\left(\left[\sum\limits_{j=t+\Delta}^{T-\Delta} \sigma(t,j;s_t)\sqrt{\Delta}\right]\Delta \right)$$

where $\cosh(x) \equiv (1/2)(e^x + e^{-x})$.

This gives the restriction on the drifts and volatilities of the forward rate process so that they are arbitrage-free and converge to the appropriate continuous-time economy.

The next subsection shows how to use this information to build trees similar to those used in the book's examples.

16.2.2 Computation of the Arbitrage-Free Term Structure Evolutions

This section shows how to use the previous expressions to compute an arbitrage-free term structure evolution.

First, we compute the evolution of the zero-coupon bond prices. Substitution of expression (16.15) into expression (16.8) gives

$$u_\Delta(t,T;s_t) = r_\Delta(t;s_t) \left[\cosh\left(\left[\sum_{j=t+\Delta}^{T-\Delta} \sigma(t,j;s_t)\sqrt{\Delta} \right] \Delta \right) \right]^{-1} e^{\left[\sum_{j=t+\Delta}^{T-\Delta} \sigma(t,j;s_t)\sqrt{\Delta} \right] \Delta}$$

(16.16a)

and

$$d_\Delta(t,T;s_t) = r_\Delta(t;s_t) \left[\left[\cosh\left(\sum_{j=t+\Delta}^{T-\Delta} \sigma(t,j;s_t)\sqrt{\Delta} \right) \right] \Delta \right]^{-1} e^{\left[-\sum_{j=t+\Delta}^{T-\Delta} \sigma(t,j;s_t)\sqrt{\Delta} \right] \Delta}.$$

(16.16b)

Using expression (5.3) of Chapter 5 for the evolution of a zero-coupon bond's price in conjunction with expression (16.16) gives:

$$P_\Delta(t+\Delta,T;s_{t+\Delta})$$

$$= \begin{cases} P_\Delta(t,T;s_t)r_\Delta(t;s_t) \left[\cosh\left(\left[\sum_{j=t+\Delta}^{T-\Delta} \sigma(t,j;s_t)\sqrt{\Delta} \right] \Delta \right) \right]^{-1} e^{\left[\sum_{j=t+\Delta}^{T-\Delta} \sigma(t,j;s_t)\sqrt{\Delta} \right] \Delta} \\ \qquad \text{if } s_{t+\Delta} = s_t u \\[2em] P_\Delta(t,T;s_t)r_\Delta(t;s_t) \left[\cosh\left(\left[\sum_{j=t+\Delta}^{T-\Delta} \sigma(t,j;s_t)\sqrt{\Delta} \right] \Delta \right) \right]^{-1} e^{\left[-\sum_{j=t+\Delta}^{T-\Delta} \sigma(t,j;s_t)\sqrt{\Delta} \right] \Delta} \\ \qquad \text{if } s_{t+\Delta} = s_t d \end{cases}$$

(16.17)

for $t < T - \Delta$ and $T \le \tau\Delta$ where both t and T are integer multiples of Δ.

Expression (16.17) gives the realization of the discrete-time bond price process for the approximation as in Figure 16.1.

Under the empirical probabilities $(1/2) + (1/2)\phi(t; s_t)$, this bond price process converges to the limiting empirical process for the bond's price.

Under the pseudo probabilities $\pi_\Delta(t; s_t) = (1/2)$, this bond price process converges to the limiting pseudo process for the bond's price. Because the pseudo economies are all that are relevant to application of the contingent claim valuation theory of Chapter 10, we never have to estimate the stochastic process $\phi(t; s_t)$.

It is shown later that $\phi(t; s_t)$ can be interpreted as a risk premium, i.e., as a measure of the excess expected return (above the spot rate) per unit of standard deviation for the zero-coupon bonds. Therefore, to apply this technology to price contingent claims, one never has to estimate a zero-coupon bond's risk premium. This is an important characteristic of the model. It is the analog of the Black-Scholes-Merton equity option pricing model for interest rates.

To construct the forward rate process evolution, from expression (16.15) we get

$$e^{\left[\mu(t, t+\Delta; s_t)\Delta\right]\Delta} = \cosh\left(\left[\sigma\left(t, t+\Delta; s_t\right)\sqrt{\Delta}\right]\Delta\right) \tag{16.18}$$

and

$$e^{\left[\mu(t, T; s_t)\Delta\right]\Delta} = \cosh \frac{\left(\left[\sum_{j=t+\Delta}^{T} \sigma\left(t, j; s_t\right)\sqrt{\Delta}\right]\Delta\right)}{\cosh\left(\left[\sum_{j=t+\Delta}^{T-\Delta} \sigma\left(t, j; s_t\right)\sqrt{\Delta}\right]\Delta\right)}$$

for $T \ge t + 2\Delta$ where both t and T are integer multiples of Δ.

Substitution into expression (16.3) yields

$$f_\Delta(t+\Delta,T;s_{t+\Delta}) = \begin{cases} f_\Delta(t,T;s_t) \left[\dfrac{\cosh\left(\left[\sum\limits_{j=t+\Delta}^{T} \sigma(t,j;s_t)\sqrt{\Delta}\,\right]\Delta\right)}{\cosh\left(\left[\sum\limits_{j=t+\Delta}^{T-\Delta} \sigma(t,j;s_t)\sqrt{\Delta}\,\right]\Delta\right)} \right] e^{\left[-\sigma(t,T;s_t)\sqrt{\Delta}\,\right]\Delta} \\[2em] \qquad\qquad\qquad\qquad \text{if } s_{t+\Delta} = s_t u \\[2em] f_\Delta(t,T;s_t) \left[\dfrac{\cosh\left(\left[\sum\limits_{j=t+\Delta}^{T} \sigma(t,j;s_t)\sqrt{\Delta}\,\right]\Delta\right)}{\cosh\left(\left[\sum\limits_{j=t+\Delta}^{T-\Delta} \sigma(t,j;s_t)\sqrt{\Delta}\,\right]\Delta\right)} \right] e^{\left[+\sigma(t,T;s_t)\sqrt{\Delta}\,\right]\Delta} \\[2em] \qquad\qquad\qquad\qquad \text{if } s_{t+\Delta} = s_t d \end{cases}$$

$$(16.19)$$

for $\tau\Delta \geq T + \Delta$ and $T - \Delta \geq t$ where both t and T are integer multiples of Δ, and where as a notational convenience we define

$$\cosh\left(\left[\sum_{j=t+\Delta}^{t} \sigma(t,j;s_t)\sqrt{\Delta}\,\right]\Delta\right) \equiv 1.$$

We include this last identity so that expression (16.19) can be written in one line. Otherwise, because of the denominator involving the cosh function, we would have to write out two expressions: one when $T = t + \Delta$ and one when $T \geq t + 2\Delta$.

Thus, expression (16.19) gives the evolution of the discrete-time forward rate curve as in Figure 16.1.

Under the *empirical probabilities* $(1/2) + (1/2)\phi(t;s_t)$, this process converges to the empirical continuous-time process for the forward rates.

Under the *pseudo probabilities* $\pi_\Delta(t;s_t) = (1/2)$, this converges to the pseudo continuous-time process for the forward rates. The computation of interest rate derivative values is done using the pseudo probabilities.

Note that under the pseudo probabilities, a specification of the volatility structure of forward rates,

$$
\begin{bmatrix}
\sigma(t, t + \Delta; s_t) \\
\sigma(t, t + 2\Delta; s_t) \\
\vdots \\
\sigma(t, \tau\Delta - \Delta; s_t)
\end{bmatrix}
$$

for all $0 \leq t \leq \tau - \Delta$ and s_t is sufficient to determine the evolution of the forward rate curve. The risk premium process $\phi(t, T; s_t)$ does not appear in expression (16.19). Again, this is an important attribute of the model, which makes its implementation practical.

Two functional forms of the volatility function $\sigma(t, T; s_t)$ have received special attention in the literature.

Case 1: Deterministic volatility function

The first case is that in which the volatility $\sigma(t, T; s_t)$ is a deterministic function, independent of the state s_t. This restriction on the volatility function implies that forward rates can go negative. This case includes as special cases Ho and Lee's [8] model ($\sigma(t, T; s_t)$ is a constant) and a discrete-time approximation in the HJM framework to Vasicek's [11] model ($\sigma(t, T; s_t) = \xi e^{-\eta(T-t)}$ for ξ, $\eta > 0$ constants).

Case 2: Nearly proportional volatility function

The volatility is $\sigma(t, T; s_t) = \eta(t, T)\min(\log f(t, T), \ M)$, where $\eta(t, T)$ is a deterministic function and $M > 0$ is a large positive constant. In other words, in the second case $\sigma(t, T; s_t)$ is approximately proportional to the current value of the continuously compounded forward rate $\log f(t, T)$. The proportionality factor is $\eta(t, T)$. This proportionality implies that forward rates are always nonnegative. Nonnegativity of forward rates is a condition usually required in models because negative interest rates for zero-coupon bonds are inconsistent with the existence of cash currency, which can be stored costlessly at zero interest rates.

In this case, the larger the forward rate, the larger the volatility. If the forward rate becomes too large, however, the volatility is bounded by $\eta(t, T)M$. This upper bound guarantees that forward rates do not explode with positive probability. This is a necessary technical condition based on the limit economies in Figure 16.1 (see Heath, Jarrow, and Morton [6] for details).

Example: Construction of Figure 5.6

To construct Figure 5.6, we used expression (16.19), with $\Delta \equiv 1$ and the following forward rates:

$$f(0,0) = f(0,1) = f(0,2) = f(0,3) = 1.02.$$

We used the volatility function given in case 2 with the proportionality coefficient $\eta(t,T) \equiv \eta(T-t)$ depending only on time to maturity and having the values

$$\eta(1) = 0.11765,$$

$$\eta(2) = 0.08825,$$

and

$$\eta(3) = 0.06865.$$

We set $M = 1,000,000$. This completes the example. □

16.2.3 The Continuous-Time Limit

For purposes of empirical estimation and computation, we have constructed the discrete-time economies to converge to a continuous-time limit. It is instructive to study the continuous-time limits of the one-factor economy analyzed in the previous sections. This section is not used in the remainder of the text, and it can be skipped on a first reading.

The first step in analyzing the continuous-time economy is to study continuous compounding as the limit of the discretely compounded rates used in the previous sections. Intuitively, the *continuously compounded forward rate*, $\tilde{f}(t,T)$, is that rate such that for small time intervals Δ, the following condition holds:

$$f_\Delta(t,T) \approx e^{\tilde{f}(t,T)\Delta}. \tag{16.20}$$

The left side of expression (16.20) is the forward rate over $[T, T+\Delta]$ as seen at time t. This forward rate is one plus a percentage. The right side gives the appreciation obtained from the time t continuously compounded forward rate $\tilde{f}(t,T)$ per unit time, compounding for the Δ units of time from T to $T+\Delta$. The continuously compounded forward rate $\tilde{f}(t,T)$ is a percentage, expressed as a number between -1 and 1. Expression (16.20) is only an approximation.

Formally (assuming the limit exists), the continuously compounded forward rate is defined by

$$\tilde{f}(t,T) \equiv \lim_{\Delta \to 0} \log\left(f_\Delta\left(t,T\right)\right)/\Delta. \tag{16.21}$$

Expression (16.21) implies that

$$f_\Delta(t,T) = e^{\int_T^{T+\Delta} \tilde{f}(t,v)dv}. \tag{16.22}$$

Expression (16.22) is the formal version of expression (16.20). When the time interval Δ is small, expression (16.20) is approximately true.

We can rewrite the zero-coupon bond prices in terms of the continuously compounded forward rates. Indeed, it can be shown that

$$P_\Delta(t,T) = e^{-\int_t^T \tilde{f}(t,v)dv}. \tag{16.23}$$

Derivation of Expression (16.23)

Choose T so that T/Δ is an integer. From the definition of the bond price in terms of forward rates we have

$$P_\Delta(t,T) = \left[\prod_{j=t}^{T/\Delta-1} f_\Delta(t,j\Delta)\right]^{-1}$$

$$= \left[\prod_{j=t}^{T/\Delta-1} e^{\int_j^{j+\Delta} \tilde{f}(t,v)dv}\right]^{-1}$$

by the definition of

$$\tilde{f}(t,v) = e^{-\sum_{j=t}^{T/\Delta-1}\left[\int_j^{j+\Delta} \tilde{f}(t,v)dv\right]} = e^{-\int_t^T \tilde{f}(t,v)dv}$$

This completes the proof. ☐

In the one-factor case under expressions (16.2) and (16.15), given suitable restrictions upon $\mu(t,T)$, $\sigma(t,T)$, and $\phi(t)$, the random process

$$\left[\frac{\log\left(f_\Delta(t+\Delta,T)\right)}{\Delta} - \frac{\log\left(f_\Delta(t,T)\right)}{\Delta}\right] \tag{16.24}$$

converges* as the time step $\Delta \to 0$ to the random process given by the following:

$$\tilde{f}(t,T) - \tilde{f}(0,T) = \int_0^t \mu^*(v,T)dv + \int_0^t \sigma(v,T)dW^*(v) \tag{16.25a}$$

under the empirical probabilities, where $\{W^*(t): t \; \varepsilon \; [0,\tau]\}$ is a Brownian motion, initialized at zero, and

$$\tilde{f}(t,T) - \tilde{f}(0,T) = \int_0^t \mu(v,T)dv + \int_0^t \sigma(v,T)dW(v) \tag{16.25b}$$

under the pseudo probabilities, where $\{W(t): t \; \varepsilon \; [0, \tau]\}$ is a Brownian motion, initialized at zero.

Further, it can be shown (see Heath, Jarrow, and Morton [6]) that

$$dW(v) = dW^*(v) - \phi(v)dv \tag{16.25c}$$

and

$$\int_0^t \mu(v,T)dv = \int_0^t \sigma(v,T)\left[\int_v^T \sigma(v,y)dy\right]dv. \tag{16.25d}$$

The first term on the right side of expression (16.25a) is an ordinary integral from first-year calculus. The second term is a stochastic integral, whose definition we leave for outside reading (see Protter [9]). A complete understanding of these integrals is not necessary for an understanding of the remainder of the text.

---

* Formally, the random process given in (16.24) converges weakly to that in expression (16.25). The definition of weak convergence is rather technical and can be found in Billingsley [2].

Expressions (16.25a) and (16.25b) are the direct result of expressions (16.6) and (16.12), respectively. They give the limiting processes, which are characterized by their drifts and volatilities.

Expression (16.25c) in conjunction with the two previous expressions shows that

$$\mu*(t,T) = \mu(t,T) - \sigma(t,T)\phi(t). \tag{16.26}$$

This relates the drift of forward rates in the empirical economy ($\mu*(t,T)$) to the drift of forward rates in the pseudo economy ($\mu(t,T)$).

Expression (16.25d) is the limiting form of expression (16.15). It is the no-arbitrage restriction written in terms of the volatilities of the forward rate process in the pseudo economy (the proof is in Heath, Jarrow, and Morton [6]). Combined with expression (16.26), the no-arbitrage restriction expression (16.25d) is equivalent to

$$\frac{E_t(dP(t,T)/P(t,T)) - r(t)dt}{\sqrt{\mathrm{Var}_t(dP(t,T)/P(t,T))}} = -\phi(t)dt \tag{16.27}$$

for all t, T, where $dP(t,T)$ is the instantaneous change in the T-maturity zero-coupon bond's price over $[t, t+dt]$.

The proof of this expression is contained in Heath, Jarrow, and Morton [6]. This result gives $-\phi(t)$ the interpretation of being a *risk premium*, i.e., the excess expected return (above the spot rate) per unit of standard deviation for the zero-coupon bonds. The arbitrage-free restriction is therefore equivalent to the statement that all zero-coupon bonds must have the same excess expected return per unit of risk. This is the continuous-time analog of the no-arbitrage condition involving equalities of the pseudo probabilities used in Chapter 10.

Two examples of expression (16.25) are useful in applications.

Case 1: Deterministic volatility functions
The first example is that in which the volatility function is a deterministic function, i.e., $\sigma(t,T)$ is nonrandom.

In this case, the limiting random variable $\tilde{f}(t,T) - \tilde{f}(0,T)$ can be shown to be normally distributed with

$$\text{mean} \quad \int_0^t \mu(v,T)dv \qquad (16.28)$$

and

$$\text{variance} \quad \int_0^t \sigma^2(v,T)dv. \qquad (16.29)$$

This implies, from expression (16.25), that the zero-coupon bond's price $P(t,T)$ is lognormality distributed. This lognormality enables one to compute analytic expressions for various types of options (see Heath, Jarrow, and Morton [6]). Expression (16.25) with the restrictions implied by expression (16.28) is called a *Gaussian economy*.

When $\sigma(t,T)$ is a constant, independent of t and T, we get a continuous-time limit of the Ho and Lee model [8]. When $\sigma(t,T) = \xi e^{-\eta(T-t)}$ for ξ, η constants, we get in the HJM framework a version of Vasicek [10].

Case 2: Nearly proportional volatility functions

A second example useful in applications is that in which the volatility function satisfies the condition $\sigma(t,T) = \eta(t,T)\min(\tilde{f}(t,T), M)$ where $\eta(t,T)$ is a deterministic function and M is a large, positive constant. In this case it can be shown that the limiting process for $\tilde{f}(t,T)$ is positive for sure. The bound M is included to keep the forward rate process from exploding in finite time; see Heath, Jarrow, and Morton [6] for details. No known distribution for $\tilde{f}(t,T)$ is available, and the limiting random variable is best approximated via expression (16.19).

Cases 1 and 2 are the limiting forms of the two cases for the volatility functions discussed earlier in this chapter.

16.3 TWO-FACTOR ECONOMY

This section extends the previous analysis to a two-factor economy. Because we are interested in computing contingent claim values as in Chapter 10, we give only the characterization for the pseudo economies. This corresponds to the lower part of Figure 16.1.

Consider the two-factor economy as described in Chapter 5. The forward rate process can be characterized as:

$$f_\Delta\left(t+\Delta,T;s_{t+\Delta}\right)=\begin{cases}\alpha_\Delta\left(t,T;s_t\right)f_\Delta\left(t,T;s_t\right) & \text{if } s_{t+\Delta}=s_tu \\ \quad\text{with pseudo probability } 1/4 \\ \gamma_\Delta\left(t,T;s_t\right)f_\Delta\left(t,T;s_t\right) & \text{if } s_{t+\Delta}=s_tm \\ \quad\text{with pseudo probability } 1/4 \\ \beta_\Delta\left(t,T;s_t\right)f_\Delta\left(t,T;s_t\right) & \text{if } s_{t+\Delta}=s_td \\ \quad\text{with pseudo probability } 1/2. \end{cases} \quad (16.30)$$

where $\tau\Delta-\Delta\geq T\geq t+\Delta$, and where t and T are integer multiples of Δ.

Let us reparameterize expression (16.30) in terms of three new stochastic processes $\mu(t,T;s_t)$, $\sigma_1(t,T;s_t)$, and $\sigma_2(t,T;s_t)$ as follows:

$$\alpha_\Delta\left(t,T;s_t\right)=e^{\left[\mu(t,T;s_t)\Delta-\sigma_1(t,T;s_t)\sqrt{\Delta}-\sqrt{2}\,\sigma_2(t,T;s_t)\sqrt{\Delta}\right]\Delta}$$

$$\gamma_\Delta\left(t,T;s_t\right)=e^{\left[\mu(t,T;s_t)\Delta-\sigma_1(t,T;s_t)\sqrt{\Delta}+\sqrt{2}\,\sigma_2(t,T;s_t)\sqrt{\Delta}\right]\Delta} \quad (16.31)$$

$$\beta_\Delta\left(t,T;s_t\right)=e^{\left[\mu(t,T;s_t)\Delta+\sigma_1(t,T;s_t)\sqrt{\Delta}\right]\Delta}$$

Substitution of expression (16.31) into expression (16.30) gives

$$f_\Delta\left(t+\Delta,T;s_{t+\Delta}\right)=\begin{cases}f_\Delta\left(t,T;s_t\right)e^{\left[\mu(t,T;s_t)\Delta-\sigma_1(t,T;s_t)\sqrt{\Delta}-\sqrt{2}\,\sigma_2(t,T;s_t)\sqrt{\Delta}\right]\Delta} & \text{if } s_{t+\Delta}=s_tu \\ \quad\text{with probability } 1/4 \\ f_\Delta\left(t,T;s_t\right)e^{\left[\mu(t,T;s_t)\Delta-\sigma_1(t,T;s_t)\sqrt{\Delta}+\sqrt{2}\,\sigma_2(t,T;s_t)\sqrt{\Delta}\right]\Delta} & \text{if } s_{t+\Delta}=s_tm \\ \quad\text{with probability } 1/4 \\ f_\Delta\left(t,T;s_t\right)e^{\left[\mu(t,T;s_t)\Delta+\sigma_1(t,T;s_t)\sqrt{\Delta}\right]\Delta} & \text{if } s_{t+\Delta}=s_td \\ \quad\text{with probability } 1/2. \end{cases}$$

$$(16.32)$$

Thus, the stochastic processes $\mu(t,T;s_t)$, $\sigma_1(t,T;s_t)$, and $\sigma_2(t,T;s_t)$ can be interpreted as the drift and the volatilities $(\sigma_1(t,T;s_t), \sigma_2(t,T;s_t))$ for the process $\log f_\Delta(t+\Delta,T;s_{t+\Delta}) - \log f_\Delta(t,T;s_t)$ with $\sigma_i(t,T;s_t)$ for $i = 1, 2$ being the volatilities for the first and second factors. Indeed, a straightforward calculation shows that

$$\tilde{E}_t\left\{\frac{\log f_\Delta(t+\Delta,T)}{\Delta} - \frac{\log f_\Delta(t,T)}{\Delta}\right\} = \mu(t,T)\Delta \qquad (16.33)$$

and

$$\tilde{V}ar_t\left\{\frac{\log f_\Delta(t+\Delta,T)}{\Delta} - \frac{\log f_\Delta(t,T)}{\Delta}\right\} = \sigma_1^2(t,T)\Delta + \sigma_2^2(t,T)\Delta. \qquad (16.34)$$

under the pseudo probabilities given in expression (16.32).

16.3.1 Arbitrage-Free Restrictions

We next study the restrictions that the existence of the pseudo probabilities implies about the reparameterization. For computational efficiency, we set the pseudo probabilities in expression (16.32) equal to the following:

$$\pi_\Delta^u(t;s_t) \equiv 1/4,$$

$$\pi_\Delta^m(t;s_t) \equiv 1/4,$$

and

$$1 - \pi_\Delta^u(t;s_t) - \pi_\Delta^m(t;s_t) \equiv 1/2 \quad \text{for all } s_t \text{ and } t.$$

From expression (5.19) in Chapter 5 we get

$$u_\Delta(t,T;s_t) = r_\Delta(t;s_t)e^{\left[-\sum_{j=t+\Delta}^{T-\Delta}\mu(t,j;s_t)\Delta + \sum_{j=t+\Delta}^{T-\Delta}\sigma_1(t,j;s_t)\sqrt{\Delta} + \sqrt{2}\sum_{j=t+\Delta}^{T-\Delta}\sigma_2(t,j;s_t)\sqrt{\Delta}\right]\Delta}$$

$$m_\Delta(t,T;s_t) = r_\Delta(t;s_t)e^{\left[-\sum_{j=t+\Delta}^{T-\Delta}\mu(t,j;s_t)\Delta + \sum_{j=t+\Delta}^{T-\Delta}\sigma_1(t,j;s_t)\sqrt{\Delta} - \sqrt{2}\sum_{j=t+\Delta}^{T-\Delta}\sigma_2(t,j;s_t)\sqrt{\Delta}\right]\Delta}$$

$$\qquad (16.35)$$

$$d_\Delta(t,T;s_t) = r_\Delta(t;s_t)e^{\left[-\sum_{j=t+\Delta}^{T-\Delta}\mu(t,j;s_t)\Delta - \sum_{j=t+\Delta}^{T-\Delta}\sigma_1(t,j;s_t)\sqrt{\Delta}\right]\Delta}$$

Recall the martingale condition:

$$\frac{P_\Delta(t,T;s_t)}{\cdot B_\Delta(t;s_{t-1})} = \frac{P_\Delta(t,T;s_t)}{B_\Delta(t;s_{t-1})r_\Delta(t;s_t)}\begin{bmatrix}(1/4)u_\Delta(t,T;s_t)+(1/4)m_\Delta(t,T;s_t)\\+(1/2)d_\Delta(t,T;s_t)\end{bmatrix}.$$

(16.36)

Substitution of expression (16.35) into expression (16.36) gives, after some algebra, the no-arbitrage restriction:

$$e^{\left[\sum_{j=t+\Delta}^{T-\Delta}\mu(t,j;s_t)\Delta\right]\Delta} = (1/2)e^{\left[\sum_{j=t+\Delta}^{T-\Delta}\sigma_1(t,j;s_t)\sqrt{\Delta}\right]\Delta}\left((1/2)e^{\sqrt{2}\left[\sum_{j=t+\Delta}^{T-\Delta}\sigma_2(t,j;s_t)\sqrt{\Delta}\right]\Delta}\right.$$
$$\left.-\sqrt{2}\left[\sum_{j=t+\Delta}^{T-\Delta}\sigma_2(t,j;s_t)\sqrt{\Delta}\right]\Delta\right)$$
$$+(1/2)e^{\left[-\sum_{j=t+\Delta}^{T-\Delta}\sigma_1(t,j;s_t)\sqrt{\Delta}\right]\Delta}$$

(16.37)

for $t \le T-2\Delta$ and $T \le \tau\Delta - \Delta$, where t and T are integer multiples of Δ.

16.3.2 Computation of the Arbitrage-Free Term Structure Evolutions

We show how to compute the arbitrage-free evolution of the forward rate curve. Using expression (16.37), we get

$$\left[\sum_{j=t+\Delta}^{T-\Delta}\mu(t,j)\Delta\right]\Delta = \log\left((1/2)e^{\left[\sum_{j=t+\Delta}^{T-\Delta}\sigma_1(t,j)\sqrt{\Delta}\right]\Delta}\left[(1/2)e^{\sqrt{2}\left[\sum_{j=t+\Delta}^{T-\Delta}\sigma_2(t,j)\sqrt{\Delta}\right]\Delta}\right.\right.$$
$$\left.+(1/2)e^{-\sqrt{2}\left[\sum_{j=t+\Delta}^{T-\Delta}\sigma_2(t,j)\sqrt{\Delta}\right]\Delta}\right]$$
$$\left.+(1/2)e^{-\left[\sum_{j=t+\Delta}^{T-\Delta}\sigma_1(t,j)\sqrt{\Delta}\right]\Delta}\right)$$

(16.38)

This system can be solved recursively, as

$$\left[\mu(t,t+\Delta)\Delta\right]\Delta = \log\left(\left(1/2\right)e^{\left[\sigma_1(t,t+\Delta)\sqrt{\Delta}\right]\Delta}\left[\begin{array}{c}(1/2)e^{\sqrt{2}\left[\sigma_2(t,t+\Delta)\sqrt{\Delta}\right]\Delta}\\+(1/2)e^{-\sqrt{2}\left[\sigma_2(t,t+\Delta)\sqrt{\Delta}\right]\Delta}\end{array}\right]+(1/2)e^{-\left[\sigma_1(t,t+\Delta)\sqrt{\Delta}\right]\Delta}\right) \quad (16.39a)$$

and

$$\left[\mu(t,K)\Delta\right]\Delta = \log\left((1/2)e^{\left[\sum\limits_{j=t+\Delta}^{K}\sigma_1(t,j)\sqrt{\Delta}\right]\Delta}\left[\begin{array}{c}(1/2)e^{\sqrt{2}\left[\sum\limits_{j=t+\Delta}^{K}\sigma_2(t,j)\sqrt{\Delta}\right]\Delta}\\+(1/2)e^{-\sqrt{2}\left[\sum\limits_{j=t+\Delta}^{K}\sigma_2(t,j)\sqrt{\Delta}\right]\Delta}\end{array}\right]+(1/2)e^{-\left[\sum\limits_{j=t+\Delta}^{K}\sigma_1(t,j)\sqrt{\Delta}\right]\Delta}\right)$$

$$-\sum_{j=t+\Delta}^{K-\Delta}\left[\mu(t,j)\Delta\right]\Delta \quad for \; \tau\Delta-\Delta\geq K\geq t+2\Delta$$

$$(16.39b)$$

Given two vectors of volatilities:

$$\begin{bmatrix}\sigma_1(t,t+\Delta;s_t)\\\sigma_1(t,t+2\Delta;s_t)\\\vdots\\\sigma_1(t,\tau\Delta-\Delta;s_t)\end{bmatrix} \quad and \quad \begin{bmatrix}\sigma_2(t,t+\Delta;s_t)\\\sigma_2(t,t+2\Delta;s_t)\\\vdots\\\sigma_2(t,\tau\Delta-\Delta;s_t)\end{bmatrix}$$

expressions (16.39a) and (16.39b) can be used in conjunction with expression (16.32) to generate an evolution of forward rates for the pseudo economy.

To get the evolution of the zero-coupon bond price process, one uses the evolution of the forward rates just computed plus the definition of a bond's price. These evolutions are all that are needed to compute contingent claim values as in Chapter 10.

16.4 MULTIPLE-FACTOR ECONOMIES

The previous analysis is easily extended to $N \geq 3$-factor economies. The basic equations (16.30) are augmented to include additional states, with additional volatilities. Generalized versions of expressions (16.31)–(16.37) follow through the existence of the pseudo probabilities. As before, using knowledge of the volatility parameters alone and not the drifts can determine the forward rate curve evolution. This is a key attribute of the model.

To illustrate the basic pattern for $N \geq 3$, we give the relevant equations for the three-factor case. As we are only interested in computing contingent claim values as in Chapter 10, we only provide the equations for the discrete-time pseudo economy. This is the lower part of Figure 16.1.

First, the forward rate process, under the pseudo probabilities, evolves according to expression (16.40):

$$f_\Delta\left(t+\Delta,T;s_{t+\Delta}\right) = f_\Delta\left(t,T;s_t\right)e^{\mu(t,T;s_t)\Delta^2}
\begin{cases}
e^{-\left[\sigma_1(t,T;s_t)\sqrt{\Delta}-\sqrt{2}\sigma_2(t,T;s_t)\sqrt{\Delta}-2\sigma_1(t,T;s_t)\sqrt{\Delta}\right]\Delta} \\
\qquad\text{with probability } 1/8 \\
e^{-\left[\sigma_1(t,T;s_t)\sqrt{\Delta}-\sqrt{2}\sigma_2(t,T;s_t)\sqrt{\Delta}+2\sigma_1(t,T;s_t)\sqrt{\Delta}\right]\Delta} \\
\qquad\text{with probability } 1/8 \\
e^{-\left[\sigma_1(t,T;s_t)\sqrt{\Delta}+\sqrt{2}\sigma_2(t,T;s_t)\sqrt{\Delta}\right]\Delta} \\
\qquad\text{with probability } 1/4 \\
e^{+\left[\sigma_1(t,T;s_t)\sqrt{\Delta}\right]\Delta} \\
\qquad\text{with probability } 1/2
\end{cases}$$

$$(16.40)$$

where $\tau\Delta - \Delta \geq T \geq t+\Delta$, and where t and T are integer multiples of Δ.

Under the pseudo probabilities, a straightforward calculation shows that

$$\tilde{E}_t\left\{\frac{\log f_\Delta(t+\Delta,T)}{\Delta} - \frac{\log f_\Delta(t,T)}{\Delta}\right\} = \mu(t,T)\Delta \qquad (16.41a)$$

$$\tilde{\mathrm{Var}}_t\left\{\frac{\log f_\Delta(t+\Delta,T)}{\Delta} - \frac{\log f_\Delta(t,T)}{\Delta}\right\} = \sigma_1(t,T)^2\Delta + \sigma_2(t,T)^2\Delta + \sigma_3(t,T)^2\Delta.$$

$$(16.41b)$$

Under (16.40), the no-arbitrage condition is

$$e^{\sum_{j=t+\Delta}^{T-\Delta}[\mu(t,j;s_t)]\Delta} = \left(\frac{1}{2}\right)e^{+\Sigma_1}\left(\left(\frac{1}{2}\right)e^{-\Sigma_2}\left(\left(\frac{1}{2}\right)e^{-\Sigma_3}+\left(\frac{1}{2}\right)e^{+\Sigma_3}\right)+\left(\frac{1}{2}\right)e^{+\Sigma_2}\right)+\left(\frac{1}{2}\right)e^{-\Sigma_1}$$

$$(16.42)$$

where

$$\Sigma_1 \equiv \left[\sum_{j=t+\Delta}^{T}\sigma_1\left(t,j;s_t\right)\sqrt{\Delta}\right]\Delta$$

$$\Sigma_2 \equiv \left[\sqrt{2}\sum_{j=t+\Delta}^{T}\sigma_2\left(t,j;s_t\right)\sqrt{\Delta}\right]\Delta$$

$$\Sigma_3 \equiv \left[2\sum_{j=t+\Delta}^{T}\sigma_3\left(t,j;s_t\right)\sqrt{\Delta}\right]\Delta$$

for $t \leq T + 2\Delta$ and $T \leq \tau\Delta - \Delta$, where t and T are integer multiples of Δ.

Expression (16.42) can be solved recursively, just as in expression (16.39) for $\mu(t,T;s_t)$. Given this value, expression (16.40) provides the equations for computing the arbitrage-free evolution of the forward rate curve. The inputs needed are the vectors of volatilities:

$$\begin{bmatrix} \sigma_1(t,t+\Delta;s_t) \\ \vdots \\ \sigma_1(t,\tau\Delta-\Delta;s_t) \end{bmatrix}, \begin{bmatrix} \sigma_2(t,t+\Delta;s_t) \\ \vdots \\ \sigma_2(t,\tau\Delta-\Delta;s_t) \end{bmatrix}, \begin{bmatrix} \sigma_3(t,t+\Delta;s_t) \\ \vdots \\ \sigma_3(t,\tau\Delta-\Delta;s_t) \end{bmatrix}.$$

16.5 COMPUTATIONAL ISSUES

This section briefly discusses the computational issues involved in implementing the one-, two-, or three-factor model on a computer. Three techniques are discussed, and references are provided: *(i)* bushy trees, *(ii)* lattice computations, and *(iii)* Monte Carlo simulation.

16.5.1 Bushy Trees

The procedure provided for computing forward rate curve evolutions in expression (16.19) for the one-factor case, expression (16.39) for the two-factor case, or expression (16.42) for the three-factor case is often called a *bushy tree*. It is called a bushy tree because the number of branches on the tree expands exponentially as the number of time steps increases. For example, in the one-factor case, the number of branches (nodes) at time t equals 2^t. For the two-factor case, the number of nodes at time t equals 3^t, and so forth.

For large numbers of time steps, depending upon what computational tricks are employed, the computing time becomes large. Consequently, it is often incorrectly believed that contingent claim valuation cannot be done using bushy trees. This belief is incorrect because a large number of time steps are not always essential for obtaining good approximations.

For European options or American options with six or seven decision nodes (of economic importance), bushy trees provide very accurate values with step sizes of only 12–14. This is because the branches spread out very quickly, giving a fine grid of values at the last date in the tree. From a numerical integration perspective (recall valuation is equivalent to computing an expected value), the approximating grid at the last date will be quite accurate.

For exotic options with multiple cash flow times (say ≥ 14) or long-dated American options with many decision nodes of economic importance (say ≥ 14), bushy trees provide a less attractive, time-intensive computational procedure.

Of course, as the computing technology improves, these concerns with bushy trees become less and less of a problem. For path-dependent options, such as index-amortizing swaps, bushy trees and Monte Carlo simulation are often the preferred approaches. This occurs because each path through the tree must be recorded to determine a value, and the lattice or partial differential equation approach does not record this information. For a more complete discussion of these issues, see Heath, Jarrow, and Morton [7].

16.5.2 Lattices

Special cases of the one-factor model allow for more efficient computation. These are the cases in which the tree recombines at various nodes; for example, the values of forward rates after an up followed by a down are the same as after a down followed by up.

For the one-factor economy, the tree recombines when the volatility function $\sigma(t,T)$ is a constant, independent of either time or the maturity date. Furthermore, under a time transformation, case 1 (the deterministic volatility function) can also be shown to recombine; see Amin [1].

One-factor lattice approaches work well for most contingent claims, except those that are path dependent, e.g., index-amortizing swaps. This is because a lattice does not remember the path taken through the tree, but only the current node. For path-dependent options, bushy trees and Monte Carlo simulation are often better approaches.

16.5.3 Monte Carlo Simulation

As stated earlier, contingent claims valuation reduces to calculating an expected value given the arbitrage-free evolution of the term structure of interest rates. Monte Carlo techniques are well suited for such computations. These techniques appear to be especially useful for multiple-factor models (greater than 3), in which computations using bushy trees are time-consuming. A good reference for this technique is Glasserman [3].

REFERENCES

1. Amin, K., 1991. "On the Computation of Continuous Time Options Prices Using Discrete Approximations." *Journal of Financial and Quantitative Analysis* 26, 477–496.
2. Billingsley, P., 1968. *Convergence of Probability Measures.* John Wiley & Sons, New York.
3. Glasserman, P., 2003. *Monte Carlo Methods in Financial Engineering*, Springer, New York.
4. He, H., 1990. "Convergence from Discrete-to-Continuous-Time Contingent Claims Prices." *Review of Financial Studies* 3 (4), 523–546.
5. Heath, D., R. Jarrow, and A. Morton, 1990. "Contingent Claim Valuation with a Random Evolution of Interest Rates." *Review of Futures Markets* 54–76.
6. Heath, D., R. Jarrow, and A. Morton, 1992. "Bond Pricing and the Term Structure of Interest Rates: A New Methodology for Contingent Claims Valuation." *Econometrica* 60 (1), 77–105.
7. Heath, D., R. Jarrow, and A. Morton, 1992. "Easier Done Than Said." *Risk Magazine* 5 (9), 77–80.
8. Ho, T. S., and S. Lee, 1986. "Term Structure Movements and Pricing Interest Rate Contingent Claims." *Journal of Finance* 41, 1011–1028.
9. Protter, P., 1990. *Stochastic Integration and Differential Equations.* Springer-Verlag, New York.
10. Vasicek, O., 1977. "An Equilibrium Characterization of the Term Structure." *Journal of Financial Economics* 5, 177–188.

CHAPTER 17

Parameter Estimation

THE PREVIOUS CHAPTERS TAKE the input parameters, the initial forward rate curve, and the volatility function(s) as given. From these inputs, interest rate derivatives are priced and hedged. This chapter studies how to obtain these inputs from observable market prices of zero-coupon bonds, coupon bonds, and various other interest rate options. This chapter does not exhaust the possible approaches to this problem. Rather, it provides a first-pass analysis constructed to illustrate the issues involved. In any particular implementation, these techniques will almost certainly need to be modified, refined, and extended.

This chapter is divided into four sections. The first shows how to strip zero-coupon bond prices from coupon bonds. The second shows how to obtain the initial forward rate curve from the zero-coupon bonds, and the third studies volatility function estimation. The fourth section illustrates these techniques by applying them to weekly observations of U.S. Treasury security prices. This application is purely pedagogical.

17.1 COUPON BOND STRIPPING

This section discusses how to obtain the zero-coupon bond prices implicit in observed coupon bond prices. As discussed in Chapter 4, except for floating rate notes, Treasury securities with maturities greater than a year (notes and bonds) are all coupon bearing. Although Treasury strips trade, they are not as liquid as Treasury bills, notes, and bonds. Consequently, coupon-bearing bonds usually provide the best source of data for the determination of the underlying zero-coupon bond prices. Of these securities, the on-the-run Treasury securities have the most liquid markets and provide the most accurate prices. The procedure for inferring the underlying zero-coupon bond prices we call *coupon bond stripping.*

In theory, looking at market prices, we can observe a collection of coupon-bond prices, denoted by

$$\mathcal{B}_j(0) \quad \text{for } j = 1, \ldots, n.$$

This set could include some Treasury bills or strips.

From Chapter 11, we know that the arbitrage-free price of the coupon bond $\mathcal{B}_j(0)$ with coupons C_j principal L_j and maturity T_j can be written as:

$$\mathcal{B}_j(0) = \sum_{t=1}^{T_j} C_j P(0,t) + L_j P(0,T_j). \tag{17.1}$$

For most bonds the index summation dates, $t = 1, \ldots, T_j$ correspond to 6-month intervals (the time between coupon payments).

To "strip out the zero-coupon bonds" means to solve this set of linear equations (expression [17.1] for all j) for the zero-coupon bond prices ($P(0,T)$ for all T).

Depending upon the set of coupon bonds included, this system could have no solutions, one solution, or many solutions. The system has no solutions if there are arbitrage opportunities present, because then the equality in expression (17.1) is violated. The system has one solution (generically) if there are just enough bonds to infer the zeros, and there are no arbitrage opportunities. Last, the system has many solutions (generically) if there are no arbitrage opportunities and there are less coupon bonds than there are zeros to be estimated.

The best method to employ, therefore, is to choose an error-minimizing procedure. This method works under all these circumstances. For example,

Choose $P(0,t)$ for $0 \leq t \leq \max\{T_j : j = 1, \ldots, n\}$ to

$$\text{minimize} \sum_{j=1}^{n} \left[\mathcal{B}_j(0) - \left(\sum_{t=1}^{T_j} C_j P(0,t) + L_j P(0,T_j) \right) \right]^2. \tag{17.2}$$

The error is the difference between the coupon bond's price and the arbitrage-free value, i.e.,

$$\mathcal{B}_j(0) - \left(\sum_{t=0}^{T_j} C_j P(0,t) + L_j P(0,T_j) \right).$$

Using this procedure, the best zero-coupon bond prices, by definition, are the ones that minimize this sum of squared errors. This is a quadratic programming problem for which standard software is readily available. The solution to expression (17.2) provides a vector of zero-coupon bond prices for use in the next section. We illustrate this procedure first with a hypothetical example, then with real data in a subsequent section.

Example: Coupon Bond Stripping

This example shows how to strip zero-coupon bond prices from a set of coupon bond prices. Suppose that we go to our fixed income broker and he gives us the following prices for five different coupon bonds.

Bond Price $\mathcal{B}(0)$	Coupon C	Maturity T	Face Value L
100.2451	2.25	1	100
101.9415	3	2	100
100	2	3	100
101.9038	2.5	4	100
98.8215	1.75	5	100

The coupon bonds have different coupon payments and maturity dates, but equal face values. One can solve expression (17.2) for the underlying zero-coupon bond prices. The solution is:

T	P(0,T)
1	.980392
2	.961168
3	.942322
4	.923845
5	.905730

In this special case, the sum of squared errors is zero because there is an exact set of zero-coupon bond prices that generate the coupon bond prices. The reader is encouraged to check that these zero-coupon bond prices do indeed give back the original coupon bond prices from which they were generated. This completes the example. □

17.2 THE INITIAL FORWARD RATE CURVE

This section studies estimation of the initial continuously compounded forward rate curve $\tilde{f}(0,T)$ for all T measured on a *per-year* basis. These forward rates are determined from the set of zero-coupon bond prices $P(0,T)$ for all T determined as the solution to expression (17.2).

From the procedure given in expression (17.2), the resulting zero-coupon bond prices will have maturities with discrete spacings (perhaps 6 months apart). Let us represent the price observations at time 0 by the $m \times 1$ vector

$$\begin{bmatrix} P(0,1) \\ P(0,2) \\ \vdots \\ P(0,m) \end{bmatrix}. \tag{17.3}$$

The difficulty encountered here is that the number of observed zero-coupon bond prices each day (m) are insufficient to determine the continuously compounded forward rates of all maturities (a continuous curve). There are missing zero-coupon bond price observations. The price vector (17.3) is insufficient information to price interest rate derivatives, because interest rate derivative cash flows often occur on days other than the payment dates for the coupon bonds included in our estimation procedure.

There are many methods for getting around the missing zero-coupon bond price observations. We will discuss the simplest (but perhaps most robust) approach. This approach assumes constant forward rates over the missing maturities. As such, it parameterizes a continuous curve with a finite number of parameters. The parameters can be determined from the estimated zero-coupon bond prices.

To understand this approach, we start with the definition of the continuously compounded forward rates.

$$\frac{P(0,t)}{P(0,t+1)} = e^{\int_t^{t+1} \tilde{f}(0,v)dv}. \tag{17.4}$$

Assuming that $\tilde{f}(0,v)$ is constant between time t and $t+1$, we can rewrite this as:

$$\frac{P(0,t)}{P(0,t+1)} = e^{\tilde{f}(0,t)1}$$

where

$$\tilde{f}(0,v) = \tilde{f}(0,t) \quad \text{for } t \le v < t+1. \tag{17.5}$$

Expression (17.5) is easily solved for the forward rates, given the zero-coupon bond price vector in expression (17.3). The solution is:

$$\tilde{f}(0,t) = \log(P(0,t)/P(0,t+1)). \tag{17.6}$$

This approximates the forward rate curve with a piecewise constant step function. This completes the estimation of the initial forward rate curve. We illustrate this computation first with a hypothetical example, then in a subsequent section with actual market data.

Example: Continuously Compounded, Piecewise Constant Forward Rates

This example illustrates the computation of the piecewise constant continuously compounded forward rates. Suppose that from the coupon bond stripping procedure, we obtain the following zero-coupon bond prices:

T	$P(0,T)$
1	.980392
2	.961168
3	.942322
4	.923845
5	.905730

Following the formula given in expression (17.6), we generate the following continuously compounded forward rates:

T	$\tilde{f}(0,T)$
1	.0198
2	.0198
3	.0198
4	.0198
5	.0198

These zero-coupon prices generate the same forward rates for all maturities. The forward rate graph would be a flat (constant) line. This completes the example. ☐

Instead of piecewise constant forward rate curves, one could use spline techniques. The most popular of these use cubic splines – piecewise joined polynomials of degree three. For a complete discussion of forward rate curve smoothing procedures, see Jarrow [2].

17.3 VOLATILITY FUNCTION ESTIMATION

This section studies two distinct approaches for estimating the volatility function(s)* $\sigma_j(t,T)$ for $j = 1, \dots, N$: (i) historic volatility estimation and (ii) implicit volatility estimation. Historic volatility estimation uses time-series observations of past forward rates (generated in the last section) to estimate these volatility functions. Implicit volatility estimation uses current market prices of various interest rate derivatives, and it inverts the computed price formulas to obtain the volatility functions such that the computed prices best match market prices. For this reason, implicit volatility estimation is sometimes called *curve-fitting*. Since the techniques discussed are independent of the number of factors selected, we analyze the general case of N factors.

For N factors, one must observe at least N different continuously compounded forward rates, represented by the $(N \times 1)$ vector at time t

$$
\begin{bmatrix}
\tilde{f}(t,t+1) \\
\tilde{f}(t,t+2) \\
\vdots \\
\tilde{f}(t,t+N)
\end{bmatrix}.
\tag{17.7}
$$

The basic $(N \times 1)$ vector equation for forward rates that underlies the volatility estimation procedure is expression (16.25a) of Chapter 16, rewritten here as

---

* This method allows up to N factors given N forward rates. The hope, of course, is that only a small number of factors can explain most of the variation in forward rates.

$$
\begin{bmatrix} \tilde{f}(t+\Delta,t+1) \\ \vdots \\ \tilde{f}(t+\Delta,t+N) \end{bmatrix} \approx \begin{bmatrix} \tilde{f}(t,t+1) \\ \vdots \\ \tilde{f}(t,t+N) \end{bmatrix} + \begin{bmatrix} \mu^*(t,t+1) \\ \vdots \\ \mu^*(t,t+N) \end{bmatrix} \Delta + \begin{bmatrix} \sum_{j=1}^{N} \sigma_j(t,t+1)\Delta W_j(t) \\ \vdots \\ \sum_{j=1}^{N} \sigma_j(t,t+N)\Delta W_j(t) \end{bmatrix}
$$

$$(17.8)$$

where

$$
\begin{bmatrix} \Delta W_1(t) \\ \vdots \\ \Delta W_N(t) \end{bmatrix}
$$

is an $N \times 1$ vector that is approximately normally distributed with mean 0 and covariance matrix $I\Delta$ where I is the $N \times N$ identity matrix.

It is important to stress that this evolution is under the actual or empirical probabilities. This is a discrete-time approximation to the continuous-time process of expression (16.25a).

17.3.1 Historic Volatilities

We illustrate the historic estimation procedure for the two cases of volatility functions studied in Chapter 16.

Case 1: Deterministic volatility functions:
$\sigma_j(t,T) = \sigma_j(T-t)$ is a deterministic function of $T-t$ for all $j = 1, \dots, N$.
$\mu^*(t,T) = \mu^*(T-t)$ is a deterministic function of $T-t$.

Case 2: (Nearly) proportional volatility functions*:

$$
\sigma_j(t,T) = \sigma_j(T-t)\min(\tilde{f}(t,T), M)
$$

---

* In Chapter 16 we used $\eta_j(T-t)$ instead of $\sigma_j(T-t)$ in case 2. We change the notation in this chapter to facilitate the subsequent exposition.

where $\sigma_j(T-t)$ is a deterministic function of $T-t$ for $j = 1, \dots, N$.

$$\mu^*(t,T) = \mu^*(T-t)\min\left(\tilde{f}(t,T), M\right)$$

where $\mu^*(T-t)$ is a deterministic function of $T-t$ and M is a large, positive constant.

For case 1, define the vector stochastic process $x(t)$, an $N \times 1$ vector, as

$$\begin{bmatrix} x_1(t) \\ \vdots \\ x_N(t) \end{bmatrix} = \begin{bmatrix} \tilde{f}(t+\Delta, t+1) - \tilde{f}(t, t+1) \\ \vdots \\ \tilde{f}(t+\Delta, t+N) - \tilde{f}(t, t+N) \end{bmatrix} \tag{17.9a}$$

and for case 2, define the vector stochastic process $x(t)$, an $N \times 1$ vector, by

$$x_j(t) = \begin{cases} \left[\tilde{f}(t+\Delta, t+j) - \tilde{f}(t, t+j)\right] \Big/ \tilde{f}(t, t+j) & \text{if } \tilde{f}(t,j) \leq M \\ \left[\tilde{f}(t+\Delta, t+j) - \tilde{f}(t, t+j)\right] \Big/ M & \text{if } \tilde{f}(t,t+j) > M \end{cases} \tag{17.9b}$$

for $j = 1, \dots, N$.

In both cases, $x(t)$ is a time-homogenous normally distributed random process with an $N \times 1$ mean vector

$$\mu^* = \begin{pmatrix} \mu^*(1) \\ \vdots \\ \mu^*(N) \end{pmatrix}$$

and an $N \times N$ covariance matrix Σ whose (i, j)-th element is

$$\sum_{k=1}^{N} \sigma_k(i)\sigma_k(j).$$

The additional restrictions in cases 1 and 2 were imposed in order to obtain this time-homogenous normally distributed random process. This structure enables us to apply standard principal component analysis (see Jolliffe [3]) to estimate the unknown volatility functions (vectors).

Using a time-series of K observations of $x(t)$ (using either expression (17.9a) or (17.9b)), we can obtain the $N \times N$ sample covariance matrix $\hat{\Sigma}$. Because this matrix is positive semi-definite, it can be decomposed as

$$\hat{\Sigma} = ALA' \tag{17.10}$$

where the $N \times N$ matrix $A = (a_1, \ldots, a_N)$ gives the N eigenvectors a_i for $i = 1, \ldots, N$ of $\hat{\Sigma}$ and the $N \times N$ diagonal matrix $L = \mathrm{diag}(\iota_1, \ldots, \iota_N)$ provides the N eigenvalues ι_i for $i = 1, \ldots, N$. The prime denotes transpose.

This decomposition gives the estimates of the N volatility functions as

$$\begin{bmatrix} \sigma_i(1) \\ \vdots \\ \sigma_i(N) \end{bmatrix} = a_i \sqrt{\iota_i} \quad \text{for } i = 1, \ldots, N. \tag{17.11}$$

A demonstration that this identification yields expression (17.8) is presented in the appendix to this chapter.

Sampling distributions are available for these estimates (see Jolliffe [3, chapter 3]). For a one-factor model, we set $N = 1$ and use (17.11) for the volatility function (vector). Sample estimates of these volatility function vectors are provided in the next section of this text.

17.3.2 Implicit Volatilities

The idea behind implicit volatility estimation is to use market prices from traded interest rate derivatives to estimate the volatility vectors. This is done by finding those volatility vectors such that a collection of computed interest rate derivative prices best match observed market prices. This technique is sometimes called curve-fitting. We discuss two methods for this estimation.

The first method imposes no restrictions on the $N \times N$ volatility matrix

$$
\begin{bmatrix}
\sigma_1(t,t+1; s_t) & \cdots & \sigma_N(t,t+1; s_t) \\
\\
\vdots & & \vdots \\
\\
\sigma_1(t,t+N; s_t) & & \sigma_N(t,t+N; s_t)
\end{bmatrix}.
$$

This matrix can depend on the time t and history s_t. The procedure finds those $N \times N$ values of this matrix that best match market prices. An error-minimizing procedure is often utilized.

The second method reduces the number of parameters to be estimated by imposing restricted functional forms on the volatility functions, e.g.,

$$
\sigma_j(t,T) = \sigma_j e^{-\lambda_j(T-t)}.
$$

Then only the parameters in these functional forms (σ_j, λ_j) for $j = 1, \ldots, N$ need to be determined implicitly from market prices. In this case, there are only $2N$ parameters to estimate.

17.4 APPLICATION TO COUPON BOND DATA

This section applies the previous parameter estimation procedures to actual U.S. Treasury bond price data. The data is from the U.S. Department of the Treasury (https://www.treasury.gov/resource-center/data-chart-ce nter/interest-rates/), and it corresponds to weekly Treasury par-bond yields for securities with maturities 1 month, 3 months, 6 months, 1 year, 2 years, 3 years, 5 years, 7 years, 10 years, 20 years, and 30 years over the time period January 2007–December 2018.

A typical week's data is contained in Table 17.1. These numbers are for December 12, 2018. The first column in Table 17.1 corresponds to the Treasury security's price. Note that for securities with maturities greater than a year, notes, and bonds, they are priced at par. This is due to the definition of a par-bond yield. For securities with maturities a year or less (bills), these are discount bonds and the price is per a 100 par. For example, the 1-month Treasury bill has a price of 99.81 dollars. The second column gives the time-to-maturity. For the first security, this is 1 month. The third column of Figure 17.1 provides the security's coupon rate, which for Treasury bills is zero and for the coupon bearing bonds, it is equal to the bond's yield. As seen, the 2-year coupon-bearing Treasury security

TABLE 17.1   Treasury Security Data for December 12, 2018

Price	Time-to-Maturity	Coupon Rate	Yield
99.81	1 Month	0	2.3
99.39	3 Months	0	2.43
98.73	6 Months	0	2.56
97.34	1 Year	0	2.7
100	2 Years	2.77	2.77
100	3 Years	2.78	2.78
100	5 Years	2.77	2.77
100	7 Years	2.84	2.84
100	10 Years	2.91	2.91
100	20 Years	3.04	3.04
100	30 Years	3.15	3.15

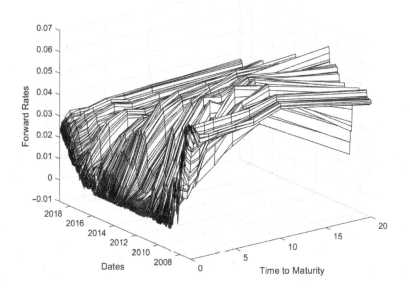

FIGURE 17.1   Forward Rate Curve Evolutions over January 2007–December 2018.

(a note) has a coupon rate of 2.77 percent. Last, the fourth column gives the security's par-bond yield. This is 2.30 percent for the Treasury security in the first row.

17.4.1 Coupon Bond Stripping and Forward Rate Estimation

The first task in the estimation procedure is to strip out the zero-coupon bond prices underlying the coupon bond data. Unfortunately, a complication arises in applying the procedure of expression (17.2). Looking at expression (17.2), we see that expression (17.2) requires that a zero-coupon bond price is estimated for each coupon payment date for each of the coupon bonds under consideration. This is a lot of zero-coupon bonds. For example, in Table 17.1, the coupon bond with the longest time to maturity remaining has 30 years left in its life (the last row). This implies that there are at 60 different coupon payment dates for this bond alone, and at least 60 zero-coupon bond prices will need to be estimated using expression (17.2). As there are only 11 bonds included in our data set for this date, there will be a non-unique solution to the minimization problem in expression (17.2). This non-uniqueness of the zero-coupon bond prices is a problem.

To obtain a unique solution, we can combine the zero-coupon bond price stripping procedure and the forward rate estimation procedure into one operation. To see why this works, consider expression (17.5) relating zero-coupon bond prices to (continuously compounded) forward rates. In this expression, the forward rates are assumed to be constant over various maturity time intervals. For our analysis, we assume that the forward rates are constant over the following (unequal) maturity intervals: 0–1 month, 1–3 months, 3–6 months, 6 months to 1 year, 1–2 years, 2–years, 3–5 years, 5–7 years, 7–10 years, 10–20 years, and 20–30 years. This gives 11 different forward rates.

Then, we substitute the forward rate expression (17.5) into the minimization problem of expression (17.2). This transforms the minimization problem from choosing zero-coupon bond prices to choosing the 11 different constant maturity forward rates $\tilde{f}(0,\overline{T})$ for time-to-maturities $\overline{T} \in \{0,.083,.25,.5,1,2,3,5,7,10,20\}$ to minimize the sum of squared errors given in expression (17.2). As there are 11 different forward rates, and 11 coupon bond price observations, the minimization problem now has a unique solution.

Given the forward rates from this minimization problem, expression (17.5) can now be used again, but in the reverse direction, to obtain the zero-coupon bond prices.

We applied this estimation procedure to our bond data. For the bond data contained in Table 17.1, Table 17.2 contains the estimated forward rates and zero-coupon bond prices obtained for December 12, 2018. The 0-month forward rate (the spot rate) is .0235. The forward rates increase up to 6 months, decrease to 2 years, and then increase again until year 20.

TABLE 17.2 Forward Rates and Zero-Coupon Bond Prices on December 12, 2018

Maturity	Forward Rate	Maturity	Zero-Coupon Band Price
0–1 Month	.0235	1 Month	.9980
1–3 Month	.0241	3 Month	.9940
3–6 Month	.0267	6 Month	.9874
6 Month–1 Year	.0275	1 Year	.9739
1–2 Year	.0263	2 Year	.9487
2–3 Year	.0260	3 Year	.9243
3–5 Year	.0265	5 Year	.8766
5–7 Year	.0297	7 Year	.8262
7–10 Year	.0301	10 Year	.7548
10–20 Year	.0312	20 Year	.5522
20–30 Year	.0340	30 Year	.3931

The zero-coupon bond prices start with .9980 for the 1-month maturity and decline to .3931 for the 30-year maturity.

Figure 17.1 contains a time series graph of the estimated forward rates over the entire time period studied, i.e., January 2007–December 2018. There are 626 weekly observations during this time period. In this graph, the forward rate curves are piecewise constant. As seen, the forward rate curves evolve across time in a non-parallel fashion.

These forward rate observations are the inputs necessary to do the parameter estimation for the forward rate's stochastic process.

17.4.2 Volatility Function Estimation – Principal Components Analysis

This section uses the forward rates obtained in the previous section to estimate the volatility functions used in the HJM model. The principal components analysis method is employed.

For this analysis, we use (absolute) changes in the forward rates. This corresponds to expression (17.9a) in the text. The appendix to this chapter shows that under the assumption of a piecewise constant forward rate curve using changes in constant maturity forward rate rates over a small time interval Δ gives the same drifts and volatilities as changes in the fixed-maturity date forward rates (as defined in this book).

Figure 17.2 contains the histograms of weekly changes in the 11 different constant maturity forward rates over this sample period. A normal distribution is superimposed on each of these histograms to give a sense for the quality of the underlying normality assumption. The normal approximation appears reasonable.

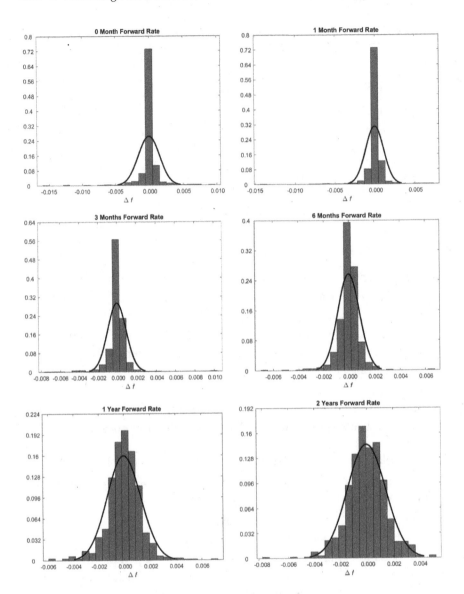

FIGURE 17.2   Histogram of Weekly Changes in Forward Rates from January 2007–December 2018.

The means, standard deviations, and correlations between the weekly changes in the various constant maturity forward rates are contained in Table 17.3. The mean weekly change in the 0-month forward rate over this sample period is –.00003792. This is less than one-third of one basis point (.0001). The standard deviation of the weekly change in the 0-month forward rate is .0015662. Similar numbers are given for the remaining

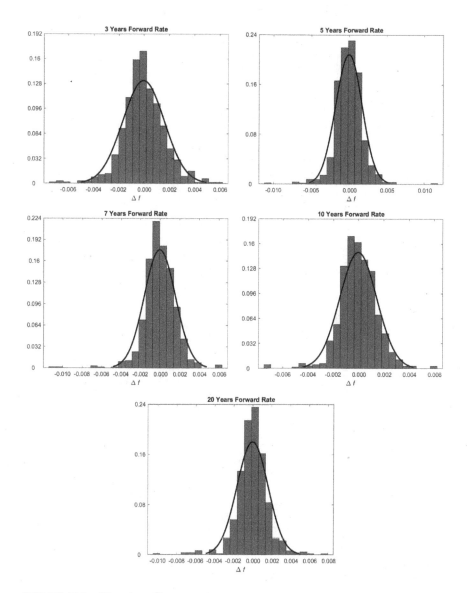

FIGURE 17.2 (Continued)

maturities. We see from this table that weekly changes in forward rates are not that large or variable.

Examining the correlation matrix, we see that the correlation between the weekly changes in the various constant maturity forward rates are more correlated for similar maturity forward rates. Weekly changes in the 0-month forward rate are positively correlated with changes in all the other maturity forward rates, with the correlation decreasing as the

TABLE 17.3 Correlation Matrix, Means, and Standard Deviations for the Changes in Weekly Forward Rates from January 2007–December 2018

	0 Month	1 Month	3 Month	6 Month	1 Year	2 Year	3 Year	5 Year	7 Year	10 year	20 year	Mean	Standard Deviation
0 Month	1	.5757	.1525	.2786	.0550	.1419	.0750	.0456	.0048	.0344	.0458	-.00003792	.0015662
1 Month	.5757	1	.4743	.3300	.1245	.1922	.0660	.0242	-.0172	.0308	.0393	-.00004416	.00112278
3 Month	.1525	.4743	1	.4559	.3643	.3922	.3000	.2133	.1829	.1832	.1930	-.00004048	.00098041
6 Month	.2786	.3300	.4559	1	.5539	.5575	.3998	.2759	.1282	.1994	.1591	-.00003456	.00085601
1 Year	.0550	.1245	.3643	.5539	1	.7735	.6679	.4384	.3184	.3353	.2351	-.00002941	.00132348
2 Year	.1419	.1922	.3922	.5575	.7735	1	.7729	.6620	.4956	.5176	.3797	-.00003035	.00145529
3 Year	.0750	.0660	.3000	.3998	.6679	.7729	1	.8335	.6574	.6860	.4620	-.09902980	.00166245
5 Year	.0456	.0242	.2133	.2759	.4384	.6620	.8335	1	.7232	.7449	.5542	-.00002686	.00176435
7 Year	.0048	-.0172	.1829	.1282	.3184	.4956	.6574	.7232	1	.8510	.6840	-.00002570	.00160374
10 Year	.0344	.0308	.1832	.1944	.3353	.5176	.6860	.7449	.8510	1	.7571	-.00003127	.00147419
20 Year	.0458	.0393	.1930	.1591	.2351	.3797	.4620	.5542	.6840	.7571	1	-.00001366	.00163785

maturity increases. Similarly, weekly changes in the 20-year forward rate are positively correlated with all the other forward rates, with the correlation largest for the 10-year forward rate and decreasing as the maturity of the forward rate decreases. These correlations show that a parallel shifting yield curve model for weekly changes in forward rates is inappropriate. Indeed, if there were a parallel shifting forward rate curve, all maturity forward rates would have identical correlations, which equal one. And, this is clearly not consistent with the correlations in Table 17.3. Such a model underlies the traditional approach to risk management (duration and convexity) discussed in Chapter 3.

The covariance matrix underlying these correlations is the basis for a principal components analysis using the deterministic volatility assumption (Case 1) underlying expression (17.9a). Standard statistical software can be used to generate the principal components of the covariance matrix. The volatility functions are then obtained using expression (17.11). These computations, based on weekly changes in forward rates, generate volatility functions normalized on a per week basis. To transform these volatilities to a per year basis, these weekly estimates were multiplied by $\sqrt{52}$, to adjust for the number of weeks in a year. The estimates of these per year volatility functions are contained in Table 17.4.

In this case, since there are 11 different constant maturity forward rates, there are 11 potential factors and 11 volatility functions estimated. The percentage variances of the changes in weekly forward rates explained by the 11 volatility functions are also provided in Table 17.4. As seen, the first three factors account for 77.5 percent of the total variance in forward rate changes.

The first three volatility functions are graphed in Figure 17.3. The first factor is roughly a parallel shift, except in the middle range. The second factor accounts for isolated movements in mid-term rates. Finally, the third factor emphasizes the relation between the short and long rates.

The volatility functions in Table 17.4 are the inputs required in Chapter 16 to generate the forward rate evolutions used for pricing and hedging. This completes the principal components approach to volatility estimation.

17.4.3 Volatility Function Estimation – A One-Factor Model with Exponentially Declining Volatility

This section applies a second procedure useful for estimating HJM volatility functions. This alternative procedure utilizes additional structure

TABLE 17.4 Volatility Functions and Percentage of Variance Explained

	1	2	3	4	5	6	7	8	9	10	11
0 Month	-.000410	-.000158	.000446	-.000139	-.000823	-.000206	-.001197	-.001222	-.001963	.002303	.000173
1 Month	.000642	.000688	-.000472	-.000189	.002528	-.000016	.001888	.000735	-.000964	.001682	.000143
3 Month	-.001047	-.000815	.001214	-.000511	-.002255	.000020	.002411	.001778	.000124	.000958	.000422
6 Month	.001523	.001064	-.003204	.001112	-.001929	.000413	-.000183	.000925	.000479	.001148	.000467
1 Year	-.001712	.002194	.000363	-.001297	.000553	-.001044	-.001669	.001288	.001987	.001226	.001117
2 Year	.000364	-.002660	0-.001198	.003702	.001210	-.000011	-.000954	.000637	.001823	.001552	.001974
3 Year	.003033	-.002099	-.000151	-.004092	-.000088	.000331	.000183	-.001747	.001767	.000637	.002944
5 Year	-.002118	.003161	.000508	.001655	-.000301	.004051	.002268	-.003778	.000478	-.000608	.003796
7 Year	.003364	.003256	.002324	.002225	-.001468	-.007876	.001473	-.000624	-.002570	-.002821	.004544
10 Year	-.007043	-.004946	-.007292	-.001056	.001608	-.002925	.000310	.000352	-.003119	-.002619	.005012
20 Year	.004161	.001182	.004248	-.002197	.000803	.010781	-.007978	.013107	-.010817	-.005929	.008507
Percentage Variance Explained	52.7	15.4	11.4	5.8	3.9	3.0	1.9	1.8	1.6	1.3	1.1

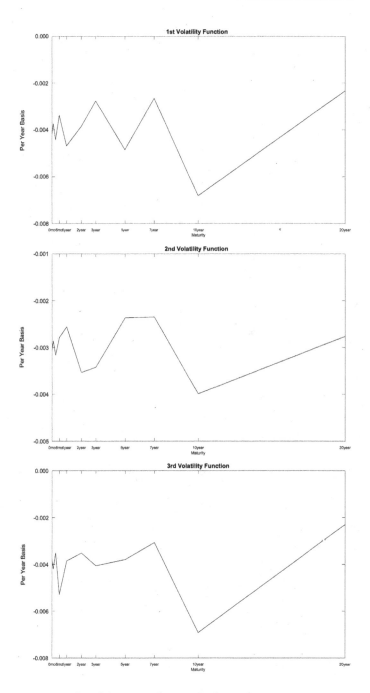

FIGURE 17.3   Graphs of the First Three Volatility Functions.

obtained from specifying a fixed number of factors and a specific functional form for each volatility function. In this section, we study a one-factor model with an exponentially declining volatility of the form:

$$\sigma(t,T) = \sigma e^{-\lambda(T-t)}. \tag{17.12}$$

This is a special case of the model estimated using principal components analysis in the previous section.

Using the continuous time model of Chapter 16, it can be shown (see Heath, Jarrow, Morton [1]) that the variance of the zero-coupon bond prices over the time interval $[t, t+\Delta]$ satisfies the following identity:

$$\text{var}\left(\log P(t+\Delta,T) - \log P(t,T) - r(t)\Delta\right) = \sigma^2 \left(e^{-\lambda(T-t)} - 1\right)^2 \Delta \big/ \lambda^2. \tag{17.13}$$

This identity forms the basis of an estimation procedure.

The first step is to estimate the variance on the left side of this expression. Using the time-series observations of zero-coupon bond prices generated by stripping the coupon bonds we can compute this sample variance. To do this, we set $\Delta = 1/52$ (1 week) and consider the maturities $T = 3$ months, 6 months, 1 year, 2 years, 3 years, 5 years, 7 years, 10 years, 20 years, and 30 years. The observation period is the same as before from January 2007 to December 2018. The 1-month zero-coupon bond price is omitted because its return is the spot rate by construction.

For each of these maturities, we compute the sample variance, denoted by v_T, over the entire observation period (January 2007–December 2018). These sample variances are contained in Table 17.5. The sample variances increase as the maturity of the zero-coupon bond increases. The 3-month sample variance is .0001228, while the 30-year sample variance is .0260424.

We then run the following (cross-sectional) nonlinear regression across the different maturities to estimate the parameters (λ, σ):

$$v_T = \sigma^2 \left(e^{-\lambda T} - 1\right)^2 \Delta \big/ \lambda^2 + e_T \quad \text{for all } T \tag{17.14}$$

where the error terms e_T are assumed to be independent and identically distributed random variables with zero means and constant variances.

The results of this estimation are also contained in Table 17.5. The estimated $\sigma = .0762$ and the estimated $\lambda = .0547$. The parameters, combined

TABLE 17.5   Estimates of the Sample Variance and Volatility
Parameters over January 2007–December 2018

Time to Maturity T	Sample Variance v_T
3 Months	.0001228
6 Months	.0002787
1 Year	.0004339
2 Years	.0007334
3 Years	.0018002
5 Years	.0030789
7 Years	.0059627
10 years	.0089158
20 Years	.0126329
30 Years	.0260424
$\sigma = .0762$	$\lambda = .0547$

with expression (17.13), provide the volatility inputs needed in the one-factor HJM model to price and hedge interest rate options.

These volatility inputs are contained in Table 17.6 and graphed in Figure 17.4. The volatility function declines exponentially in the forward rate's maturity. It is interesting to compare Figure 17.4 with the first volatility function graph in Figure 17.3. One can view Figure 17.4 as a smoothed version of this graph. The advantage of the volatility function given in expression (17.12) over the principal components–based volatility function is that expression (17.12) facilitates the derivation of analytic formulas for various standard interest rate options (see Heath, Jarrow, and Morton [1]).

TABLE 17.6   Volatility Function Values for the
Parameters $\sigma = .0762$ and $\lambda = .0547$

Time to Maturity $(T-t)$	$\sigma(t, T) = \sigma e^{-\lambda(T-t)}$
3 Months	.0751234
6 Months	.074103
1 Year	.0721038
2 Years	.0682656
3 Years	.0646318
5 Years	.0579341
7 Years	.0519305
10 Years	.0440712
20 Years	.0255033
30 Years	.0147583

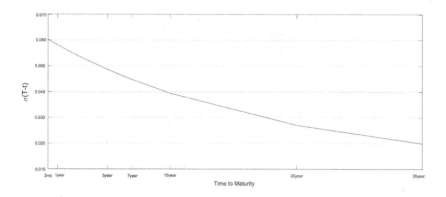

FIGURE 17.4 Graph of the Volatility Function $s(T-t)$ for the Parameters $\sigma = .0762$ and $\lambda = .0547$.

APPENDIX

Mathematical Demonstration That the Principal Components Decomposition Yields Expression (17.8)

This demonstration comes from Joliffe [6, chapters 1–3].

Define

$$X \equiv [x(1) \quad \dots \quad x(K)]$$
$$N \times K \quad N \times 1 \qquad N \times 1$$

where K is the number of time-series observations. Principal components analysis finds the $N \times N$ orthogonal matrix A (orthogonal means that $A'A = AA' = I$) such that

$$X' = Z'A$$
$$K \times N \quad K \times N \quad K \times N$$

where $Z' \equiv [z(1), \dots, z(K)]'$, $z(j)$ is an $N \times 1$ vector, and the sample covariance matrix satisfies

$$\hat{\text{cov}}(X) = \hat{\Sigma} = ALA'.$$

But, $X'A' = Z'AA'$ implies

$$Z' = X'A'$$

where

$$\hat{\text{cov}}(Z) = A'\hat{\text{cov}}(X)A$$

$$= A'ALA'A$$

$$= L.$$

Define

$$\underset{N \times 1}{w(t)} \equiv \begin{pmatrix} 1/\sqrt{\iota_1} & & \\ & \ddots & \\ & & 1/\sqrt{\iota_N} \end{pmatrix} z(t) \text{ and}$$

$$N \times N$$

$$\underset{N \times K}{W} \equiv [w(1) \dots w(K)].$$

Then

$$\underset{N \times 1}{z(t)} \equiv \begin{pmatrix} \sqrt{\iota_1} & & \\ & \ddots & \\ & & \sqrt{\iota_N} \end{pmatrix} w(t)$$

$$\hat{\text{cov}}(W) = \begin{pmatrix} 1/\sqrt{\iota_1} & & \\ & \ddots & \\ & & 1/\sqrt{\iota_n} \end{pmatrix} \hat{\text{cov}}(Z) \begin{pmatrix} 1/\sqrt{\iota_1} & & \\ & \ddots & \\ & & 1/\sqrt{\iota_n} \end{pmatrix}$$

$$= \begin{pmatrix} 1/\sqrt{\iota_1} & & \\ & \ddots & \\ & & 1/\sqrt{\iota_n} \end{pmatrix} L \begin{pmatrix} 1/\sqrt{\iota_1} & & \\ & \ddots & \\ & & 1/\sqrt{\iota_n} \end{pmatrix}$$

$$= I.$$

Thus,

$$
\underset{K \times N}{X'} = \underset{K \times N}{W'} \begin{pmatrix} \sqrt{\iota_1} & & \\ & \ddots & \\ & & \sqrt{\iota_n} \end{pmatrix} A
$$

$$
\underset{N \times K}{X} = A' \begin{pmatrix} \sqrt{\iota_1} & & \\ & \ddots & \\ & & \sqrt{\iota_n} \end{pmatrix} \underset{N \times K}{W}
$$

$$
x(t) = A' \begin{pmatrix} \sqrt{\iota_1} & & \\ & \ddots & \\ & & \sqrt{\iota_n} \end{pmatrix} w(t)
$$

because $\hat{\text{cov}}(W) = I$

This last expression is the matrix form of expression (17.8). The vector $w(t)$ has a nonzero mean vector. This completes the demonstration.

Theorem. *For piecewise constant forward rates, constant maturity forward rates and fixed-maturity date forward rates (as defined in this book) have the same drifts and volatilities for Δ small.*

Proof

As defined previously, $\tilde{f}(t,T)$ corresponds to a continuously compounded fixed-maturity date forward rate where T is some fixed date in the future, say January 1, 2025.

Let us introduce the new notation for *constant maturity* forward rates $\hat{f}(t,\bar{T})$ where $\bar{T}$ corresponds to time-to-maturity. The fixed future date is $t + \bar{T}$.

We want to compute $\tilde{f}(t + \Delta, T) - \tilde{f}(t, T)$.

We have that $\tilde{f}(t,T) = \hat{f}(t, T - t)$ and
$\tilde{f}(t + \Delta, T) = \hat{f}(t + \Delta, T - t - \Delta)$.

Substitution yields

$$\tilde{f}(t+\Delta,T)-\tilde{f}(t,T)=\hat{f}(t+\Delta,T-t-\Delta)-\hat{f}(t,T-t).$$

For a piecewise constant forward rate curve, we have that $\hat{f}(t+\Delta,T-t-\Delta)=\hat{f}(t+\Delta,T-t)$ for small Δ. Substitution yields

$$\tilde{f}(t+\Delta,T)-\tilde{f}(t,T)=\hat{f}(t+\Delta,\bar{T})-\hat{f}(t,\bar{T})$$

where $\bar{T}=T-t$ is the time-to-maturity. This completes the proof. □

REFERENCES

1. Heath, D., R. Jarrow, and A. Morton, 1992. "Bond Pricing and the Term Structure of Interest Rates: A New Methodology for Contingent Claims Valuation." *Econometrica* 60(1), 77–105.
2. Jarrow, R., 2014. "Forward Rate Curve Smoothing," *Annual Review of Financial Economics* 6, 443–458.
3. Jolliffe, I. T., 1986. *Principal Component Analysis*. Springer-Verlag, New York.

CHAPTER 18

Extensions

A MAJOR ADVANTAGE OF THE HJM term structure model presented in this textbook is that it is easily extended to incorporate additional term structures. The introduction of additional term structures is the generalization needed to price and hedge foreign currency derivatives, credit derivatives, and commodity options. This chapter briefly discusses each of these generalizations, providing references for subsequent reading.

18.1 FOREIGN CURRENCY DERIVATIVES

To price and hedge foreign currency derivatives, one needs a spot exchange rate of foreign into domestic currency and two zero-coupon bond price curves: *(i)* one for the domestic currency and *(ii)* one for the foreign currency.

The method for building an arbitrage-free evolution of these term structures proceeds in a fashion identical to that given in Chapters 4–10. The only complication is that in constructing the tree, two price vectors arc included at each node. One vector is for the domestic currency zero-coupon curve (just as before), and one vector is for the foreign currency zero-coupon curve with the spot exchange rate appended. The arbitrage-free conditions correspond to the existence of pseudo probabilities that make all dollar-denominated and dollar-translated securities (foreign zero-coupon bonds) martingales after normalization by the domestic money market account. Market completeness corresponds to the uniqueness of these pseudo probabilities. Pricing and hedging is done via the risk-neutral valuation procedure.

The only difficulty in applying these extensions in practice is that the computation time increases as more term structures are introduced into the model. Efficient numerical procedures become an important issue. References for this extension include Amin and Jarrow [2] and Amin and Bodurtha [1].

18.2 CREDIT DERIVATIVES AND COUNTERPARTY RISK

An important extension of the default-free term structure model to multiple term structures is when one includes securities with different levels of bankruptcy risk. The pricing and hedging of corporate debt and the pricing and hedging of swaps with counterparty risk are two prime examples.

The easiest way to analyze this pricing problem is to transform it into a foreign currency derivative problem and then to use the methods for pricing and hedging foreign currency derivatives (with obvious modifications).

To see the foreign currency analogy, consider two term structures of zero-coupon bonds: *(i)* the default-free term structure and *(ii)* the term structure for a risky firm. Call the risky firm XYZ. XYZ's zero-coupon bonds provide only a *promised dollar* payoff at future dates. The promised dollar is paid only if XYZ is not bankrupt at the payoff date.

One can think of XYZ zero-coupon debt differently. Consider XYZ zero-coupon bonds as first paying off in a hypothetical (foreign) currency, called XYZ dollars. That is, each XYZ zero-coupon bond pays one XYZ dollar for sure at its maturity. In XYZ dollars, XYZ debt can be considered default-free. But XYZ dollars need to be converted into actual dollars for analysis. The conversion rate (or spot exchange rate from XYZ dollars to dollars) is the payoff ratio at the zero-coupon bond's maturity. If XYZ is not bankrupt, the payoff ratio is unity. If it is bankrupt, less than the promised dollar is received.

Given this foreign currency analogy, the pricing and hedging problem for foreign currency derivatives can now be applied. This analogy also applies to counterparty risk. A counterparty to a contract only *promises* to make a payment, and when the contract provisions come due, payment is made only if the counterparty is not in default. Thus, this problem is identical to the one already discussed. Recommended references are Jarrow [4], Jarrow and Turnbull [7], Jarrow, Lando, and Turnbull [5], Jarrow and Yu [6], and Lando [8].

18.3 COMMODITY DERIVATIVES

The final extension studied is the pricing of commodity derivatives. Examples include oil futures, options on oil futures, precious metal futures, and options on precious metals. Again, this pricing and hedging problem has two term structures. The first is the same as that already studied, that is the term structure of default-free zero-coupon bonds. The second term structure is the term structure of commodity futures prices for future delivery. Given these two term structures, the analysis proceeds in a fashion similar to that of Chapters 4–10. The only difference is that in constructing the tree, two price vectors are included at each node. One is for the default-free zero-coupon bond prices, and the second is for the commodity futures prices for future delivery.

The arbitrage-free conditions correspond to the existence of pseudo probabilities that make the zero-coupon bond prices normalized by the money market account martingales and that make the futures prices martingales (see Chapter 13 for the motivation of this last condition). Market completeness corresponds to the uniqueness of these pseudo probabilities. Pricing and hedging are done using the risk-neutral valuation procedure. This extension can be found in Carr and Jarrow [3].

REFERENCES

1. Amin, K., and J. Bodurtha, 1995. "Discrete Time Valuation of American Options with Stochastic Interest Rates." *Review of Financial Studies* 8 (1), 193–234.
2. Amin, K., and R. Jarrow, 1991. "Pricing Foreign Currency Options under Stochastic Interest Rates." *Journal of International Money and Finance* 10, 310–329.
3. Carr, P., and R. Jarrow, 1995. "A Discrete Time Synthesis of Derivative Security Valuation Using a Term Structure of Futures Prices." In W. Ziemba, R. Jarrow, and V. Maksimovic, eds., *Finance: Handbook in Operations Research and Management Science*, North Holland, Amsterdam.
4. Jarrow, R., 2001. "Default Parameter Estimation using Market Prices." *Financial Analysts Journal* October, 1–118.
5. Jarrow, R., D. Lando, and S. Turnbull, 1997. "A Markov Model for the Term Structure of Credit Risk Spreads." *The Review of Financial Studies* 10 (2), 481–523.
6. Jarrow, R., and F. Yu, 2001. "Counterparty Risk and the Pricing of Defaultable Securities." *Journal of Finance* 56 (5), 1765–1799.
7. Jarrow, R., and S. Turnbull, 1995. "Pricing Derivatives on Financial Securities Subject to Credit Risk." *Journal of Finance* 50 (1), 53–85.
8. Lando, D., 1998. "On Cox Processes and Credit Risky Securities." *The Review of Derivatives Research* 2, 99–120.

Index